MACKIE.

Compact Mixers

Edition 2.1

Published by HAL LEONARD CORPORATION
7777 West Bluemound Road, P.O. Box 13819
Milwaukee, WI 53213

Printed in the United States of America

ISBN 0-634-00670-3

About the Author

Rudy Trubitt is a free-lance writer, sound designer and musician. He has written hundreds of articles for Electronic Musician, Mix, EQ, Keyboard, Pro Sound News, Gig and other magazines. He is also the author of the book "Live Sound For Musicians," available from Hal Leonard. Rudy Lives in the San Francisco bay Area with his wife Robin and daughter Riley.

As an audio engineer, he specializes in field recording, music editing, and sound effects design. One of his most unusual gigs is recording locomotives and editing the results for Lionel® toy train's onboard sound systems. Send e-mail to **ask_rudy@trubitt.com** and he'll do his best to anwer your questions. Also feel free to visit his web site for more mixer and general audio-related information at **http://www.trubitt.com.**

Acknowledgments

Illustrations: Mark Wilcox and Rudy Trubitt. Many thanks to Mackie Designs for use of their product illustrations.

Book Design: Debbi Murzyn.

Rudy's Special Thanks To: All my friends at Mackie Designs, including Paul Larson, Keith Medley, Dave Franzwa, Ron Koliha, Scott Garside, Ivan Schwartz, Ken Jager, Kyle Ritland, Kevin Johnson, Jason, Brian, Scott, Diane, Sara and the rest of the crew. Keep making those great mixers!

At Hal Leonard—Brad Smith, Jackie Muth, Mark Rattner, Sue Gedemer, Jon Eiche, Kimberly Walker, Lori Hagopian and company. Finally, to those who offered valuable comments and corrections: Dave Carr, Dan Phillips, Devon, Benny Rietveld, Amir Nevo, Michael Ladeau and Chris Dunnett. And thanks to my love Robin as well as Louie and Stella for their on-going support.

MACKIE®
Compact Mixers

By Rudy Trubitt

Edition 2.1

7777 W. Bluemound Rd. P.O. Box 13819 Milwaukee, WI 53213

Table of Contents

Preface

When I was eleven, the family got a big cassette recorder for Mother's day. It had a little plastic microphone and a switch that turned it into a tiny PA. It was my introduction to the wonders of home recording, feedback, and, after stuffing the mic into an acoustic guitar, a hint as to how I would be spending my life. What a revelation!

Although the cassette recorder was eventually replaced by more sophisticated toys, the purpose remained the same: *Make an interesting sound.*

It doesn't matter if you love twiddling knobs or consider sound equipment a necessary evil. What's important is learning to use the technology to create and to communicate. This book can help you down that path.

Where's My Mixer?

This book includes specific coverage of the following model Mackie mixers:

MS1202
1202-VLZ
1202-VLZ PRO
1402-VLZ
1402-VLZ PRO
CR-1604
1604-VLZ
1604-VLZ PRO
1642-VLZ PRO
CFX•12
CFX•16
CFX•20
SR24•4
SR32•4
PPM-406M
PPM-408M
PPM-408S
PPM-808M
DFX•6
DFX•12

This book includes specific information on quite a few Mackie mixers. Many models share particular features, so you'll find many sections of this book apply to more than one mixer.

For example, if you see a reference to "SR-series" mixers, those comments will apply to both SR24•4 and SR32•4 models. Similarly, "CFX-series" means CFX•12, CFX•16 and CFX•24 and "DFX-series" applies to both the 6 and 12 channel model DFX mixers. Along the same lines, first-generation VLZ and subsequent corresponding VLZ "PRO" models are identical from an operational standpoint. So, any reference to a "1202-VLZ" applies equally to a "1202-VLZ PRO." The same is true for 1402-VLZ and 1604-VLZ models. 1642- and 1604-VLZ are also quite similar. Any 1604-VLZ comments apply to the 1642-VLZ PRO unless otherwise noted.

While specific model Mackie mixers are used to illustrate the topics in this book, the concepts presented here apply to all mixers, regardless of their size, model or manufacturer. This means that even if "your" mixer is not directly mentioned, the information presented still applies to you!

I know many people find this a bit difficult to believe. After I wrote the first edition of this book, Mackie came out with a bunch of new mixers and I immediately started getting e-mail (drop me a line at: ask_rudy@trubitt.com) saying, "...I have a such-and-such mixer, does your book apply to me?" I'd write back saying "While my book uses a couple of specific Mackie mixers as examples, the concepts covered apply to all mixers, regardless of size, model or manufacturer."

And even though this second edition of this book, fully revised and freshly illustrated, includes many more Mackie models than my original book, there are a few Mackie mixers that this book *doesn't* include. So, at the risk of triggering a massive attack of *deja vu*, I'll say it one more time:

While this book describes specific Mackie mixers, the concepts presented apply to all mixers, regardless of their size, model or manufacturer!

Your Pal,

RUDY

CHAPTER 1

An Introduction To Mixers

Until recently, few people owned their own mixing boards. Those who did were usually experienced engineers with large, expensive mixers, or musicians with a simple "PA head," combining a *very* simple mixer with an amplifier for live performances.

Mackie crossed an invisible price/performance threshold with the release of their original CR-1604 mixing board. Suddenly, many more people could afford their own "real" mixer. New Mackie designs have put even more power in the hands of musicians and sound engineers everywhere.

Still, first-time mixer owners start with a lot of questions: "How do I hook up the rest of my equipment?" "What's the best way to get a mix?" And "What are all these knobs for?!?"

Questions like these are what this book is all about. Using Mackie's wildly popular compact mixers as examples, we'll look at how simple and slightly more complicated mixing boards work.

In addition to specific ideas to help you get the most from your Mackie, this book explains the concepts behind mixers and defines their role at the center of an audio system. That way, you'll get a good idea of how all mixing boards work, regardless of model or brand.

Finally, the most important thing you'll learn from this book is how sounds get from your mixer's inputs to its outputs. I guarantee that understanding this "signal flow" will help you every time you use any sound system!

P.S. This book is designed to complement, not replace, the succinct and entertaining manual that came with your Mackie mixer!

All Those Knobs!

Mixers (also called mixing consoles or boards) can look a bit intimidating at first glance. How will you ever remember what all those knobs are for?

Learning to use just 20 of these 83 controls is all you need to start mixing on a Mackie 1202-VLZ.

Put your fears to rest. Here's the deal: Yes, the 1202-VLZ PRO (shown above) has 83 knobs and buttons. But guess what? You only have to learn 20 of them!

The same is true of the 1604-VLZ. Its 340 knobs and buttons can look pretty overwhelming. But master just 43 of them and you've got the 1604-VLZ licked!

These surprising statements are true because although mixers have a large *number* of individual controls, most of them are *duplicates*. Once you've learned one, you'll know how lots of the others work too. So don't be alarmed—you *can* learn to use these things.

As an added bonus, mixers share a certain "seen one, seen 'em all" similarity. Most models have a lot in common, so learning a basic mixer like the 1202 today can help prepare for an encounter with a much larger mixing board tomorrow.

What Is a Mixer For?

Beneath its dizzying array of controls, a mixer actually has some important similarities to a simple home stereo receiver. A stereo receiver has controls that let you switch between different components of your hi-fi system, so you can listen to the CD player, or the phonograph, or a cassette deck. A receiver also has controls to set overall volume, the balance between left and right speakers, and some tone controls to shape the overall sound.

A mixer does many of the same things, including changing levels and tone. The most important difference between a mixer and a stereo system is that a mixer allows you to control and *combine* or mix sounds from many different sources at once. Rather than simply choose between one sound *or* another, a mixer gives you the option to combine many sounds at the same time.

This brings us to three simple functions of a mixer:

- Changing the character of an individual sound—making it louder, softer, brighter, fatter, etc.
- Combining many sounds together, and finally;
- The artistic process of blending all sources into a finished "mix," either recorded to tape or for a live audience.

Let's look at these three tasks in a little more detail:

Changing the Character of a Sound

Every sound has a few basic properties: It can be soft, loud or somewhere in between. It can be shrill, muffled, smooth, bright, dark, etc.

Within limits, a mixer has the ability to alter a sound's sound. For example, you can make it louder or softer with a touch of a dial. However, when it comes to changing a sound's basic character, there are limits. For example, let's say you have a steel-string acoustic guitar and want to record a classical guitar piece. While a mixer's tone controls (called *equalization*) can make the bright steel-string sound darker and rounder—a little more like a nylon-string guitar— no one will be fooled into thinking you've switched instruments.

There is no substitute for starting with the right sound to begin with!

Combining Many Sounds

A single recording or live performance can include dozens of individual sounds or instruments. Therefore, a crucial function of a mixing board is to serve as a traffic manager for all the individual audio signals involved in a project.

Signal routing is important because you may not want all your signals to go to the same place. For instance, you might want a vocal microphone to go through an echo or other "effects unit."

Mixing consoles offer many ways to route signals. The most familiar is the *pan* control. Similar to the *balance* knob on a hi-fi, the pan control directs a signal to the left or right side of a stereo sound system, or somewhere in between. In practice, this makes it possible to simulate the original location of each performer in the finished stereo mix.

The Final Mix

The finished product from your mixer will be a blend of sounds or musical parts, balanced just the way you want them. This "mix" typically takes two forms: It can be recorded to tape to be replayed later and perhaps duplicated and distributed to others; alternately, you may be mixing a live performance and helping to balance the sound heard by the audience over a sound system.

In either case, your mixer is the tool that allows you to create these finished pieces. Once you have mastered the basic functions of your board, you'll be ready to use it to further your creative goals.

CHAPTER 2

The Basics

This chapter is *really* important. Here, you will gain an understanding of concepts you'll return to again and again in your work with mixers and sound systems. *Here we go!*

The purpose of a mixer is to allow you to adjust the character of audio signals individually and then combine them in whatever balance suits your ear.

Individual audio signals are connected to a mixer's input and the combined result is the mixer's output. When this output (or two outputs, in the case of a stereo mixer) is connected to an amplifier and speakers, you'll hear the mix of the individual sounds you started with.

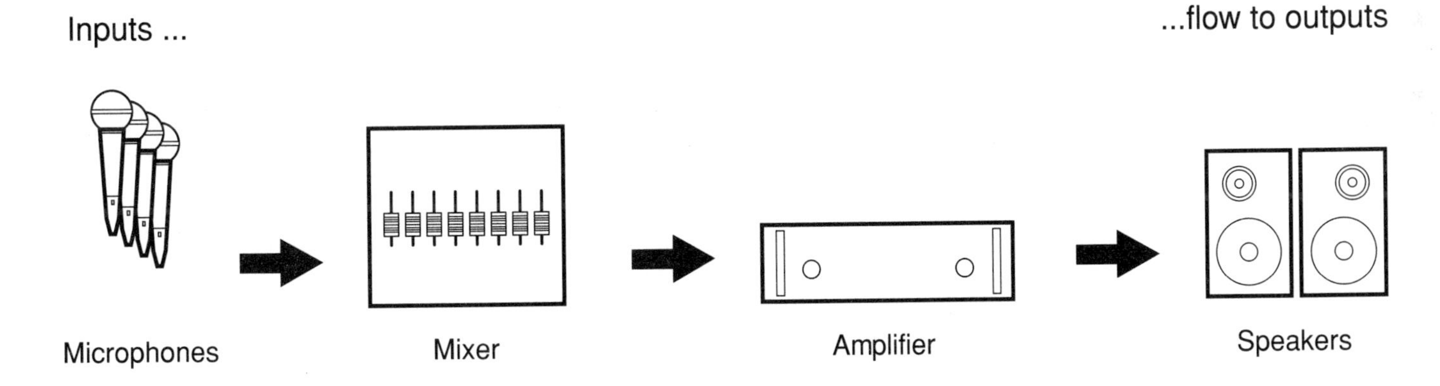

Every sound system starts with inputs and ends with outputs. When connecting devices, remember that the output of one unit should be connected to the input of the next.

The rest of this chapter will introduce you to the conceptual building blocks of a mixer. Even though it is a single piece of equipment, you'll quickly learn to recognize the functional parts built into every mixer.

These parts fall into one of two simple categories: Inputs or Outputs.

Mixer Inputs

All mixers have a number of input "channels." Each channel provides controls for an individual sound source. Typical channel controls include volume and tone (also called EQ), as well as others we'll get into later. Whenever the mood strikes, you may use the controls on any input channel to change the sound of the signal connected to it.

Input channels can be recognized as identical sets of controls, arranged in vertical strips. A four-channel mixer offers individual control over up to four different sound sources, a 16-channel mixer up to 16, and so on.

All mixers have a number of independent input channels. These allow individual control over the volume and tonal character of each signal connected to that input. On the Mackie 1202-VLZ, input channel controls are grouped into eight vertical strips, as shown here.

These sets of identical controls explain why you need only learn a fraction of the total number of knobs to understand how a whole mixer works. Each input channel usually has the same controls found on all other input channels, although as you'll see, Mackie's compact mixers don't exactly follow this general rule.

Individual Input Channel Strip

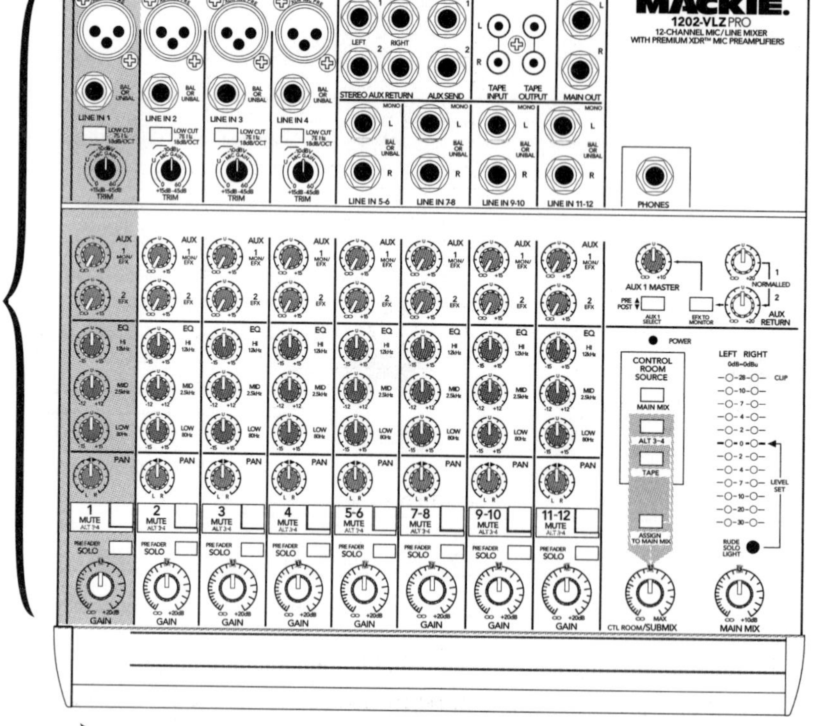

Mixer Input Section

Note that you'll occasionally hear engineers refer to "channel strips" when describing the input channels of a mixer. A little tip: To avoid looking like a greenhorn, never use the word "tracks" when referring to mixers. Tape recorders have *tracks*, and mixers have *channels* or *inputs*.

Mixer Outputs

In order to get the sounds from your mixer to the rest of your sound system, mixer inputs must eventually lead to mixer outputs.

Mixer outputs have controls of their own, separate from those of the input channels. This *output section* is usually found on the right side of the mixer (it's in the center of some very large mixers). You can identify a mixer's output section at a glance, because its controls look different than the repeating columns of input channels.

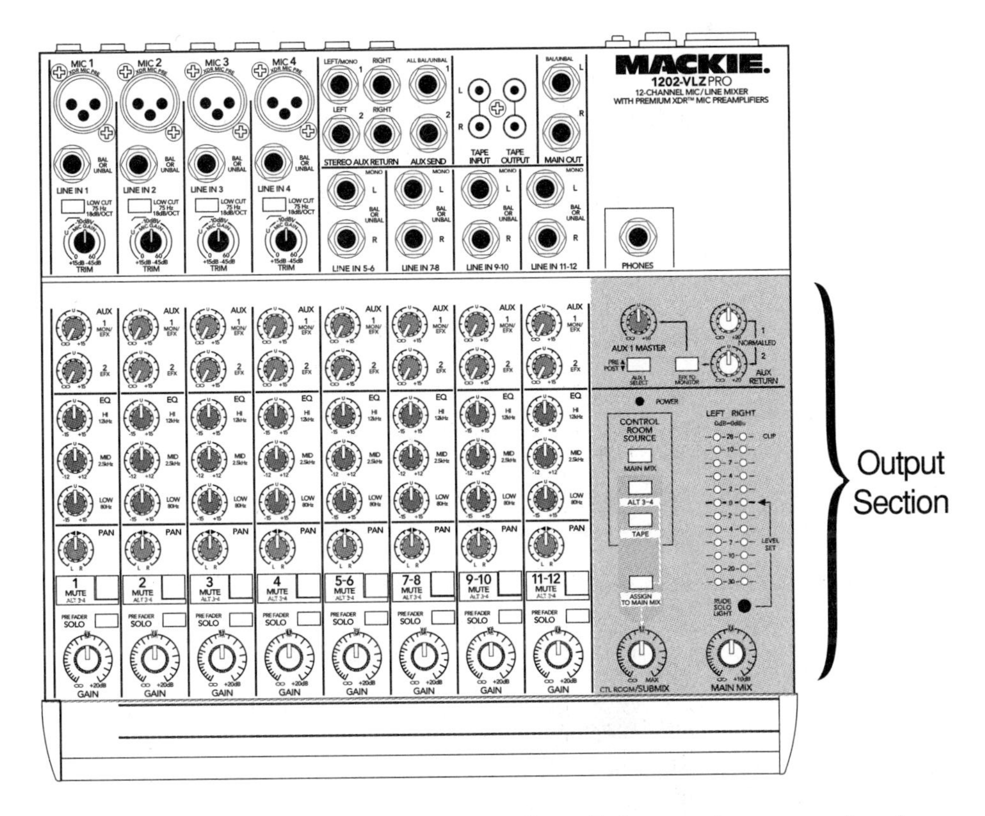

Audio signals pass through individual input channels and then are combined into the mixer's output. This output is typically connected to an external amplifier and speakers or to a tape recorder. Controls in the output section usually apply to the entire mix.

In general, controls in the output section affect all the signals connected to the mixer, while controls on individual channel strips always apply *only* to that particular channel's input.

The most basic of the output section controls is the *master volume*. This knob provides a single point where the entire mix can be turned up to "11" or completely off. *If the master volume is all the way down, you won't hear anything out of the main outputs* (although you may hear something through your headphones).

Headphone outputs are typically found in the master section as well. Plugging in a pair of headphones (and adjusting the headphone level control) lets you hear the main output of your mixer. On many mixers, you can hear individual channels through the headphones by pressing their *SOLO* buttons.

The master section is also where you'll find *Meters*, which give you a visual indication of how loud different parts of your mix are. Many Mackies also include a "control room" section in their output area. This lets you adjust the signal source and overall level of your studio monitor speakers, without affecting the levels being sent to your master mixdown recording through your mixer's main left and right outputs.

Signal Flow: The Key to Audio Understanding

Here is the most important topic in this book!

There are lots of ways to hook up mixers. You'll find plenty of them later in this book. However, no cookbook approach will ever meet your needs exactly, because the variations in equipment and their uses are limitless. Instead, understanding how signals get in and out of your mixer is the best way to learn how to set up a sound system for *your* needs.

As you know, a mixer has inputs and outputs for audio signals. You can actually see where these inputs and outputs are by looking at the labeled connectors on your mixer's top or back panel.

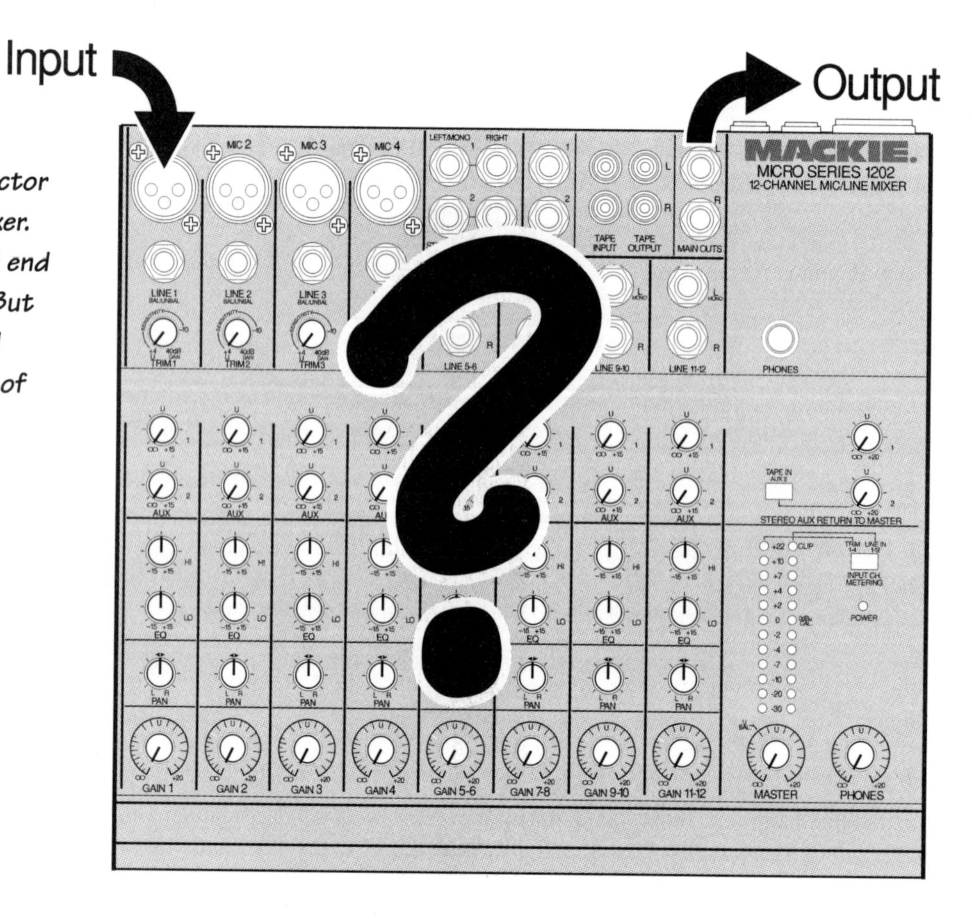

The arrows point to an input and output connector of a Mackie MS1202 mixer. These are the start and end point of a signal's trip. But what path does a signal take through the inside of the mixer?

However, the exact route a signal takes from a given input to a given output is less obvious. Grasping this flow of audio signals from in to out is the key to understanding audio systems. This *signal flow* or *signal path* might not sound like much, but—trust me—it's everything.

If you can envision a mental image of where each signal comes from, where it goes, how it gets there and what happens along the way, you'll be able to configure, operate and trouble-shoot practically any sound system.

Signal flow can be illustrated by diagrams in which each functional part of a mixing board is a separate symbol, with lines showing the audio connections from one to the next.

When studying a signal flow diagram, find the input first then trace the connections towards the output. By convention, signal flow is usually laid out like English sentences, reading from left to right.

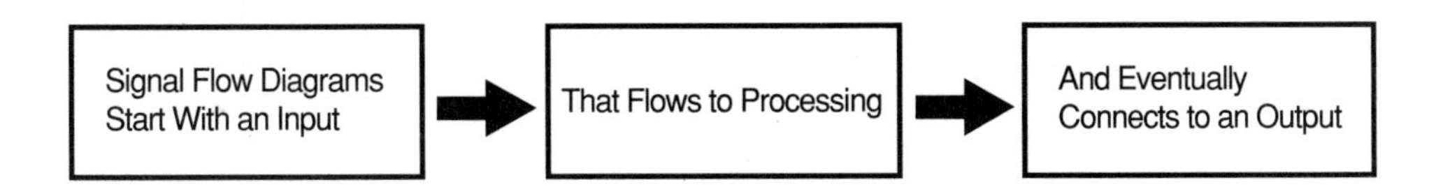

An example signal flow diagram showing the progression from input, through processing, to output.

By tracing this signal path, you'll see all the portions of the mixer that the signal travels through. Then you'll know which knobs, switches and connectors control that particular signal.

Sometimes the signal path may get a little convoluted (both on paper and within the mixer itself). It helps to remember that audio signals *always* flow from the output of one device to the input of the next, never the other way around.

After tracing a signal flow block diagram (or just running though it mentally, as you'll eventually learn to do), you will have a clear idea of what mixing and signal routing options are available to you in a given situation.

The following section will introduce you to a few simple signal flow diagrams. I'm assuming you are at least a little familiar with a few basic mixer elements. We'll consider examples that use inputs, outputs, volume, pan and tone controls. (If you're not entirely clear on the purpose of each, you'll find descriptions of them in Chapters 2 and 3.)

On with the examples!

In this signal flow example, a mic is connected to an input. Then, it passes through a volume control and then to an output which drives a pair of headphones.

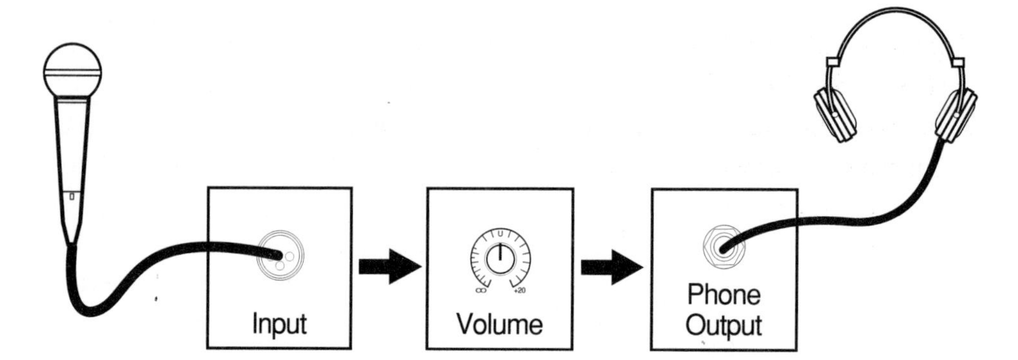

Here, a microphone is connected to an input. The input signal is then connected to a volume control, which in turn is connected to an output which feeds a pair of headphones. With a little imagination, you might guess that operating the *volume* control will change the amount of signal that reaches the output. And you'd be right!

Remembering that input flows to output helps you understand the relationship between blocks: Any control "upstream" or before another control affects the signal that is available to later blocks.

Signal flow is a little like water in a river. If a farmer diverts part of the river into a field, there's less water for other farms downstream. If someone dumps mud into the river, everyone downstream gets muddy water, unless someone filters it.

Therefore, in this example, turning the volume control down all the way means no signal makes it to the output, even if there's still signal present at the input connector.

Now, a slightly more sophisticated example:

The addition of EQ to this signal flow diagram adds flexibility to this hypothetical sound system.

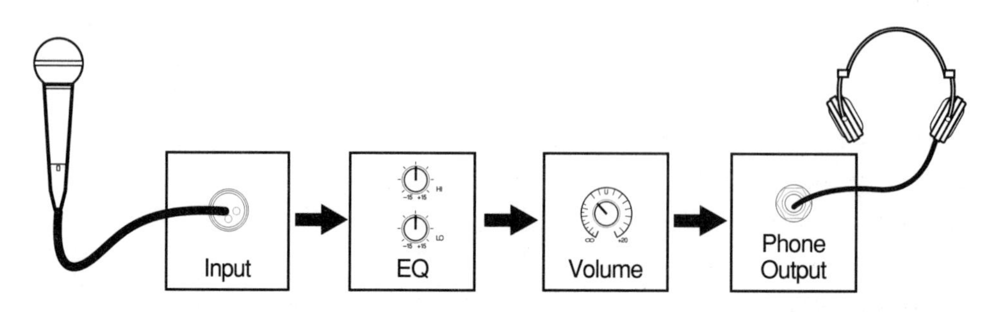

Here, we add simple bass/treble (or low/high) tone controls to our signal path. Again, these EQ controls will be familiar to anyone with a hi-fi system. (You'll learn more about EQ in Chapter 3.)

In this case, the input signal passes through the tone controls where you could make it darker or brighter, then the volume knob and finally to the output.

While block diagrams are usually less "pictographic" than the examples on these pages, the purpose of all such diagrams is the same: to show a signal's path from input to output.

Both previous examples deal with one input and one output. But as you know, mixers need multiple inputs to combine different sounds. The example below uses the signal path from the previous illustration, but instead of one input and output, we now have *four* inputs and *one* output.

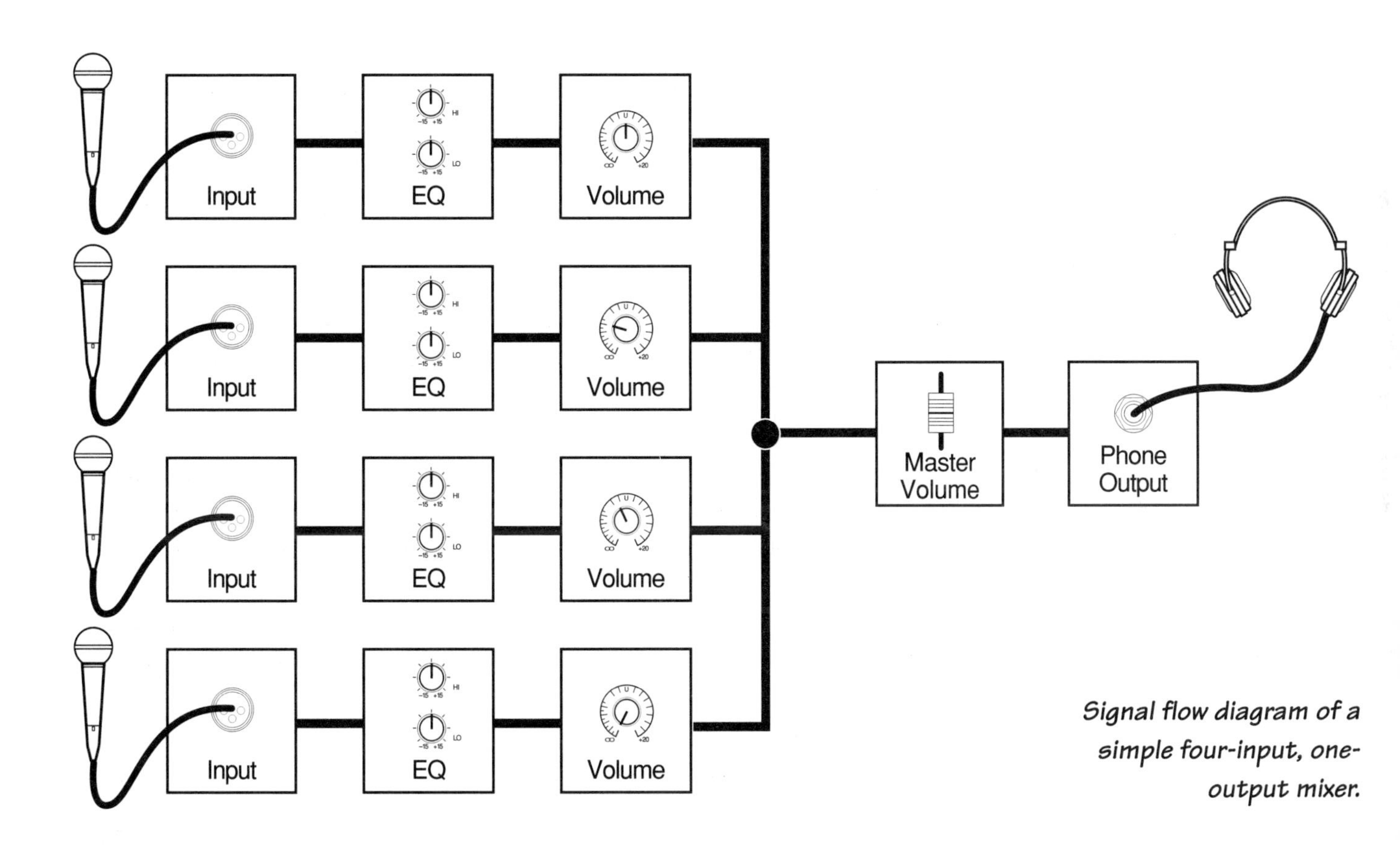

Signal flow diagram of a simple four-input, one-output mixer.

You should already understand the signal flow within each of the inputs. What's new is the master volume and common output. Note how the signal from each input is combined to one output. This happens internally in all mixers—it's in this circuit (called a *mix bus*) that individual audio signals are combined. This mixed signal then goes through one more volume control, the master volume, and then to the output.

A Signal Flow Pop Quiz

As you become familiar with signal-path diagrams, you'll begin to be able to answer various "what-if" questions. Try and answer this one: What's the easiest way to increase the output from our little example mixer without changing the relative balance between input channels? Take a look at the previous diagram and decide...

OK, time's up, pencils down.

You could increase the level of each input channel's volume control, (which we'll soon be calling the channel volume fader), but this would be imprecise and time consuming especially on a mixer with more than four channels!.

Instead, the master volume control should be used to bring the overall volume up or down, which requires turning *just one knob*. Since the signal passing through the master is the sum of all individual inputs, you can be positive that the relative level between inputs won't change.

This sort of deduction is *the key to understanding how each and every audio system works*. Start with an idea and then check your mental image of the system's signal path to see if it can be implemented. When necessary, refer to a paper block diagram, like the one in your Mackie manual (or in Chapter 14 of this book). **The key to solving every problem starts with an understanding of the signal flow involved.**

Introduction to Buses

Our little four-input mixer example should help you remember that signal flow isn't a collection of individual paths or parallel lines that never meet. Instead, a mixer has different intersections where individual input signals blend together, like tributaries flowing into a single river.

This section introduces you to the concepts and terminology behind this key feature. Busing is perhaps the most difficult concept in this *Basics* chapter. If it doesn't make complete sense the first time, read it again—it's really important.

As you know, one critical requirement of every mixer is the ability to combine multiple signals—in other words, to mix! All of a mixer's inputs can appear at the main stereo left and right outs, assuming that the settings of each channel's level and pan controls permit it.

Inside the mixer, there is a point where all the individual input signals can blend together. This point is called a *mix bus*. Other names used for the mix bus are *mix output*, *send*, *output bus* and *stereo bus*. The term *bus* comes from the world of electrical power systems, where a "bus bar" carries the electricity for several connected circuits.

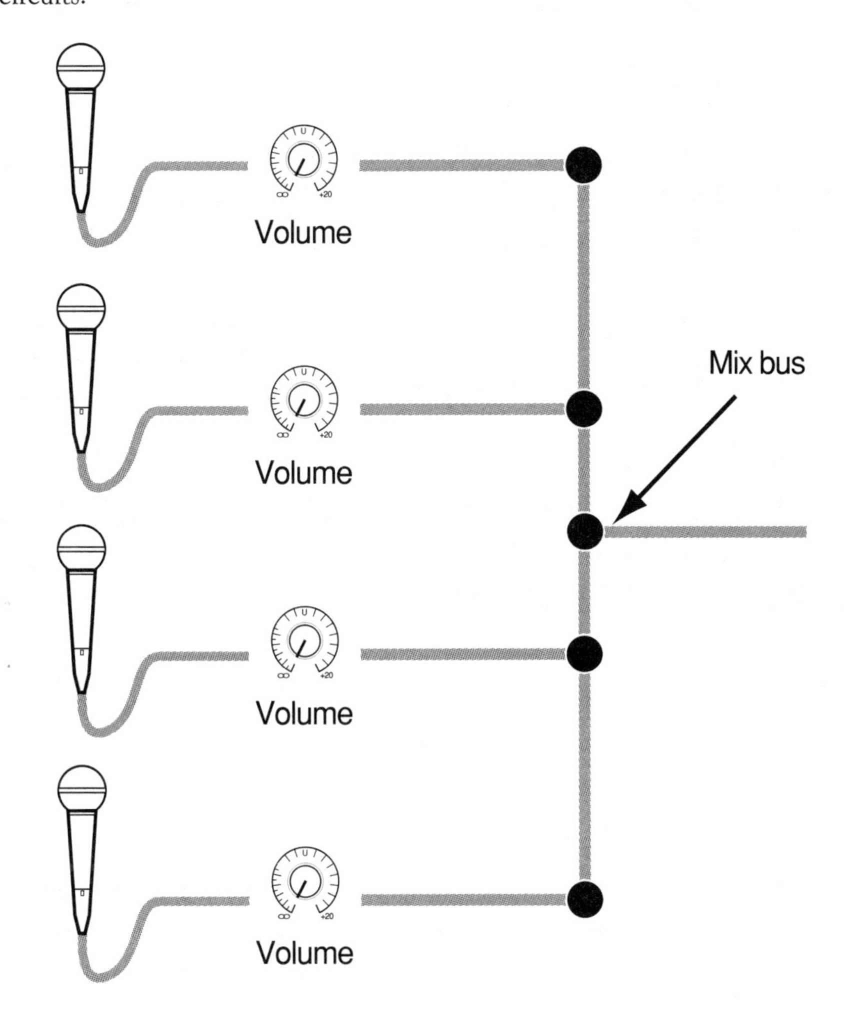

The black dots identify points along the mix bus, where individual inputs are combined to become a common output. While not shown in this illustration, stereo mixers have a separate mix bus for left and right outputs.

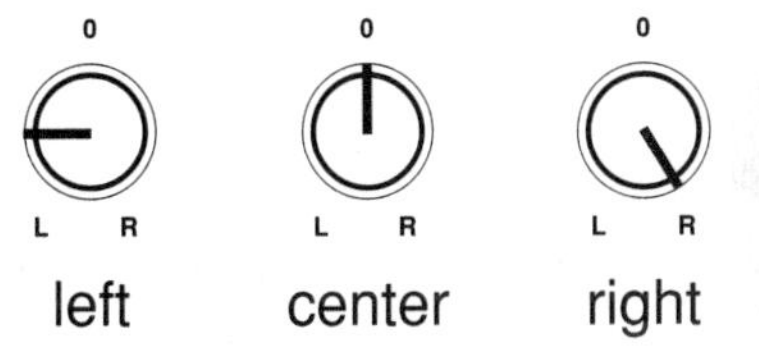

When a pan control is centered, the sound that is heard should appear to be coming from a point between the left and right speakers. When turned more to the left, for instance, the sound will appear to be coming more from the left speaker.

Most Mackies are stereo mixers, which means there is a separate point internally where the *left* signals are combined and another point for the *right* signals. Each of these internal points flows to the stereo Left and Right output connectors on your mixer.

Any given input channel can be sent to the left or right mix bus or both, depending on the position of the Pan control. This ability to route signals to specific outputs is a key element of a mixer's usefulness.

There is another set of buses in your mixer in addition to the stereo outputs: the *aux sends*. Although these look different on the surface and are used differently, they are doing the same thing inside.

When you turn up an aux send knob on a channel, part of that channel's audio signal gets sent to an aux bus, where it is mixed with the signals added from other channels. Then, this combination of signals makes its way to the aux send output.

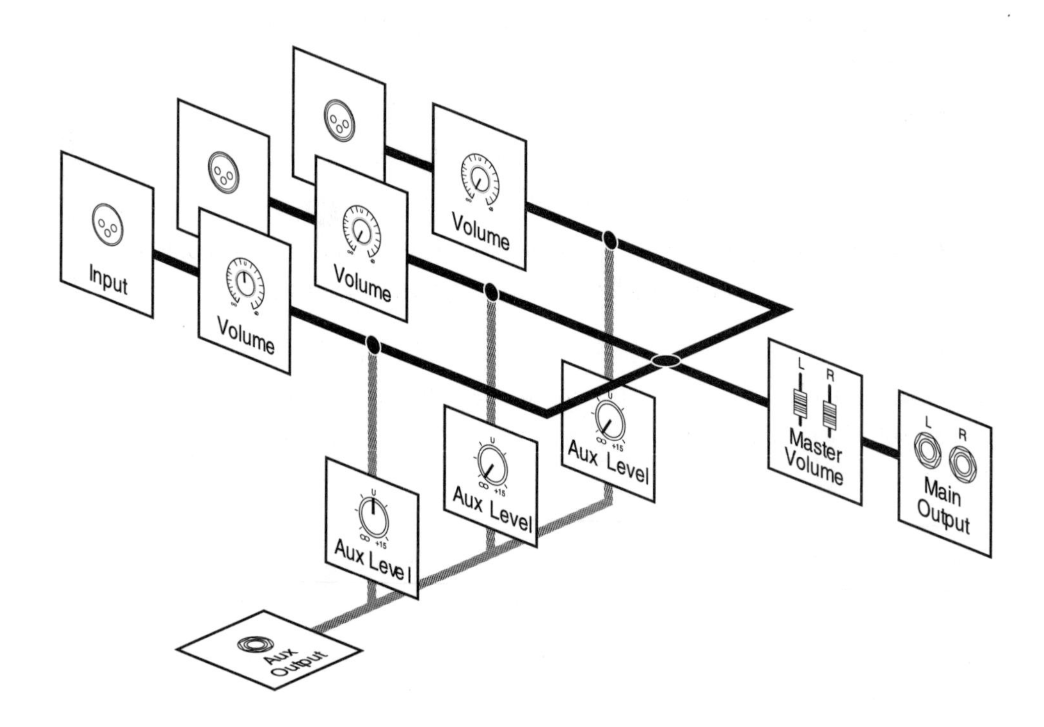

In this conceptual illustration, each input channel feeds the main mixer output, but can also feed the Aux Output if its Aux Level knob is turned up. For clarity, parts of each input channel (like EQ and Pan controls) are not shown.

Because different controls set the amount of each input that gets sent to each different bus, you can have a completely different combination of signals, also called a different mix, on each bus. So, even the classic MS1202 is capable of creating four different mixes—Left and Right and Aux 1 and 2. Usually, aux outputs are used to feed signals to effects units, or create a headphone or stage monitor mix to help musicians play at their best.

However, I hasten to add that aux mixes are not always independent. Often, each channel's level control limits the amount of signal available to that channel's aux sends. So, if a channel is turned all the way down, you may not be able to get its sound into the aux buses.

Finally, some Mackies have "assignable buses," which give you yet another way to route a given input channel to one or more pairs of stereo outputs (see illustration of facing page). The 160-VLZ, CFX and SR-series mixers include this feature, and the classic CR-1604, 1202-VLZ and 1402-VLZ models have a simplified version of it as well.

Aux sends and assignable buses are important enough topics to warrant their own chapters a bit later on. For now, the important points to remember are:

- Each mixer output corresponds to a point in the mixer where signals from multiple input channels are combined. This point is called an output bus.
- Not all inputs share the same routing options. The input channel strips are the most flexible. Tape inputs and effect returns are often only able to be routed to the main stereo bus.
- Understanding how your mixer controls signal mixing and routing is a critical step in making your entire sound system work for you.

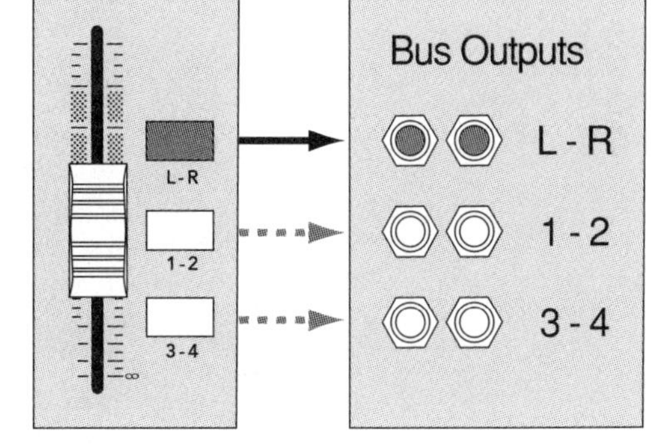

On some, but not all Mackies, "bus assign switches," labeled L-R, 1-2 and 3-4, route an individual channel's signal to one or more stereo outputs. In this case, the example channel's signal will come out of the main left-right outputs. Don't worry, there's a whole chapter explaining this a bit later. I just wanted to give you a little heads-up now.

If you've made it this far, you're doing great. Even if some of this is still murky, don't panic—these concepts take a little time to sink in. There's also no substitute for hands-on experimenting; stop reading and play around with your mixer any time.

In a moment, we'll apply these concepts to *your* mixing board. But before that happens, it's time to try your patience with *just one more little lecture...*

Up to this point, we've talked about "audio signals" in slightly vague terms. You may be happy to take their existence on faith, but I recommend reading this section, despite the ominous warning above.

In keeping with the spirit of this book, we're going to cover the concepts behind audio signals without getting into any rigorous techno-speak. There are plenty of good books for that—refer to them for further study.

Speaking of books, if you've read other books on sound equipment, you'll recognize the little "talk" I'm about to give. It's not the mind-numbing lecture about how well-behaved your classmates are (why can't you be more like them?), and it's not the embarrassing warning from the nice officer about obeying traffic signs.

It's a tiresome speech called *"The Nature of Sound."*

The Nature of Sound

Thanks to a scary movie from the late seventies, we all remember that "*In space, no one can hear you scream.*" This is because in space, there is no air. Although a bit of peace and quiet might be welcome, especially if you live next to my house, lack of air may be a bigger problem than getting relief from that *infernal racket.*

But how are air and sound related? Here's the scoop: Something—anything—vibrates. As it does, these vibrations push back and forth against the atmosphere. This pushing and pulling motion against the air sends out little ripples of changing air pressure, much like the ripples that spread out after you toss a pebble into a pond.

These ripples, called *sound waves*, radiate out from the sound source (*not* equally in all directions, but that's a topic for a whole book). If your ears happen to be in the field of a sound's influence, you "hear" the little changes in air pressure in great detail.

Faster vibrations make higher pitched sounds, while slower ones sound lower. These vibrations can be staggeringly complex, as is our ability to perceive them. We can hear little flutters taking place up to 10,000 or even 20,000 times a second! And, the more intense the change in air pressure becomes (the greater the back-and-forth motion of the vibrating object), the louder the sound will seem to be.

I think it's a miracle that hearing works at all—imagine telling the difference between a clarinet and an oboe hidden from view, simply by measuring little ripples in air pressure. Preposterous! But you do it every day.

Now you are eligible to receive (and hopefully comprehend) two exciting buzzwords, at no charge, and with no further obligation! The buzzwords are *frequency* and *amplitude*.

A sound's frequency is the number of times the air pressure wavers up and down each second. Sound frequency is measured in *cycles per second* or *Hertz*, abbreviated *Hz*. The theoretical range of human hearing is from 20 Hz to 20,000 Hz. (Using the standard shorthand for big numbers, the letter "k" implies "thousand." So, "10 kHz" means the same thing as "10,000 Hz." Sometimes people will just say "10k," although to be picky, one could respond "Ten thousand *what?*")

Amplitude is a measurement of how intense the changes in air pressure (or any back-and-forth motions) are. A mechanical example of amplitude (and frequency, for that matter) is a playground swing: the greater the distance the swing moves back and forth, the larger the amplitude of its motion. The length of time it takes to complete one back-and-forth cycle is its frequency.

With sound, the greater the change in air pressure, the louder the sound will seem to be. You may have heard the term *sound pressure level* (SPL), which is a scale used to measure the loudness of sound caused by these variations in air pressure.

Although a swing, sound waves and all such *oscillating* motions have both amplitude and frequency these properties are independent from each other. You can have low-frequency/high-amplitude sounds or high-frequency/low-amplitude sounds, etc., in any combination.

Be aware that a third component called *Phase* exists in sound waves and other "oscillating systems." A pair of playground swings moving in opposite directions is an example of differences in phase.

Another extremely important property of sound is called *resonance*. It describes the frequency at which a thing vibrates most efficiently, the so-called *frequency of resonance*.

Both phase and resonance are worthy of your attention, but are sadly beyond the scope of this book (you could at least *pretend* to be disappointed!).

But seriously, if you are into learning more about sound and acoustics, go for it—it's fascinating. On the other hand, if you're just trying to get your mixer set up so you can play music, you can function in blissful ignorance of these topics.

Sound Into Electrical Signals

We just briefly explored the realm of sound in air. But mixers work with electrical signals, not changing air pressure. Therefore, the atmospheric vibrations that make up sound must be transformed into electrical signals that your mixer can accommodate. We call these electrical impulses *audio signals*.

After running through your sound system, the audio signals must again be converted into changes in air pressure for human ears to hear the result. Converting air pressure variations to electrical signals is accomplished using microphones; speakers convert the signals back to changes in air pressure.

Inside a microphone, there's a little diaphragm that is sensitive to fluctuations in air pressure, just like your ear drum. But instead of converting these pressure changes to nerve impulses, as the ear ultimately does, a microphone creates an electrical signal that wiggles up and down, tracking the changes in air pressure. This electrical signal has nearly identical *frequency* and *amplitude* characteristics to those present in the sound waves it's picking up.

These audio signals are used throughout mixing boards and the rest of your sound system. They can be stored and replayed by audio tape recorders, and manipulated by your mixer's controls and other audio equipment.

Although all sound equipment works with audio signals, some audio signals are weaker, while others are stronger.

For example, the electrical signal that drives the hundreds of speakers at rock concerts is very powerful—hundreds or even thousands of *watts*, enough juice to run a house! On the other hand, the electrical signal produced by a microphone isn't strong enough to power a penlight.

higher strength audio signals

Speaker-Level

Line-Level

Mic-Level

lower strength audio signals

Audio signal levels generally fall into one of these ranges. Your mixer's trim control can compensate for most of these variations, but speaker-level signals should never be connected to your Mackie mixer.

The range of signals connected to mixing boards isn't quite as dramatic. Typical mixers are capable of dealing with low "mic-level" signals like those from microphones, instrument pickups and the like, as well as higher "line level" signals from equipment like tape recorders, synthesizers, effects units and so on.

Part of any mixing board's job is to bring these different signals into a roughly equal operating range. Your mixer's *trim* control is used to compensate for the varying levels of these different sources.

One last term to get acquainted with is *waveform*. A waveform is a visual representation of a sound wave, tracing its frequency and amplitude. You've probably seen an *oscilloscope* display, which is a piece of test equipment that draws waveforms from audio and other electronic signals, much like the little green squiggles showing the patient's heartbeat on countless TV medical shows.

Electrical Signals Back Into Sound

After passing through your mixer and the rest of your sound system, audio signals are turned back into sound waves by loudspeakers. Speakers (and headphones) are the reverse of microphones—they take an audio signal and convert it into similarly varying sound waves. However, this is a pretty inefficient process, so extra power is required. The *amplitude* (here's another chance to use your new buzzwords) of the audio signal is boosted by an amplifier, or *amplified*, in order to compensate for the inefficiencies of loudspeakers. This allows the reproduced sound to be as loud as, or louder, than the original sound.

Although good microphones and speakers do an excellent job of converting sound waves to and from audio signals, they do tend to change or "color" the sound. That, plus the fact that sound quality is a matter of perception and taste, explains the endless debate over the merits of different mics and speakers.

So, although this book is mostly about mixing boards, try not to lose sight of the big signal path: from sound source, through your entire sound system to the ears of your listeners. Each stage of the process is important! Ignore any part of the entire system at your peril!

CHAPTER 3

Mixer Input Concepts

This and the following seven chapters will introduce you to all the controls on your mixer, what each is for, and how to use them. All mixers, from the simplest to the most complex, share many common features. Getting a clear understanding of even one small mixer (like a 1202) can help prepare you for dealing with almost any mixer, however large.

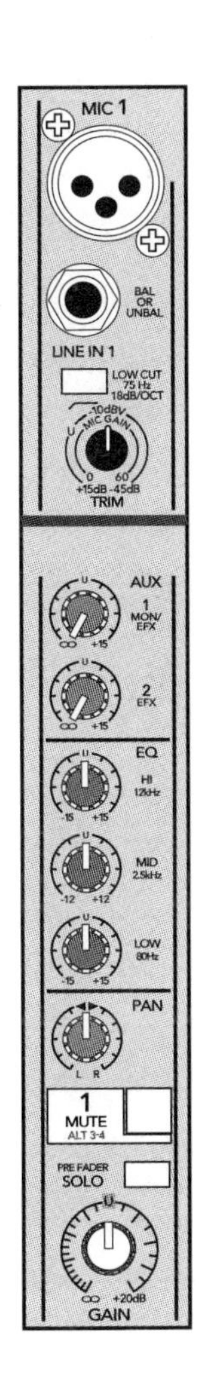

Because mixers have a lot of different controls, we're going to break things up into four nice, manageable chunks: Input Channels, Output Sections, Inserts and Aux Sends, and finally, Signal Routing and Bus Assigns.

Each of these four sections is further divided into two chapters. The first contains an explanation of each type of control found in that mixer section. Regardless of what model mixer you own, this information is for you!

Following those generic concepts is another chapter detailing how each model Mackie implements that set of mixer controls. When you reach these mixer-specific chapters, you'll probably want to just read about *your* Mackie. If you are considering a new mixer, you can browse through the description of each model to get an idea of their similarities and differences.

As was repeated ad nauseam in the previous chapter, the path audio signals follow from input to output is called signal flow. When considering signal flow, it's tempting to assume that the signal flow follows the same physical sequence of the controls on your mixer's front panel. Tempting, but wrong! In fact, the *external* physical layout and *internal* signal flow are often surprisingly different. Since your understanding of the *internal* signal flow is so important, each mixer-specific section will include a special block diagram illustrating the signal flow of one section of your mixer, in addition to a picture of how the front-panel controls are arranged.

So, take a deep breath and we'll begin. Eight chapters might seem like a lot, but you'll breeze through it, I promise.

Input Channel Road Map

Understanding how your input channels work is critical if you want to use your mixer to its full potential. While each Mackie model's signal path is a little different, a "generic" signal path through an input channel might look like this:

Mic/Line in—Trim—Low cut —Insert—EQ—Volume—Pan

There are other elements, but these seven items are the most important. The following section will walk you briefly through each. We won't go into all the gory details—those will come next—but you will get a brief idea of what happens on each step of your signal's journey through an input channel.

Before we begin, let's make one very important point. As we have discussed, signal flow is like a river. Anything that gets dumped into the water upstream affects everything downstream. The same is true for the signal path through an input channel. Any adjustments made affect all subsequent parts of the channel's signal path.

Mic and Line Inputs

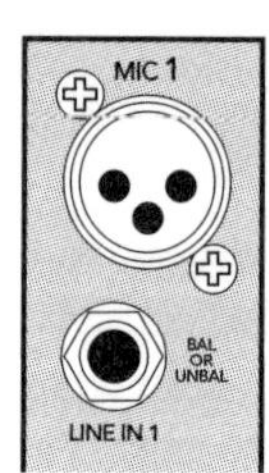

As you can see in the illustration, a typical channel begins with a pair of inputs, one XLR mic input and one 1/4" line input. In order to accommodate higher-strength signals, the line input is immediately followed by a little electronic circuit, called an "attenuator," that "pads" or reduces the level of that signal slightly.

Trim

The next stage in the input channel is a circuit that boosts the level of the input signal. The amount of boost is controlled by a knob called "Trim." PPM series mixers label this control "Input Level Set" while DFX mixers' are marked "Gain."

Your mixer's Trim control is used to set the amount of signal boost that is being added to each mic (and in many cases, line input) connected to your mixer. *Failure to set trims correctly will result in a distorted or excessively noisy sound!*

Low Cut Filter

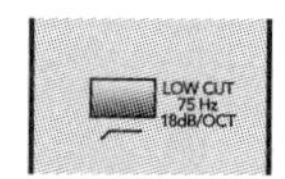

Many Mackies provide a Low cut filter (sometimes called "high pass") early in the channel's signal path. Controlled by a button on each channel, this filter cuts out the lowest of the low frequencies. You might not want to use this button on a channel connected to an upright bass, but you'd certainly want to press it in for a mic picking up a mandolin or fiddle. *Getting rid of unnecessary lows in your sound improves clarity and reduces distortion!*

The most common use of this control is on vocal microphones. When mics are hand-held, most will pick up a certain amount of "handling noise" as the performer rubs, jostles or otherwise bumps the mic in his or her hands. The most distracting components of this handling noise are very deep, low frequency tones. By engaging the low cut switch, these irritating sounds can be suppressed.

Even when the mic is placed on a mic stand, it's not immune from getting bounced around. Wooden stages are often noisy—as performers walk across the stage, their footfalls resonate across and beneath the platform. These vibrations are readily transmitted through a mic stand into a mic, and voila—more handling noise, even though no one is touching the mic!

An additional benefit of the Low cut switch on vocals is the reduction in popping "plosive" vocal noises. These are the loud thumps heard on "P" and "B" sounds, such as the phrase "put Peter's peanuts in a poly-propylene package!" While a Low cut switch won't make these sounds go away entirely, like handling noise, the low frequency components of these popping sounds are the most noticeable and distracting. The Low cut switch will tame them substantially.

The moral of the story: Start by engaging the Low cut switch on each and every vocal mic in your setup. Of course, you can certainly turn it off on any particular mic you wish.

Insert/Channel Access

Most Mackies provide "Insert" points or "Channel Access" connectors on at least some input channels. A channel insert provides you a way to connect an external signal processing device to an individual mixer input channel. Compressors and sophisticated equalizers are two devices commonly used with mixer inserts. In this case, the signal leaves the mixer channel through the insert, is processed by the external device, and then is returned back into the same input channel to continue its journey through the signal path of the mixer.

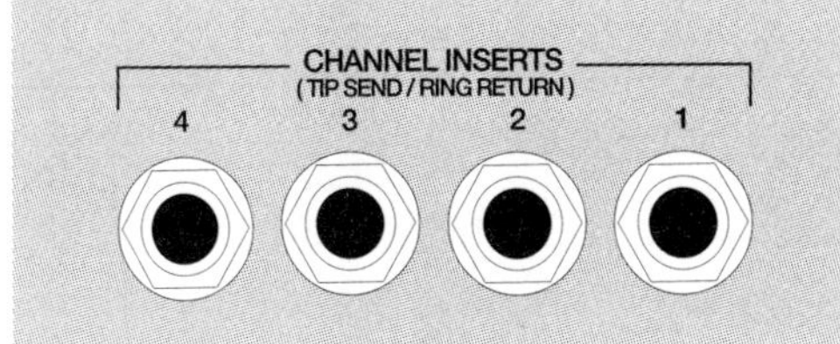

The other main purpose of an insert point is to take a given channel's signal and connect it to another device, such as the input of a multitrack recorder. In this case, the signal may exit the mixer at the insert, and...sniff, sniff...never return! OK, I'll try not to get teary-eyed—we all have to leave sometime. Actually, it's a very practical option if you want to put a single mic's sound on an individual track of a multitrack recorder. We'll talk a lot more about using inserts in Chapter 7, Auxes, Inserts and Effects.

Note that the position of the insert point in the signal path varies among different Mackie models. For instance, on the 1604-VLZ, SR- and CFX-Series mixers, the Insert comes *after* the Low cut switch. And, on the classic CR-1604, the insert comes much, much later in the signal path—*after* the channel's main volume control. But remember—all the specific details about your mixer start in the next chapter.

EQ

The next stop in the signal path is the EQ section. These controls let you contour the character of each individual input's sound, making it brighter, fuller, darker, etc. Even though the Low cut filter comes earlier in the signal path, it will interact with the channel EQ section. For instance, you can engage a channel's Low cut filter, yet boost that channel's Low knob. This can provide a nice plump tone, but leaves out the deepest lows, which helps avoid distortion.

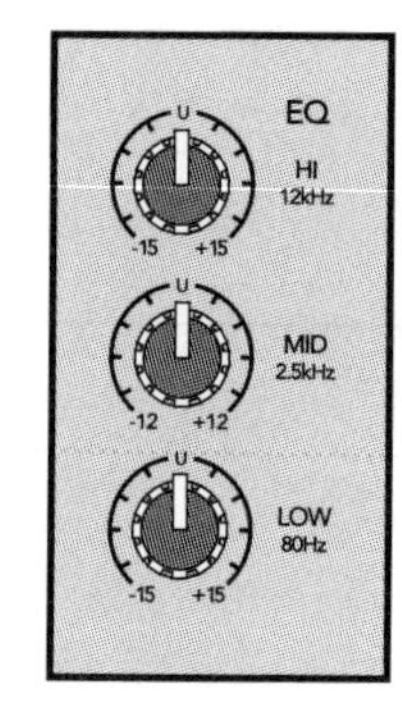

Note that the EQ comes *after* the insert point. This means that if you're using a channel's insert to connect to an external device, such as a multitrack recorder, *changing the EQ controls will have no effect on the recorded signal!* Why? Think about the signal flow—the signal that is being tapped off at the insert at a point "upstream" of the EQ. Changes to the EQ controls affect the signal at all points *after* the EQ, but have no affect on the signal *before* the EQ section of the input channel.

Channel Volume Fader

Next, our signal passes through the channel's "fader," or volume control. When you set this control at the "U," or nominal mid-point position, it isn't doing anything to the level of the sound. When you turn this control down, the sound gets softer and eventually goes away completely. Turning it up past the U adds more volume to that sound. (For you trivia buffs, the U stands for unity. The phrase "unity gain" is a technical term which means a signal has passed through a circuit with its level unchanged. Aren't you glad you asked?)

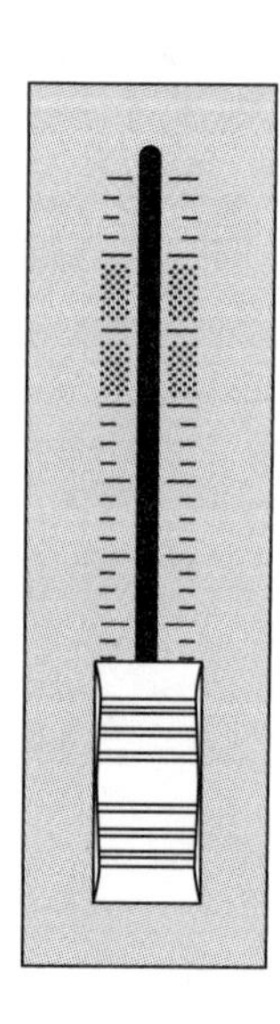

Don't confuse the function of the Trim and Channel Volume Fader controls. The fader should be used for mixing; in other words, adjust it based on your aesthetic judgment about how loud this particular sound should be. The Trim control should be set correctly once, and then changed only when necessary. A properly set Trim will avoid distortion and excessive noise in a given channel's signal path. If you set the trim too high, distortion will result near the beginning of the input channel's signal path. Adjusting the channel level control (which is near the *end* of the channel's signal path) won't make trim-induced distortion go away!

Pan

After the channel fader comes the Pan control. As you probably know, when this knob is in its center position, that channel's sound will appear to emanate from the center, or mid-point between a pair of stereo loudspeakers. Turning the knob off-center will move the apparent sound of the sound in that direction. This illusion of changing positions is achieved by simply turning down the level of the sound in one speaker while simultaneously turning it up in the other. Assuming that you are sitting in between the two speakers, your brain will perceive pan control adjustments as movement of the sound between the two speakers.

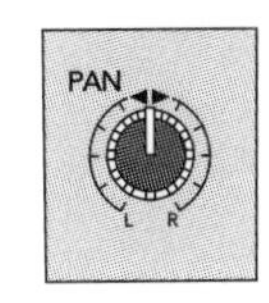

Pan is the most common example of a "routing" control. By twiddling this knob, you direct, or "route" the signal between two outputs—left and right. We'll be talking a lot more about routing in Chapter 5.

The Gory Details: Input Channels

Now that we've taken a quick sniff at the components of an input channel, we are going to dig a little bit deeper into this important topic. Get out your shovel.

Inputs

The fun with any mixer begins at its inputs! As was mentioned in the Mixer Basics chapter, mixers can accept two basic kinds of inputs: microphone and line inputs.

Before connecting anything to a mixer input, make sure the level control of the channel you are connecting is turned down to avoid an unexpected loud noise. A little buzz or pop here or there won't hurt, but very loud sounds can damage your speakers, or, worse yet, your ears!

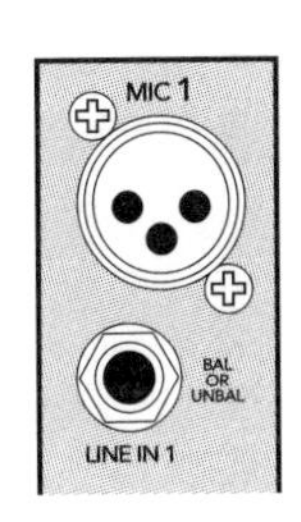

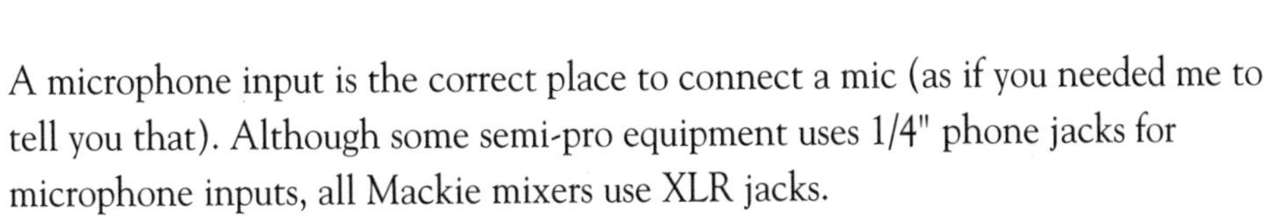

A microphone input is the correct place to connect a mic (as if you needed me to tell you that). Although some semi-pro equipment uses 1/4" phone jacks for microphone inputs, all Mackie mixers use XLR jacks.

Mackie Line inputs use 1/4" jacks. Line inputs are the appropriate jacks for connecting the outputs of analog or digital recorders, other mixer line-level outputs, electronic keyboards, effects units, drum machines and similar equipment.

Never connect amplifier outputs to mixer line inputs, since these signals are much too high in level! You should only connect line-level mixer *outputs* to amplifier *inputs*. The same is true for guitar amplifiers. If a guitar amp has a *line out* or *direct out*, it's OK to connect them to a mixer input, because a line or direct output will come before the power amp in the unit's signal path. But *never* connect a guitar amp's *speaker* output to a mixer (or any other) line input.

Microphones and Phantom Power

Phantom power is used in conjunction with a type of microphone known as a *condenser* mic. Condenser microphones need power to operate. Some can use batteries, but phantom power is a more reliable method, as batteries eventually run down.

Your Mackie can provide phantom power to mics connected to its XLR mic inputs. On most Mackies, phantom power is turned on or off for all XLR mic inputs from a switch next to the main power on/off control.

It's called phantom power because the voltage supplied to the mics travels on the mic cable itself, without any extra wires. Turning on the Phantom switch causes 48 volts of DC (direct current) to appear at all Mackie XLR Mic inputs, but not the TRS 1/4" line inputs (which means you can plug into the 1/4" line inputs without worrying about whether phantom is on or off).

Many mics don't need phantom power. For instance, members of the very common *dynamic* microphone family don't need it, nor do the relatively rare *ribbon* mics. However, it's OK to use a mix of phantom and non-phantom powered mics at once, even though this means phantom is being delivered to mics that don't need it (there are a few mics that should *never* be connected to phantom power. See your Mackie manual for details).

NOTE: If you plug equipment *other than microphones and devices that require phantom power* into Mackie XLR mic inputs, check before turning on the phantom power switch! As a general rule, connectors labeled *direct out* from the back of instrument amplifiers are fine, but XLR *line outs* from tape decks, other mixing boards or electronic musical instruments may not be. When in doubt, contact the manufacturer in question or better still, use adapters to patch these devices into 1/4" TRS line inputs.

Line Inputs

Equipment such as CD players, electronic keyboards, drum machines and audio recorders, to name a few, have line-level outputs. These output signals are significantly higher in level than those from microphones or instrument pickups. Line level outputs should be connected to line level inputs, not to mic-level inputs.

Mackie line inputs use 1/4" jacks. Some models offer balanced inputs; all will allow the connection of balanced or unbalanced sources. If you have balanced line-level outputs on any other equipment, it's a good idea to connect them to balanced inputs (using a balanced "tip-ring-sleeve" cable!). Using a balanced connection between two audio devices reduces the chances of humming, buzzing and picking up stray radio-frequency interference. The full benefit can only be realized when both the input and output offer balanced connections and when you use a balanced cable to connect them. I hasten to add that you can certainly hook up a system using unbalanced connections and do just fine. However, your system will be more susceptible to the kinds of interference balanced systems handily reject.

You'll find that while most equipment uses 1/4" connectors for line outputs, some gear uses XLR connectors for line-level outputs. When connecting equipment with XLR balanced line outputs to your Mackie mixer, you should use adapters to convert the XLR connectors to 1/4" (TRS) connectors. While it's tempting to plug the XLRs right into the XLR Mic inputs, the levels are usually too high. There is also some risk of equipment damage if your Mackie's phantom power switch is turned on when devices other than microphones are connected to XLR inputs.

The international "no" symbol reminds us not to connect a mic and line input to the same channel, as you'll have no way to control their levels independently.

Finally, although many Mackie channels have both a line and mic input, simultaneously connecting a mic and line signal to the same channel is not recommended, as you'll have no way to independently control the two sound sources.

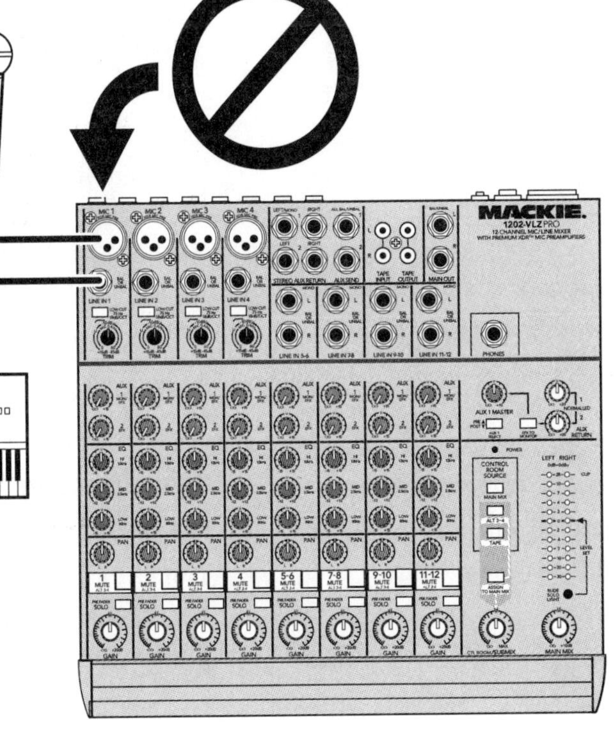

higher strength audio signals
Speaker-Level
Line-Level, Pro (+4)
Line-Level, Consumer (-10)
Mic-Level
lower strength audio signals

Remember this little chart from Chapter 2? Stop the presses! It turns out that there are two distinct types of line-level signals. Read all about them in the paragraph at right.

-10 Vs. +4 Line-Level Signals

OK, so we've talked about how the output from a microphone is at a much lower signal level than a line-output from a CD player, audio recorder, effects unit, etc. But there is one more little distinction to be aware of: There are actually two different references for line-level signals, one of which is significantly higher in level than the other.

There are two families of audio equipment: "pro" and "consumer/semi-pro." While both are "compatible" in the sense that they each process the same analog signals, there is a problem when interconnecting devices from these two families: Consumer gear operates at a much lower average level than pro gear.

If you were to take a consumer hi-fi cassette deck and play a tape that pushed the deck's meters right up to "0," where the red part of the scale started, you could measure the actual level, or voltage, of the signal at the deck's output connector. And you'd get a reading of -10 dBV (trust me).

If you took a professional recorder and did the same test, you'd find that when its meters reached the red line, the output signal level measured +4 dBu. As a result of this level mis-match, connecting both types of equipment in the same system can cause problems. If you take the output of a -10 piece of gear and patch it into something expecting a +4 level input, the sound may be too soft, even when turned up all the way. The reverse is also a problem, because the output of a +4 device may cause distortion in the input of a device designed for -10 level signals.

Some devices can operate at either signal level by including a little switch near the input or output jacks marked "-10/+4." Toggling the switch between these two positions calibrates that device's inputs or outputs to conform to one or the other operating level standard. Alternatively, there are stand-alone products designed to perform this "level-match," which allow -10 and +4 gear to operate correctly with each other. When possible, try to choose components that all use the same reference level—all +4 or all -10.

Note: -10 and +4 signals are both referred to generically as "line-level" signals, even though +4 level signals are of significantly higher amplitude.

XLR to 1/4" tip-ring-sleeve adapters should be used when connecting the XLR line outputs from another piece of gear to your Mackie's line inputs. Avoid connecting XLR line outputs to Mackie XLR mic-level inputs.

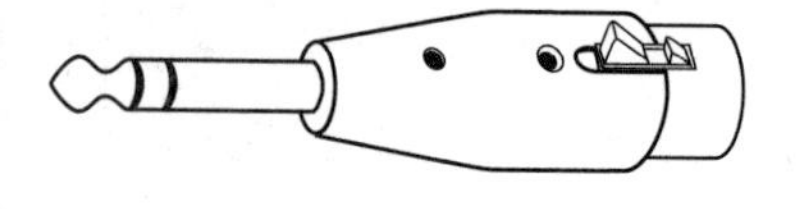

Connecting XLR Line-Level Signals to Your Mackie

When working with professional audio equipment, it's common to find XLR connectors used for line-level inputs and outputs, as well as for microphone connections. XLR line outputs are almost always +4 level signals (the difference between +4 and -10 level signals was explained in the preceding section).

When using equipment with XLR line outputs, your eyes, seeking the shortest distance between two points, will light upon the XLR inputs of your Mackie mixer channels. While it's tempting to plug XLR line outs from other equipment into Mackie XLR mic inputs, this is almost always a bad idea!! Instead, you'll be better off using adapters and plugging XLR line outputs into your mixer's TRS 1/4" line inputs.

Here's why: Microphones need a level boost to bring them up to an operating level that can match the line-level signals running around in your mixer. The circuitry behind your Mackie's XLR inputs, called the *microphone pre-amplifier,* provides this extra gain. So, if you plug a line-level signal into an XLR input, this line level signal is still getting the boost intended for microphones. This boost can be enough to cause line outputs to distort your Mackie's input channel (there are also phantom power concerns to worry about, as discussed a moment ago).

Instead, line outputs from other gear should be connected to 1/4" TRS jacks on the input channels, because these have a little circuit that reduces (or "pads" or "attenuates") the level of signals connected to it. Using an adapter from an XLR to *three conductor* TRS 1/4" connector will preserve the advantages of a balanced connection. Using an XLR-to-unbalanced 1/4" tip-sleeve (TS or "guitar cord") connector will also avoid the possibility of distortion, but won't provide a balanced connection.

A stereo input channel has left and right line inputs.

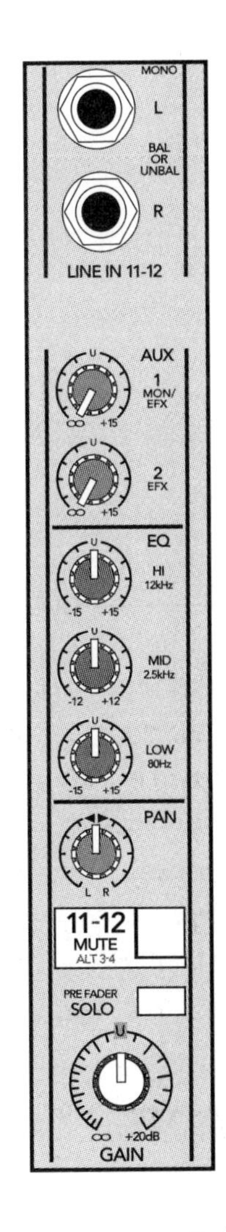

Stereo Line Inputs

Many Mackies include stereo line input channels. The need for stereo inputs has grown with the popularity of devices featuring stereo *outputs* like CD players, synthesizers, effects units and drum machines. If you have gear with stereo outs, a stereo input channel is their perfect mate.

Each stereo input channel strip can control a pair of audio signals, so while there are two inputs, there is only one set of Gain, Pan, EQ and Send controls. This reduction in knobs not only simplifies operation; it saves space and reduces its cost to boot! When a stereo input channel is used, its controls affect both channels in exactly the same way. For example, if you turn up the Gain control, both left and right inputs get equally louder. This is usually exactly what you want, since changing the sound of just half of a stereo pair may lead to odd results.

But what if you need to change the pan position of one part of a stereo signal? For instance, say you have a stereo drum machine connected to a stereo line input, and you want to move the pan position of one of the cymbals left or right. In this case, you'll have to make the adjustment on the drum machine itself—there's no way to pan an individual element of a blended stereo input from any mixing board's controls. Sorry!

The same is true if you connect two unrelated mono signals to a single stereo input channel. The two signals will appear "hard left" and "hard right" in the main stereo output. The Pan control will make one or the other inputs softer, or even turn it off completely when panning to either extreme, but the two individual sounds will always stay at the extreme left and right ends of the stereo spectrum.

Connecting a Mono Source to a Stereo Input

If you just want to connect a mono line-level signal to a stereo input channel, go right ahead. When only plugging a cord into the Left channel input, the stereo channel acts just like a mono channel, including the behavior of the Pan control. In other words, your lonely mono input won't be plastered up against the left side of the stereo image; it can be panned freely across the stereo field.

Does it seem odd that the stereo input works one way with two cables plugged in, but another way when only one connection is made? If you possess an inquisitive mind, understanding how this works will help you understand a very important piece of studio equipment, called the *patch bay*. You'll read more about patch bays at the end of Chapter 12.

Understanding Trims

The Trim control will have the *single biggest effect on overall sound quality* of any knob on your mixer. Set it correctly and all will be well. Set it too high and you'll get ugly, obvious distortion that will make your system sound "broken." Set trims too low and you'll get a lot more background hiss and noise in your final sound. (Note that Trim is called "Input Level Set" on PPM-series mixers and "Gain" on DFX-model mixers.)

As mentioned earlier, mic-level signals are much weaker (or, more accurately, have a much lower *amplitude*) than line-level ones. So, to effectively combine mic- and line-level audio signals within a mixer, they must be brought to roughly equal levels.

Why is this so? Here's an analogy: Imagine you are having a conversation with someone at a party. If they whisper, you're going to have trouble separating their voice from the background noise. On the other hand, if they yell their voice will crack (and you may want to cover your ears).

A similar situation exists within the signal path of every audio system. As you go from one section of a mixer to the next (or connect between devices in an audio system), you must set the output and input levels of each for the best compromise between noise (too little signal) and distortion (too much signal).

So, Mackie mixers have high-quality mic pre-amplifiers which add enough level boost, or *gain*, to bring a mic signal up to the board's internal operating level.

The Trim control sets the amount of gain created by the mic preamp. The Trim control is a continuously variable knob and not just a Mic/Line switch, because you will encounter a wide range of input signal levels in normal use. The overall process of optimizing signal levels throughout your mixer is often referred to as "gain structure." We'll be talking more about this soon.

Trims and Line Inputs

From our earlier discussion of channel signal flow, you know that both line and mic signals pass through the same trim control. However, there are some Mackie mixers that include line inputs without trims. For example, on the classic MS1202, trim controls are found only on inputs 1-4. On these channels, the Trim control affects either the XLR mic input, or the 1/4" line input, whichever is connected. 1202 inputs 5-12 are designed only for line-level signals and don't have trim controls. However, the 1202's gain controls have enough latitude to compensate for higher or lower than usual line-level sources.

The 1402-VLZ uses push-button switches to optimize its line inputs for -10 (consumer) or +4 (professional) line-level signals. While a switch doesn't offer the range of operation provided by a knob, these two choices are usually all you need.

The 1604 has trim controls on its mic/line inputs (1-6) as well as its line-only inputs (7-16). This trim control allows you to optimize the performance of lower line-level sources, without having to boost the fader to make up the difference.

However, a line-level input may not need to be boosted. By setting the Trim control to its "U" position (fully counter-clockwise), the signal's level is not increased. There's a name for this special condition: When a gain-providing circuit is set for zero boost, it's called a *unity gain* setting. Unity gain is an important concept we'll return to later.

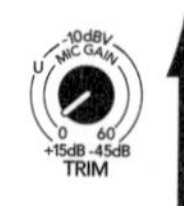

The correct trim control settings move down as signal levels move up. Pro line-level signals require low trim settings, while lower level sources, such as microphones, require higher trim settings. A loud singer will require a lower trim setting than a softer singer, assuming they are using the same model microphone.

Setting Trims

There are basically two ways to set trim controls—*quick 'n' dirty* and *The Mackie Way*. The Mackie method, which will provide the best performance, is detailed in your Mackie mixer manual. Still, there is a reason to try out the "unofficial" method, as you will soon see...

But first, the purpose of setting trims is to get a channel's input signal as loud as possible, *without distortion*. If you aren't familiar with exactly what *distortion* sounds like, setting trims is a great way to get acquainted. It's a harsh, crackling sound that gets worse as a signal gets louder.

Distortion occurs when an electronic circuit is trying to reproduce an audio signal with too high an amplitude. The result is that the peaks of the audio waveform get flattened off, or "clipped." This clipping changes, or *distorts*, the sound. The greater the clipping, the worse the sound. Proper trim settings avoid this problem.

Setting your trims properly will also ensure that you get the least amount of background hiss in your mixes, a blissful state called the "optimum signal to noise ratio."

QUICK AND DIRTY TRIM SETTING

Set your mixer's master volume control at a comfortably low level, because this procedure can get noisy. (If you use headphones, turn them down too.) First, position the mic at the source of the sound it's going to be reproducing. For instance, if you are setting the trim for your vocal mic, hold the mic up to your mouth.

Next, make the loudest possible sound you realistically expect this mic to be subjected to. While yelping into the mic, you must also be listening to the sound quality of your yelp as reproduced by your system. Having a friend listen can help.

If you hear any *distortion*, the trim is set too high. Reduce it slightly. Ignore the strange looks from others in the room and continue yelping into the mic. Conversely, if you don't hear any distortion, turn up the trim until you do. Then, back it off a little lower than the point where the signal is starting to distort. Many, but not all Mackies include a little red LED lamp that blinks when distortion is happening. If you see this light illuminate as you bark into the mic, turn the trim knob down for that channel.

OK, stop yelping! Repeat this procedure for each mixer channel, using the instrument or singer that will be connected to each input during the upcoming performance or recording.

While the official method detailed in your Mackie manual is a more robust way to go about setting trims, there's an important reason to be familiar with this quick 'n' dirty method: It teaches you to use your *ears*, rather than your eyes, when adjusting controls. It's very important to be able to recognize the sound of a signal that's just on the edge of distortion. When you're in the middle of a mix (especially a live one), you need to be able to quickly identify *which* input is distorting, and correct it. Your ears are the best tool for that job.

Be aware that even if you set each individual channel's trim to a point below clipping, it's still possible to cause distortion in the main stereo mix bus if you're running a lot of channel volume faders high. This is why the "unity gain" trim-setting procedure recommended in your Mackie manual is more conservative than the quick 'n' dirty method described here. Following Mackie's guideline will ensure that you avoid overloading the mix bus as well.

One last note about listening while twiddling knobs: It's very, very easy for your eyes to trick your brain into hearing something you really aren't. For instance, you could watch your fingers nudging up a volume fader and "hear" the sound getting louder, only to realize a moment later that you were turning up the *wrong knob!* This is an even easier mistake when making more subtle adjustments, like EQ. Believe me, pros make this mistake every day. **Get in the habit of trusting your ears.**

The "Right" Volume Setting

Everyone's familiar with volume knobs. In your car, at the TV or behind the world's biggest mixing board, volume is volume. Turn it up, the sound gets louder. Turn it down far enough and you won't hear it at all. The difference with mixing boards is that instead of one volume control, you have a dozen or more.

Each signal coming into the board has its own level adjustment. It's this *individual* control that makes mixing possible, as each sound's level is balanced relative to all others to create a particular blend or *mix* of sound.

In addition to the individual level controls on each mixer input channel, there is also a Master output volume. This control is so named because it sets an overall level of the combined sound from all the individual input channels. In practice, this means that in order to hear any sound, the master volume must be turned up, at least a little.

If you turn a volume control up far enough, it may begin to distort. How far is too far? If the trim control is set properly you will be safe from distortion if you keep your levels at or below the "U" or unity position. This position is the point at which the volume control has no effect on the signal, meaning it's not making it any softer, nor is it making it any louder than the level set by the trim control. Mackie made this important position easy to find on some mixers by adding the little click or *detent* to level controls.

Don't feel obliged to keep your level controls at their Unity position, but if most of your channel volume faders are *well* above or below this mark, you should probably change your master volume to a position that allows more faders to be set closer to unity.

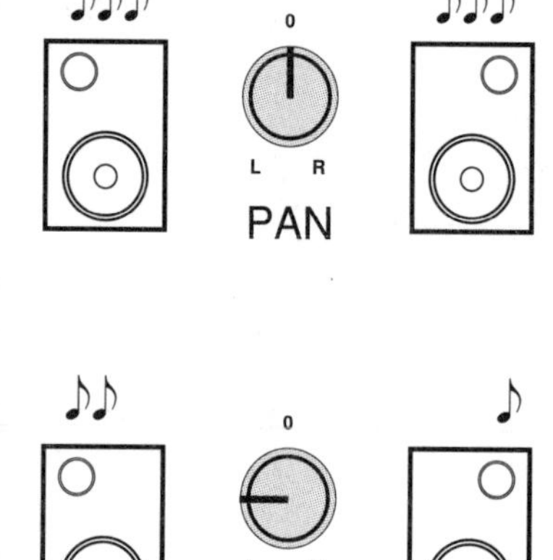

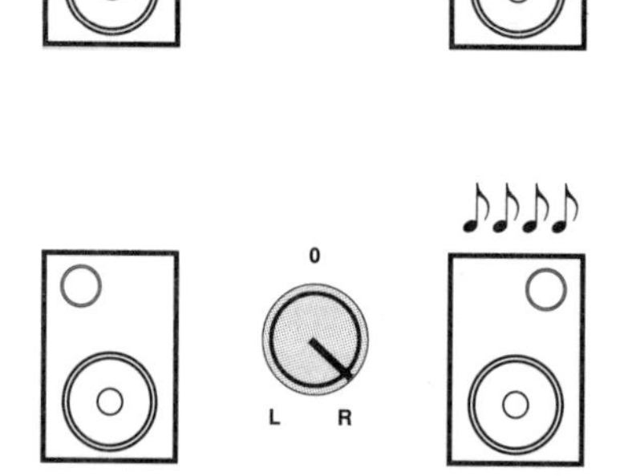

When a Pan control is turned more to the left (counter-clockwise) or right, that channel's sound will be made louder in one speaker than the other. The level difference between speakers causes the sound's apparent location to shift in the "stereo image." In the above illustration, relative level is shown by the number of musical notes above each speaker.

Alternately, re-check your trim settings, as mistakes here will force you to compensate with extreme level settings that are likely to decrease the overall fidelity of your mix. Maintaining proper trim settings and keeping your average fader levels around unity will also help you avoid the situation where you are mixing with a lot of faders at the very top or bottom of their travel. This becomes awkward when you need to turn something up or down but can't because the control is already at (or near) one extreme or the other.

The Pan Control

As you recall from our earlier discussion of signal routing and mix buses, the Pan control normally takes a single mono input and varies the amount of that signal that is routed to the left and right stereo outputs.

When you sit between the left and right speakers (or wear headphones), sounds panned to the center will appear to be coming from a point between the two speakers. When a Pan control is turned more to the left (counter-clockwise), that sound will be louder in the left speaker than in the right. The opposite is true when panning to the right. When a sound is louder in one speaker, we perceive the "source" or "position" of that sound to be coming from a point closer to that speaker's location.

To better understand how the Pan control works, let's look at a simple block diagram (facing page), which illustrates the pan control's effect on signal flow. As you can see, the Pan knob lets you direct a given input channel to either the left, right, or both outputs. This is our first example of a signal routing control. We'll cover more routing controls in a later chapter.

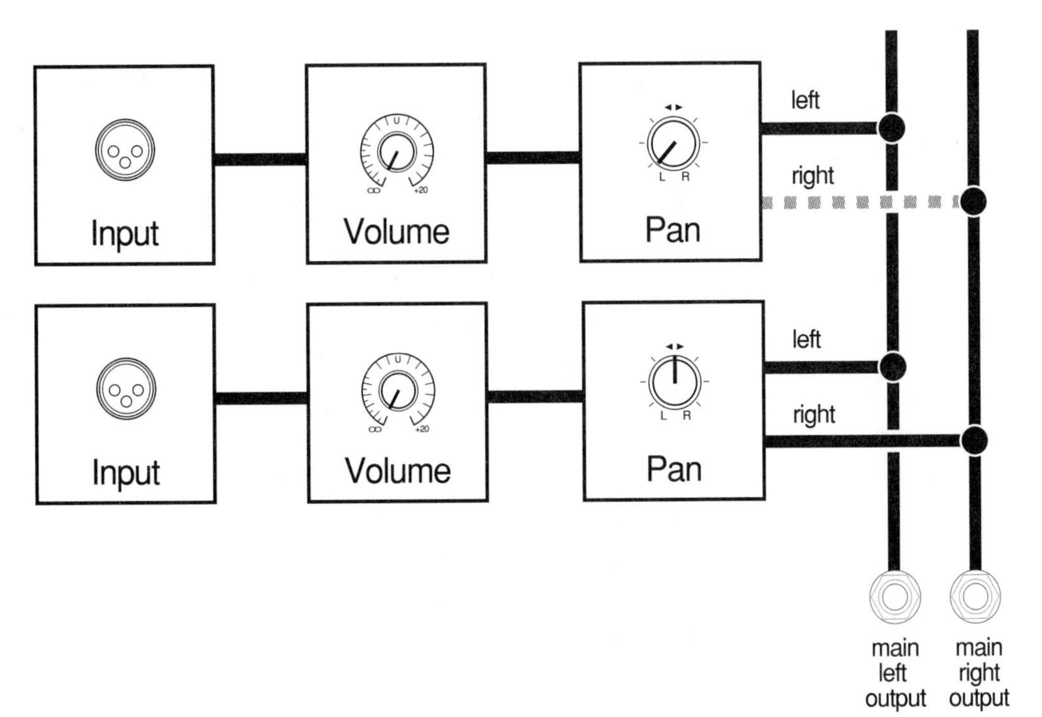

Located at or near the end of each channel's signal path, the Pan control sets that signal's balance between left and right outputs. Here, the upper input channel is panned fully-left. This prevents any of that signal from reaching the right output, shown here as a dashed line.

The Pan Control and Stereo Inputs

The Pan control on stereo input channels works a little differently. Instead of one signal to deal with, this time the Pan control has two separate signals to balance. In this case, it simply lowers the level of one channel or the other as it is turned away from its center detent position, just like a BALANCE control on a home stereo. As an aside, the classic CR-1604 mixer has four knobs actually labeled "balance" in its AUX Return section. These behave the same way as the stereo channel pans, as long as their Mono buttons aren't used. With Mono in, they behave like regular channel pans. (More information on stereo input channels appears earlier in this chapter.)

Solo

Many (but not all) Mackies feature Solo buttons on some input channels and aux returns. But what is solo, anyway? Well, it's not the button you press when the guitar player steps up to the front of the stage and goes *wheedl-ee-wheedl-ee-whoo-whoo!* Instead, the Solo button allows you to hear individual inputs by themselves; hence the term, "soloed."

In a live performance situation, soloed sounds are usually listened to using headphones—you pop on the phones, make sure the Solo/Phones slider isn't up too far, and depress one or more Solo buttons. In a studio context, you can use headphones or your main speakers to listen to solo'd sounds.

Mixers with Solo buttons may have a master Solo level control. For instance, the CR-1604 has a SOLO/PHONES fader that sets both the overall headphone and solo levels.

If you want to hear several channels at once, simply press any desired combination of Solo buttons. When you are done, push each Solo button again so it pops up, and the stereo mix will return to your headphones or monitor speakers. Note that when one or more channels are soloed, the Rude Solo Light on your mixer will blink. Only when all channels have been unsoloed will this rudeness cease. Remember that inadvertantly soloing a channel that is turned down or unused will silence your entire mixer. This can sometimes fool you into thinking that something isn't working right. The Rude Solo Light is a good thing to check if you suddenly can't figure out why your mixer isn't making a sound!

Although we haven't talked much about effects yet, here's a little hint: If you want to solo a signal *and* its effects (like reverb that is routed to the Aux return section), solo both the individual input channel you are interested in, *and* the Aux Solo button (not available on the 1202, 1402 or PPM series mixers). Note that you'll also hear the effected versions of all channels being sent to that effects unit, not just the channels you are soloing.

TYPES OF SOLO: AFL AND PFL

There are two ways to implement a solo control. They both do more or less the same thing, which is to give you a shortcut way of hearing just the signal passing through one or more solo'd channels. The key difference is this: At what point in each channel's signal path does the solo shortcut take you to? When the Solo button is pressed, are you listening to that signal at a point before the channel volume fader or after the fader? A pre-fader listen (PFL, for short) means that a solo'd channel is heard at the same volume regardless of the setting of its channel volume fader. You will also hear a PFL solo'd channel panned to the center, regardless of where that channel's pan knob is set.

In contrast, an after-fader listen (AFL) taps its signal after the fader, often after the pan control. This means you'll hear an AFL solo'd channel at the same relative level it was in the overall mix and you will also hear the sound panned left or right in the stereo field. If you try to AFL-solo a channel with its level all the way down, you won't hear anything! Note that AFL-style solo is also referred to as "solo-in-place," or SIP on some model Mackies.

Some Mackies offer both types of solo, others provide one or the other. We'll cover the specific solo implementation of each mixer in the next chapter.

Mute

Most Mackies include Mute buttons. The purpose of these controls is to shut off individual channels without disturbing the setting of their level knobs. After all, if you had set up a nice mix between half a dozen channels, and just wanted to turn off the mics temporarily, it would be a shame to have to turn all their volume controls off, thus losing the balance you had set up. To mute a channel, simply press the Mute button. To restore its operation, press the button again. It's always a good idea to Mute unused microphones. Doing so reduces sonic clutter in recordings and also minimizes the chance for feedback at live events.

On most of the smaller Mackies, the Mute button does double-duty as the ALT 3/4 assign switch. ALT outputs are covered in the routing chapter. If you don't have anything connected to the ALT output jacks of your CR-1604, 1202- or 1402-VLZ mixer, you can ignore the ALT part of the Mute button's functionality and just think of it as a plain old single-purpose Mute control.

While most Mackies offer independent mutes for each input channel, the PPM-series mixers have a single button, labeled "Break (mutes channels 1-6)." As the name suggests, this is the button to push when you leave the stage so that over-eager members of the audience can't hop up and start screaming through your microphones.

Understanding Equalization

Almost as familiar as a volume control are the "tone" knobs, the inseparable pair of bass and treble, or their alter egos, low and high.

Usually called EQ (short for frequency EQualization), these controls have a bit in common with plain volume controls. EQ controls also turn levels up and down. The difference is that they only affect a part of a sound—for instance, the lows or the highs. This is done using a special circuit that affects only a particular range of frequencies. But why would you want to affect "a range of frequencies?" Because sounds are usually made up of many different frequencies, and the balance of these frequencies determines that sound's character.

Obviously, an instrument that can play chords is able to generate a number of different frequencies at once. However, even instruments that can only play one note at a time are still capable of generating multiple frequencies simultaneously. Here's why: All sounds have *at least* one frequency. In a musical context, this primary, or *fundamental*, frequency corresponds to the pitch being played.

Additional frequencies are present as well. In pitched sounds, these *harmonics* or *overtones* appear at multiples of the fundamental. For instance, if an "A" at 440 Hz is being played, harmonics may appear at 880 (2 x 440), 1320 (3x 440), 1760 (4 x 440) Hz and so on, up to (and even beyond) the range of human hearing.

Not all sounds contain all possible harmonics. For instance, some pitched sounds have more prominent even-numbered harmonics, while others feature richer odd-numbered harmonics. Non-pitched sounds, like cymbals, have harmonics that aren't obviously related to the fundamental, which is why these sounds don't have a single recognizable pitch. But in all these cases, it's this harmonic profile that gives individual instruments their tonal character.

Even the harmonic content of the *same instrument* can change dramatically depending upon how it is played. Playing harder usually increases the level of higher harmonics, which is another way of saying that when you play harder, the sound gets brighter as well as louder. You'll also hear different harmonic contents depending on your position relative to the instrument. For instance, standing right in front of a trumpet's bell lets you hear a lot more high harmonics than standing off to the side or behind the player. This is one reason why microphone placement is such a critical and tricky issue.

Don't panic! You don't need to memorize these details—it's enough that you know harmonics exist, and that each sound is made up of its own unique combination of frequencies. Let's get back to the topic of EQ.

Low and High EQ

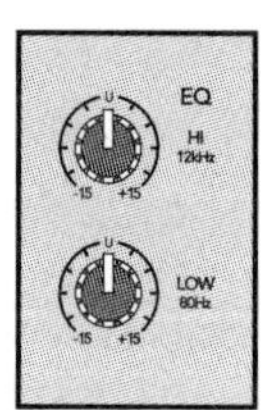

One might imagine that a control that selectively cuts or boosts a limited range of frequencies (harmonics) would allow you to alter the harmonic balance, and therefore the character of a sound. *And that is exactly what EQ does.*

To use EQ effectively, you must learn to identify, by ear, roughly what kind of frequencies are present in a particular sound. For instance, a tambourine, with its jangling metal disks, generates a lot of higher frequencies when struck. So, your Mackie's High control, which affects frequencies at 12 kHz and above, could be used to exaggerate or subdue that instrument's bright character.

Another example would be a bass drum. While bass drums do generate higher frequencies, the Low EQ, which affects frequencies at 80 Hz and below, would be highly effective at pumping up the bass drum's low frequency "kick," or perhaps cutting back the bass on an overly muffled or "rumbly" tone.

This brings up an important point: While it's easier to hear the effects of an EQ boost, you're cheating yourself out of half your EQ power if you don't use them to cut, too. These controls do nothing when left in the center detent position. Turning them up (clockwise) boosts a range of frequencies, while turning the knob counter-clockwise *cuts them.*

Also, excessive EQ boost can lead to distortion, even if your trim controls are set properly. Whenever I EQ a sound, I start by trying to identify any frequency ranges that seem too prominent, and cut them. Then, if some other part of the sound still seems too soft, it's time for the boost.

Unfortunately, this sort of detailed EQ often falls beyond the range of a simple two-band EQ circuit like the one included in the original MS1202. This is because, while the extreme lows and highs are covered by the two EQ knobs, there is a great "midrange" of frequencies between them that the MS1202 can't touch without using an external equalizer.

MID EQ

Both Mackie's High and Low EQ are called *shelving* equalizers, because they affect all frequencies above or below a particular point. Most Mackies include a third EQ control to allow independent access to these "midrange" frequencies. The MID control can add or remove the aggressive "bite" of sounds like distorted electric guitars or snare drums, bring out the fingerboard "growl" of an upright bass or subdue a shrill vocal. Because the MID control affects only a specific band of frequencies, it is called a *peaking* equalizer.

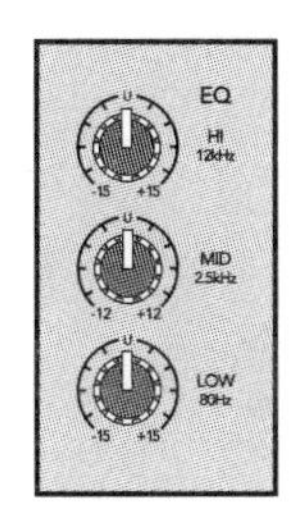

As is the case with the Low and High EQs, it's important to remember that the MID control can cut as well as boost sounds. For instance, if an acoustic guitar pickup seems too brittle or harsh, try turning down the midrange before turning up the bass.

The frequencies affected by the MID control are centered at about 2.5 kHz. This frequency range is where the "crunch" from electric guitars, the "snap" of snare drums and the sibilant sounds of speech live. It's also the part of the frequency range our hearing is most sensitive to (a topic we'll return to shortly). But first, many Mackies have a second knob that works in conjunction with the MID control. Let's check that out now.

Mid Frequency

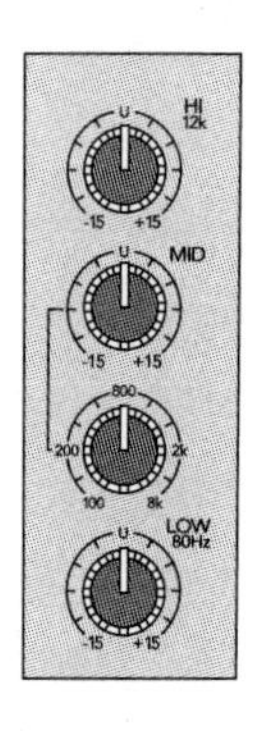

The 1604-VLZ, CFX- and SR-series mixers have a fourth knob in their channel EQ sections. This control changes the "center frequency" that the MID knob affects most strongly. In other words, rather than having a midrange control that can only cut or boost frequencies around 2.5kHz, the MID Frequency control lets you "sweep" the operation of the MID knob to higher or lower frequencies, from as low as 100Hz up to 8kHz. (A familiar example is the wah-wah pedal, which creates a midrange boost who's frequency is swept up and down by the action of a foot pedal.)

It's important to note that the frequency control only *modifies* the operation of the MID knob. If the midrange control is in its Unity (12 o'clock) position, twirling the frequency knob will have no effect on the sound. Only when the MID knob is cut or boosted will changes to the frequency control have any audible result.

Many people are intimidated by the flexibility of a swept midrange. It was hard enough deciding which of three EQ knobs to mess with! How will I choose the frequency that one of them operates at?!? Relax. If your life already requires as many decisions as you can handle, simply set the frequency control at 2.5kHz (around 3 o'clock). This is a good general-purpose default setting.

But really, it's not that hard to learn to use the frequency control. A good way to get familiar with it is to run some music from a CD player or similar source, boost the MID knob and then sweep the frequency control from low to high while listening to the music. At the low end of the frequency knob, you'll hear the bass and drums (if present) go from "thumpy" to "woofy." As you move from about 300 Hz up to 800Hz, the tone will go from boxy to nasal. In the 1.5kHz to 5kHz range the sound will be "edgy," turning to "shiny" or "airy" as you move up to 8kHz.

At either extreme of this "swept-mid" frequency range, the mid EQ will partially overlap the fixed frequency low and high EQ points. Bear this in mind if you're trying to boost the mids at 100Hz while also cutting the lows at 80Hz. It's possible this would give you just the effect you're looking for, as opposite settings of the two controls won't fully cancel each other out, but their interaction will result in a bumpy EQ curve, which may not be what you had in mind.

Some day, you'll probably run into an EQ with three knobs per band. A "parametric" equalizer has the same cut/boost and frequency knobs we've just covered. But a parametric EQ has a third knob called "bandwidth" or "Q." This broadens or narrows the "width" of the mid control, giving you the option of a broad, smooth boost across several octaves, or a super narrow peak or notch.

As a rule-of-thumb, narrow bandwidth settings are best for corrective frequency cutting—reducing feedback in a live system, notching out the fundamental of 60Hz hum or taming a resonant peak in an acoustic instrument's response. Broader Q settings are more useful for overall tonal shaping. Mackie's midrange bandwidth uses a fairly wide Q.

Some Mackies also include a "graphic equalizer" in their output sections. Since this chapter deals with mixer input sections, rather than outputs, we'll wait until Chapter 9 before we return to a discussion of graphic EQs. Feel free to skip ahead if the suspense is too much for you.

The "Right" EQ setting

While our hearing is remarkable, it's far from perfect. For example, our perception of a sound's character or frequency balance changes as the sound gets louder or softer (named the Fletcher-Munson Effect after the researchers that quantified this phenomenon).

Specifically, we tend to hear low frequencies as being relatively softer when listening at lower volume levels, *compared to the same mix played back louder over the same equipment*. (This is why "louder" always seems to sound "better." It also explains why most hi-fi systems have *loudness* buttons—it compensates for lower listening levels by boosting the lows and highs).

Because of our level-dependent perception of sound, it's often helpful to judge your EQ decision by listening to the same settings at different volume levels. Refining your EQ and level decisions after these kinds of listening experiments will help your music mixes sound good when played back on a variety of speaker systems and at varying playback volume levels.

It also turns out that our hearing is most sensitive to midrange frequencies, typically from 2 kHz to 4 kHz. Not coincidentally, acute hearing in this range is critical to understanding speech. As mentioned earlier, this is also the tonal range most affected by the fixed frequency MID knob on many Mackie mixers.

Remember that no two people hear things exactly the same way. This partly explains why getting "good" sound is such a challenging endeavor.

CHAPTER 4

Mackie Input Channels

Many mixers have a single type of input channel. In other words, although the number of individual channels vary, each is identical to all the others. However, this isn't always the case. For instance, many Mackies have mostly mono input channels, but include a few stereo channels, too. In other cases, some input channels include mic inputs, but other have only line-ins. Finally, identical looking channels (as far as their front panel layout is concerned) might have unique features, for instance, the addition of channel inserts or direct outputs that are present on some, but not all channels.

The tabs along the edges of each page will help you quickly find the information related to your model Mackie mixer.

In this chapter, we're going to look at the input sections of a number of different model Mackies. Note that we will be touching on some features (Aux sends, bus assigns and inserts, for example) that we haven't fully explained. Fear not, we'll delve into those features in following chapters.

There are two things you should learn from this chapter: The features your particular mixer's inputs include and the order in which an input signal flows through the input channel. If you don't understand what a particular control is for, you'll find a full explanation of each control type in the previous chapter.

The following specific models are covered in this book:

- MS1202
- CR-1604
- 1202-VLZ and 1202-VLZ PRO
- 1402-VLZ and 1402-VLZ PRO
- 1604-VLZ and VLZ PRO
- 1642-VLZ PRO
- SR24•4 and 32•4
- PPM Series
- CFX Series
- DFX Series **(All DFX-specific information can be found in Chapter 14)**

Even if your mixer isn't in the above list, the general information provided throughout this book still applies to you! You should also track down a copy of your mixer's manual, which will describe any unique features of your board. Somewhere near the back of the manual, you'll probably find a signal flow block diagram. Stare at this until it makes sense, or your eyes glaze over—it's the treasure map in your quest to fully understand *your* mixer.

Key to Illustrations

Before you turn to the pages describing your mixer's input channel setup, a word about how to read these examples and those in following chapters is in order. Each model (or series) of mixers has its own section. Where models in a series differ only in their number of input channels, all models will be covered in the same section. This includes the SR24•4 / SR32•4, as well as the CFX- and PPM-series mixers. On the other hand, the 1202- and 1402-VLZs, while similar, are different enough to warrant individual coverage.

Each section will open with a picture of an input channel as it appears on the mixer's front panel. A second diagram will also be shown. This block diagram represents the internal signal flow within the channel. Each labeled box corresponds to a control (or group of controls) on your mixer.

It's just as important to understand the internal signal flow as it is to remember, for instance, that the aux sends are positioned above the EQ. Knowing the signal flow of your mixer will help you troubleshoot problems and come up with creative solutions that help you get the maximum value from your mixer.

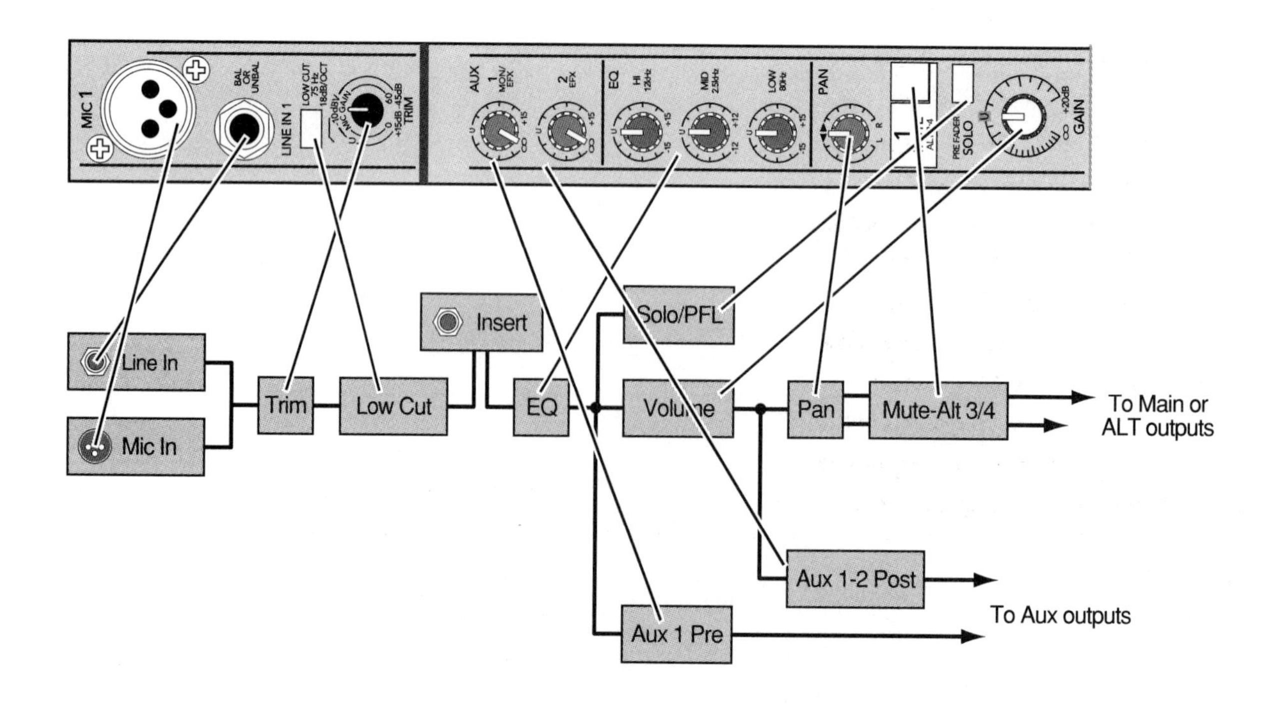

As you can see in the illustration, the physical sequence of controls on the mixer's front panel is similar, but not identical to the actual signal flow within the mixer's channel. By following the little interconnecting lines, you can see that the box labeled "Trim" in the signal flow diagram (bottom) corresponds to the trim knob called out in the mixer front panel image (top).

To avoid visual clutter, you won't see all the little lines between the two diagrams, but remember that the two drawings are different views of the same thing—a mixer input channel.

By convention, the signal flows through the block diagram from left to right, from input to output. In the few cases where the direction of signal flow in part of a diagram might be ambiguous, a little tiny arrow will point in the direction of signal flow. Remember that signal flow within the mixer is a one-way affair. There is no way for a signal to "leak" back upstream.

Finally, this section only shows the mixer input channels, which means there are other mixer sections that are *not* shown in these diagrams. You'll notice larger black arrows pointing off from various points along the right edge of the diagrams. These indicate that the signal at that point in the mixer is flowing off to another section of the mixer not shown in the illustration. We'll pick up where each of these arrows leaves off in the next few chapters, so stay tuned.

Throughout upcoming chapters, you'll find many block diagrams showing signal flow through your mixer. In the illustration at left, you can see how physical knobs on the mixer front panel relate to the labeled blocks in the diagram. These little connecting lines won't be visible in later illustrations, but they are still present, printed in very expensive (but non-toxic) imaginary ink.

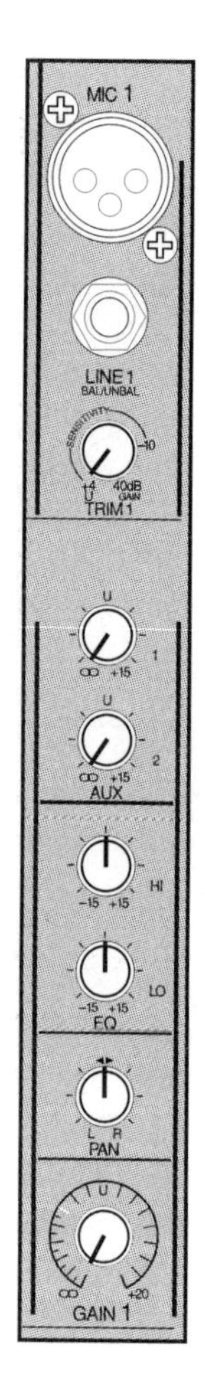

MS1202

The Mackie MS1202 is a pretty straightforward mixer with two different types of input channels. Inputs 1-4 are mono, mic/line inputs, while the next four input channel strips are stereo line inputs (inputs 5-12). Let's look at mic/line input channels 1-4 first.

MIC/LINE INPUTS (1-4)

Input channels 1-4 of the MS1202 have balanced XLR mic and 1/4" balanced line inputs. Any microphone can be plugged into one of these four XLR inputs, including those which require *phantom power* (described in Chapter 3). Although channels 1-4 of the MS1202 have both a line and mic input, simultaneously connecting a mic and line signal to the same channel is not recommended, as this leads to unpredictable results which may make you crabby.

The block diagram below shows the signal flow within the first four channels of the MS1202. A single Trim control affects the gain applied to both the line and mic input. This is followed by an *insert point*. These connectors give you access to the individual signal path within each of these four channels. You'll learn more about Inserts in Chapter 7. Note that insert points appear only on the first four input channels of this mixer; stereo inputs 5-12 don't include this feature.

Briefly, the insert jack allows you to connect, or "insert," an external signal processor into the signal path of a given channel. Compressors and equalizers are common examples. Inserting such devices provides more sophisticated control than the mixer's built-in processing can offer. If nothing is connected to an insert jack, the signal automatically flows around it; in this case, from the Trim to the Channel Volume Fader.

The MS1202 mic/line input channel (above) and corresponding signal flow diagram (right).

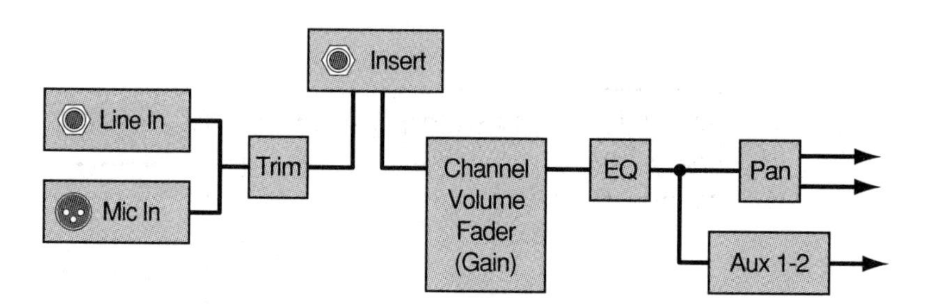

Following the Insert in the block diagram comes the Channel Volume Fader and EQ. Aux sends 1 & 2 follow the EQ and channel, and are therefore called post-fader, post-EQ sends. This type of send is ideal for use with effects, but limited when trying to create a stage monitor mix. We'll talk more about Aux sends in Chapter 7. Also following the EQ is the channel Pan control. From there, the channel's input signal (which is now stereo) moves on to the output section of the mixer.

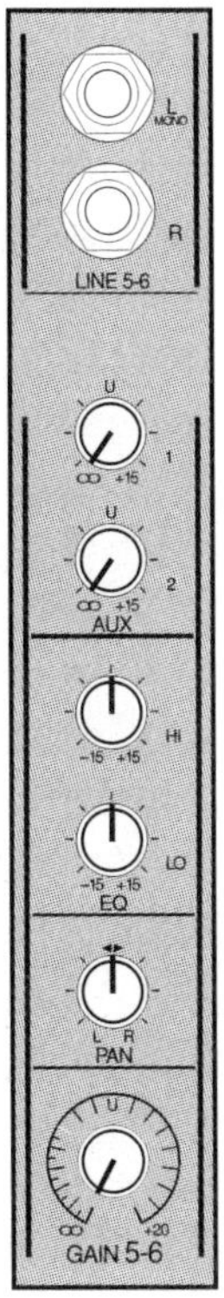

STEREO LINE INPUTS (5-12)

Inputs 5-12 are configured as four stereo unbalanced line inputs without mic input connectors. When a stereo input channel is used, its controls affect both channels in exactly the same way. For example, if you turn up the Gain control, both Left and Right inputs get equally louder. This is usually exactly what you want, since changing the sound of just half of a stereo pair may lead to odd results. If you plug a signal source into just the Left/Mono channel input, the stereo channel acts just like a mono channel.

As you can see, the block diagram of the MS1202's stereo input channels are simpler than the signal flow found in the first four channels. There is no mic input, nor is there a Trim control or Insert point. The arrangement of the Channel Volume Fader, EQ, Aux sends 1-2 and the Pan control are essentially the same, except for the fact that separate left and right signal paths are maintained from the channel input to the mixer output section (not shown in this illustration).

The only tricky item to note is that Aux 1-2 tap off a mono signal from both Left and Right inputs. This means that any effects from this channel will be applied to all sounds connected to either the left or right stereo input, regardless of the setting of the Pan control.

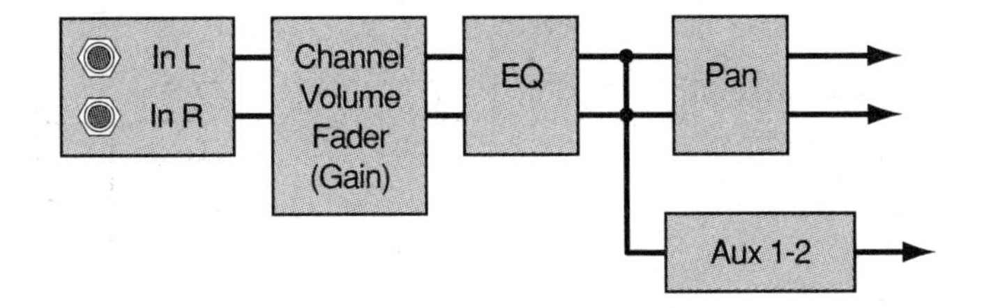

The MS1202 stereo line input channel (above) and corresponding signal flow diagram (left).

1202-VLZ & VLZ PRO

Like its predecessor the MS1202, the 1202-VLZ (and VLZ PRO) has four mono mic/line inputs and another four stereo line inputs. However, the VLZ model incorporates many additional features; we'll cover all those related to the input channels here.

While the 1202-VLZ is a physically small and economically-priced mixer, its first four channels have most of the features included on Mackie's higher-end models. Let's go through the list, following the order of the signal flow diagram shown below. Keep referring back to the block diagram as you read through this section; it's a lot easier to understand with the illustration as your guide.

The line and mic inputs both run through the same Trim control. (Connecting line level signals to the XLR mic input can result in distortion; connect line-level signals to the 1/4" Line connector instead.) Once past the Trim knob, your mixer doesn't care if the source signal came from the line or mic input; it treats them both the same.

Next stop is the Low cut switch. While you can consider the low cut a conceptual part of the EQ section, you can see it is a separate animal, both from a front panel layout and signal flow standpoint. By placing the low cut *before* the channel insert, you can prevent extreme lows from getting into any device you connect to your mixer's insert points. This can be especially helpful when using a compressor, since low frequency thumps can cause compressors to overly-squash things like vocals (you'll learn all about inserts in Chapter 7). Briefly, the insert jack allows you to connect, or "insert," an external signal processor into the signal path of a given channel. Compressors and equalizers are common examples. Inserting such devices provides more sophisticated control than the mixer's built-in processing can offer. If nothing is connected to an insert jack, the signal automatically flows around it; in this case, from the low cut switch to the EQ section.

The 1202-VLZ mic/line input channel (above) and corresponding signal flow diagram (right).

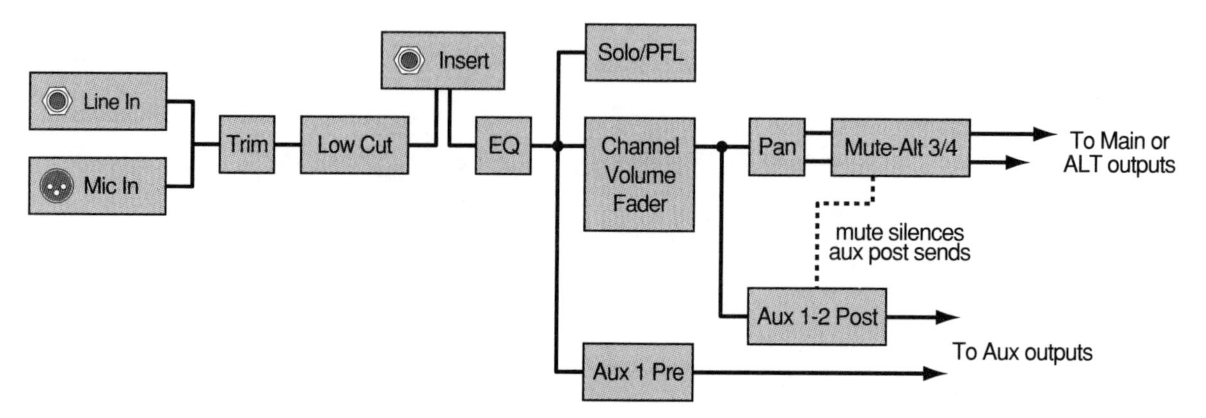

The EQ section consists of three controls, low and high shelving and a midrange peak/dip EQ. This 3-band EQ is essentially the same as found on the original CR-1604, 1402-VLZ and PPM series mixers.

Following the EQ is a major intersection in the signal flow. When you hit a channel's Solo button, this is the place in the signal flow you will hear. It's after the EQ, so your EQ settings will be heard when soloing, but it comes *before* the Fader and Pan controls, so the solo signal will ignore the settings of these knobs. Solo will also ignore the position of the Mute-ALT 3/4 switch.

After the EQ is also the point in the circuit from which Aux 1 taps it's signal, *if* the Aux 1 select button is in its "Mon/PRE" up position. In this case, Aux 1 is a *pre-fader, post-EQ* send. The Mon/PRE switch is a global control affecting all channels. There is no way for some channels to use Aux 1 in pre-fader and others in post-fader modes.

Finally, the output of the EQ section feeds the Channel Volume Fader and continues on through the rest of the input channel. As the signal leaves the Channel Volume Fader, it again splits in two directions. Aux 2 always takes its signal from this point in the circuit, making it a *post-fader, post-EQ* send. Aux 1 also takes its feed from this point *if* the Aux 1 select button is in its "POST," down position.

The signal leaving the Channel Volume Fader also feeds into the Pan control, and then the Mute-ALT 3/4 switch. You already know what a mute does; we'll talk about the functionality of the ALT 3/4 part of the equation in Chapter 5.

A special note: Pressing the Mute-ALT 3/4 button on any channel also mutes that channel's post-fader Aux sends (shown by a dotted line in the illustration). The only exception to this is aux 1 when set for pre-fader operation. In this case, the Mute button has no effect on the signal sent through aux 1.

(See next page for 1202-VLZ stereo input channel)

1202-VLZ (continued)

STEREO LINE INPUTS (5-12)

The remaining four input channel strips each control a stereo pair of inputs. As you can see in the block diagram, the stereo input channels are simpler than the signal flow found in the first four channels. There is no mic input, nor is there a Trim control, Low cut or Insert point. The arrangement of the Channel Volume Fader, EQ, Solo Aux sends 1-2 and the Pan control are essentially the same, except for the fact that separate left and right signal paths are maintained from the channel input to the mixer output section. Note that all 1202-VLZ channel line inputs are balanced connectors, which support the connection of balanced or unbalanced equipment.

The only tricky item to note is that Aux 1-2 tap off a mono signal from both left and right inputs. This means that any effects from this channel will be applied to all sounds in the stereo input, regardless of their placement in the stereo image.

As was described in the mic/line input channel section on the previous page, pressing the Mute-ALT 3/4 button on any channel also mutes that channel's post-fader aux sends (shown by a dotted line in the illustration).

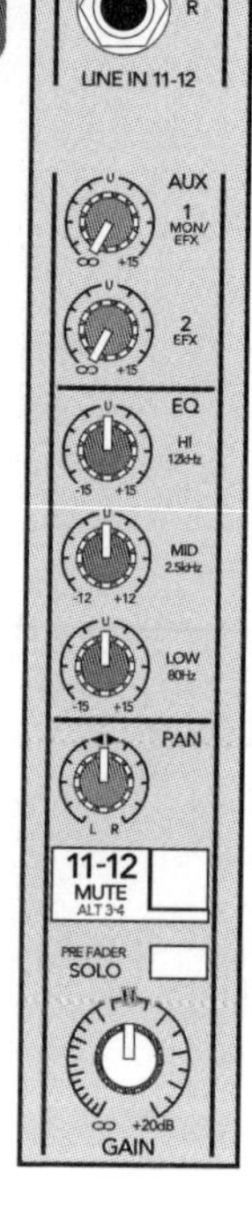

The 1202-VLZ stereo line input channel (above) and corresponding signal flow diagram (right).

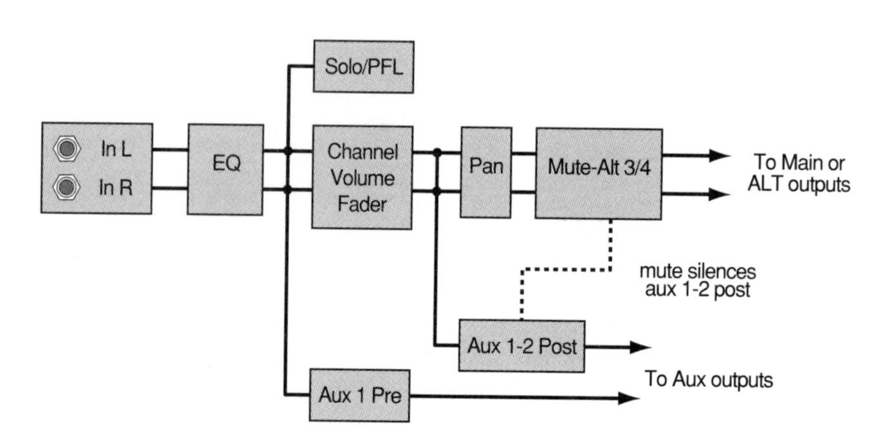

1402-VLZ & VLZ PRO

The 1402-VLZ was designed to fill the gap between the 1202 and 1604 mixers. From a features standpoint, however, it has more in common with the 1202-VLZ than the larger 1604-VLZ model. The main difference between 1402 and 1202-VLZ is the addition of two mic/line input channels and sliders, rather than knobs for the Channel Volume Fader controls.

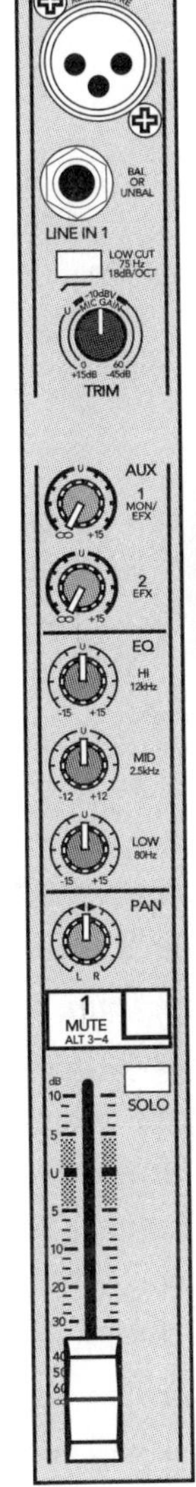

MIC LINE INPUTS (1-6)

This section describes the features of the first six input channels. Let's go through them one at a time, following the order of the signal flow diagram shown below. Keep referring back to the block diagram as you read through this section; it's a lot easier to understand with the illustration as your guide.

The line and mic inputs both run through the same Trim control. (Connecting line level signals to the XLR mic input can result in distortion; connect line-level signals to the 1/4" Line connector instead.) Once past the Trim knob, your mixer doesn't care if the source signal came from the line or mic input; it treats them both the same.

Next stop is the Low cut switch. While you can consider the low cut a conceptual part of the EQ section, you can see it is a separate animal, both from a front panel layout and signal flow standpoint. By placing the low cut *before* the channel insert, you can prevent extreme lows from getting into any device you connect to your mixer's insert points. This can be especially helpful when inserting a compressor, since low frequency thumps can cause compressors to overly-squash things like vocals (more about inserts coming up in Chapter 7). Briefly, the insert jack allows you to connect, or "insert," an external signal processor into the signal path of a given channel. Compressors and equalizers are common examples. Inserting such devices provides more sophisticated control than the mixer's built-in processing can offer. If nothing is connected to an insert jack, the signal automatically flows around it; in this case, from the Low cut switch to the EQ.

The 1402-VLZ mic/line input channel (above) and corresponding signal flow diagram (left).

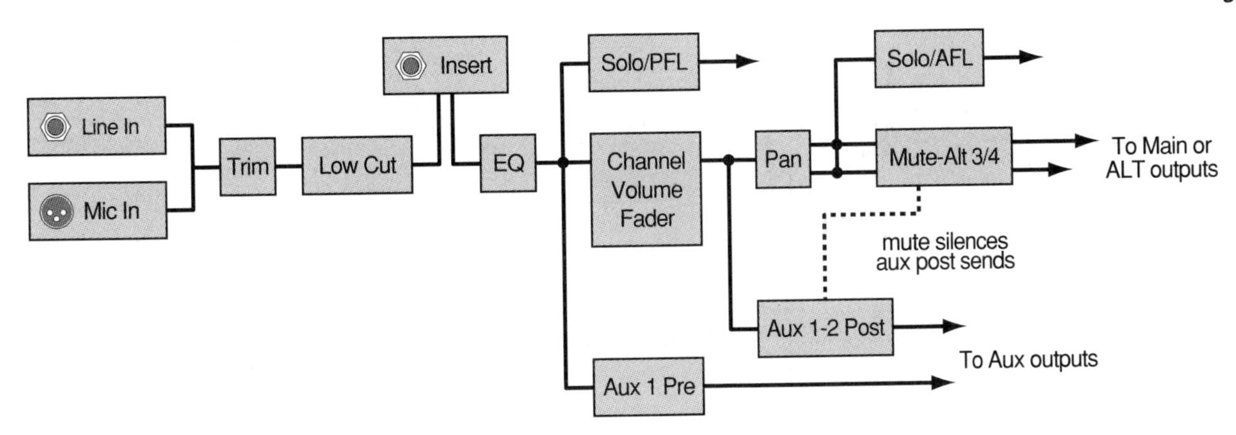

After the Insert comes the EQ section. This consists of three controls, low and high shelving and a midrange peak/dip EQ. This 3-band EQ is essentially the same as found on the original CR-1604, 1202-VLZ and PPM series mixers.

Following the EQ is a major intersection in the signal flow. When you hit a channel's Solo button, this is the place in the signal flow you will hear. It's after the EQ, so your EQ settings will be heard when soloing, but it comes *before* the fader and pan controls, so the solo signal will ignore the settings of these knobs. Solo will also ignore the position of the Mute-ALT 3/4 switch.

After the EQ is also the point in the circuit from which Aux 1 taps its signal *if* the Aux 1 Select button is in its "PRE," or up position. In this case, Aux 1 is a *pre-fader, post-EQ* send.

Finally, the output of the EQ section feeds the Channel Volume Fader and continues on through the rest of the input channel. As the signal leaves the Channel Volume Fader, it again splits in two directions. Aux 2 always takes its signal from this point in the circuit, making it a *post-fader, post-EQ* send. Aux 1 also takes its feed from this point *if* the Aux 1 Select button is in its "POST," or down position.

Here's the 1402-VLZ mic-line input channel signal flow diagram, reprinted from the previous page. This duplicate illustration will reduce (but not eliminate) the chance of repetitive stress injury (RSI) that might otherwise occur from turning back to the preceding page while reading the text on this page. Fight RSI! Avoid unnecessary page-turning! Thank you.

The signal leaving the Channel Volume Fader also feeds into the Pan control, followed by the tap for the AFL-mode solo. Next comes the Mute-ALT 3/4 switch. You already know what a Mute does; we'll talk about the functionality of the ALT 3/4 part of the equation in Chapter 5.

A special note: Pressing the Mute-ALT 3/4 button on any channel also mutes that channel's post-fader Aux sends (shown by a dotted line in the illustration). The only exception to this is Aux 1 when set for pre-fader operation. In this case, the Mute button has no effect on the signal sent through Aux 1.

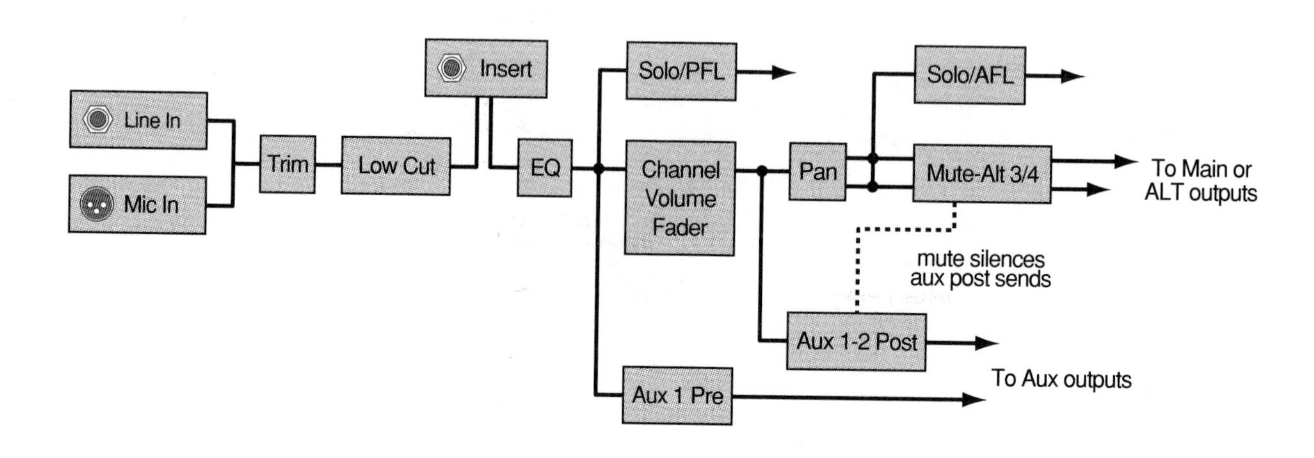

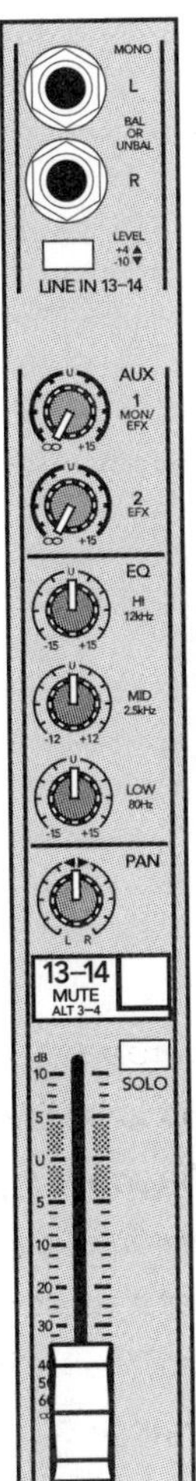

STEREO LINE INPUTS (7-14)

The remaining four input channel strips each control a stereo pair of inputs. As you can see in the block diagram, the stereo input channels are simpler than the signal flow found in the first four channels. There is no mic input, nor is there a low cut or insert point.

The Trim control of the first six channels has been replaced with a switch that selects either -10 or +4 line level operation. Choose the -10 position for consumer hi-fi equipment and most electronic musical instruments. The +4 position should be used when the line output of professional gear (pro tape decks, high-end signal processing, mic preamplifier outputs, etc.). Note that all 1402-VLZ channel line inputs are balanced connectors, which support the connection of balanced or unbalanced equipment.

The arrangement of the Channel Volume Fader, EQ, Aux sends 1-2 and the Pan control are essentially the same, except for the fact that separate left and right signal paths are maintained from the channel input to the mixer output section.

The only tricky item to note is that Aux 1-2 tap off a mono signal from both left and right inputs. This means that any effects from this channel will be applied to all sounds in the stereo input, regardless of their placement in the stereo image.

As was described in the mic/line input channel section on the previous page, pressing the Mute-ALT 3/4 button on any channel also mutes that channel's Aux post-fader sends (shown by a dotted line in the illustration).

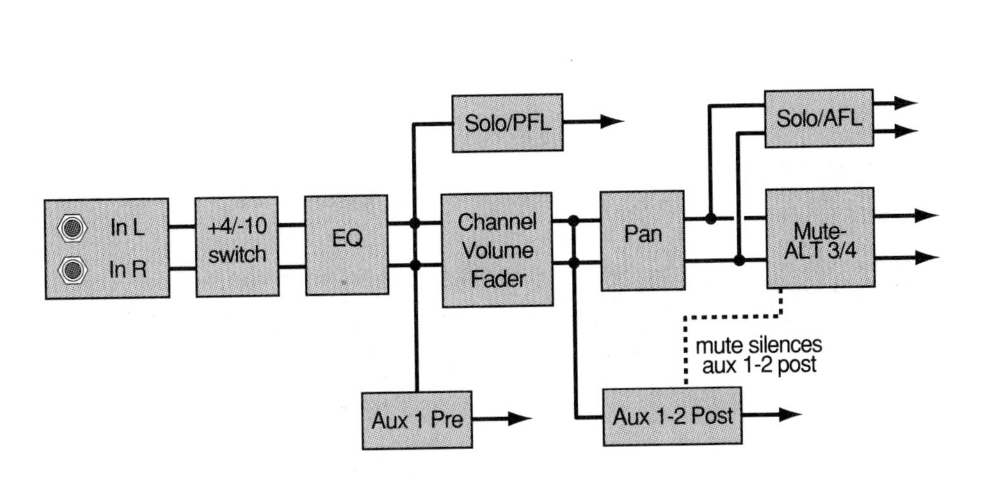

The 1402-VLZ stereo input channel (above) and corresponding signal flow diagram (left).

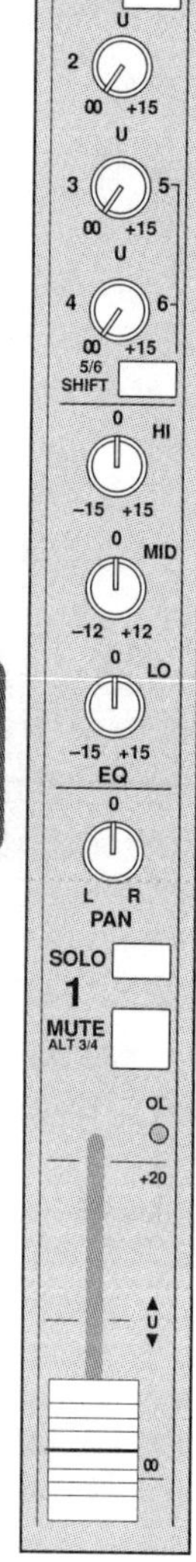

The CR-1604 input channel (above) and corresponding signal flow diagram (below).

CR-1604

While not the company's first mixer (that distinction belongs to the rarely seen LM-1602), the CR-1604 was the model that put Mackie on the map. It's packed with interesting features, some of which appear on many of the mixers that followed it. It also has a few unique features and eccentricities, as you'll soon see.

INPUT CHANNEL CONFIGURATIONS

The "16" in "CR-1604" stands for this mixer's 16 inputs. As seen from the front panel, all 16 look identical. However, there are actually three different types of CR-1604 input channels.

- Channels 1-6: Balanced XLR mic and line inputs, insert points (channel access)
- Channels 7-8: Unbalanced line inputs, insert points (channel access)
- Channels 9-16: Unbalanced line inputs

Despite these differences, the internal signal flow of all 16 input channels is very similar, so we'll discuss them as a group.

The first six inputs have XLR inputs for microphones. Any microphone can be plugged into the CR-1604, including those which require phantom power. Each channel with a mic input also has a balanced line input connector. However, simultaneously connecting a mic and line input to the same channel is not recommended, as this will lead to unpredictable results.

The line and mic inputs (Ch 1-6) both run through the same Trim control. (Connecting line level signals to the XLR mic input can result in distortion; connect line-level signals to the 1/4" Line connector instead.) Once past the Trim knob, your mixer doesn't care if the source signal came from the line or mic input; it treats them both the same.

Upon leaving the trim section, the signal path can split off to the Aux 1 Monitor send, but only if that channel's Aux 1 MONITOR button is pressed. When engaged, this makes the Aux 1 monitor send *pre-EQ, pre-fader*, which is ideally suited for stage monitors in a live sound setup. Interestingly, Aux 1 Monitor has its own output on the rear panel, separate from the "regular" Aux 1 out. In a sense, this gives the CR-1604 seven aux sends, rather than six. No other

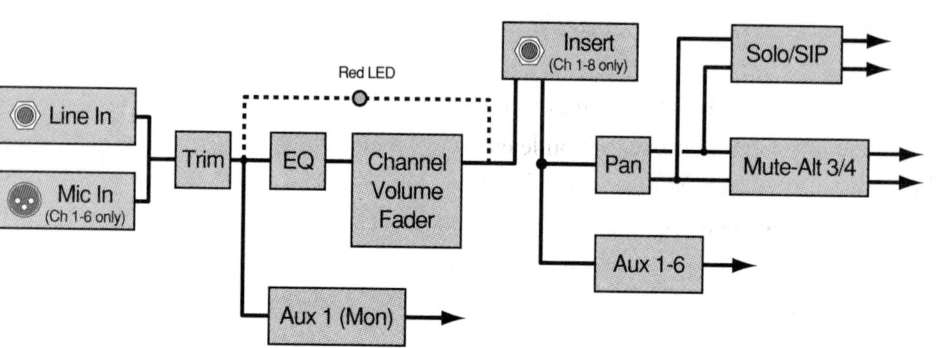

compact Mackie mixer to date includes this unusual and useful feature, so rejoice, proud owner of a classic CR-1604!

Regardless of the Aux 1 Monitor button setting, the signal leaving the trim control also flows into the EQ section. This consists of three controls, low and high shelving and a midrange peak/dip EQ. This 3-band EQ is essentially the same as found on the later 1202-VLZ, 1402-VLZ, CFX and PPM series mixers.

After the EQ is the Channel Volume Fader. The little dotted line shows the points sensed by the little red clip LED. By monitoring these two places in the circuit, the light can go off when distortion occurs at the trim stage, or is caused by excessive boost in the EQ or Channel Volume Fader sections.

Following the Channel Volume Fader on channels 1-8 comes the insert point, which is labeled "Channel Access" on this mixer (we'll be looking at inserts in-depth in Chapter 7). The decision to position Inserts *after* the Channel Volume Fader is an unconventional one, with both good and bad consequences. On the down side, it makes it very awkward to use compressors and other "dynamics processors" on a CR-1604 insert. We'll talk about possible work-arounds in Chapter 8. However, a post-fader insert is ideally suited for use as a direct output to feed a multitrack audio recorder. The CR-1604 is the only Mackie mixer to date to have a post-fader insert. Owners of a classic CR-1604: rejoice or despair as needed! If nothing is connected to an insert jack, the signal automatically flows around it; in this case from the Channel Volume Fader to the Pan control.

Following the Insert (channels 1-8) or Channel Volume Fader (channels 9-16), the signal splits again. Here is the point where all the Aux sends tap their signal, except for Aux 1 on channels where the Monitor button is pressed. As you can see, these Aux sends are *post EQ, post-fader*.

At this point, the signal also goes to the channel Pan control. Following the output of the Pan knob, our mono input is now a stereo signal. When a channel is soloed, you'll hear the signal "soloed in place," which means its left-right position will be maintained. Note that a solo that taps into the signal path after the fader is often called AFL, for "After-Fader Listen."

The last stop after the Pan control is the Mute-ALT 3/4 switch, which routes that channel's signal between the main stereo and ALT outputs. If you leave the ALT outputs unconnected, the button functions as a Mute control. We'll talk more about the importance of these signal routing controls in Chapter 5.

1604-VLZ, VLZ PRO & 1642-VLZ PRO

The 1604-VLZ is a major update on the original CR-1604. More powerful EQ, a control room output section, more mic inputs and "real" output buses are all welcome additions. As you can see from both the channel strip illustration and the corresponding signal flow diagram, the 1604-VLZ and 1642-VLZ PRO have a lot going on.

The line and mic inputs both run through the same Trim control. It's not a good idea to use both mic and line input connectors at once, because you have no way to adjust their levels individually. Also, connecting line level signals to the XLR mic input can result in distortion; connect line-level signals to the 1/4" Line connector instead.

Next stop is the Insert point, which we'll cover in Chapter 7. Briefly, this jack allows you to connect, or "insert," an external signal processor into the signal path of a given channel. Compressors and equalizers are common examples. Inserting such devices provides more sophisticated control than the mixer's built-in processing can offer. If nothing is connected to an insert jack, the signal automatically flows around it; in this case, from the trim to the low cut switch.

Following the Low cut switch, several things happen. First, the signal passes on to the EQ section (more about that in a moment). This is also the point where Aux 1-2 monitor sends get their signals. Each channel has a Pre switch. Pressing it makes the aux 1 and 2 tap their signals here, after the Insert and Low cut, but before the EQ, Mute and Fader. Since each channel has its own Pre button, you can set some channels aux 1 and 2 to Pre (*pre-fader*) and others to "not-Pre" (*post-fader*). Finally, the channel's green "signal present" LED indicator senses the level at this point in the circuit, and blinks to indicate the presence of audio passing through the channel.

The 1604-VLZ and 1642-VLZ PRO input channel (above) and corresponding signal flow diagram.

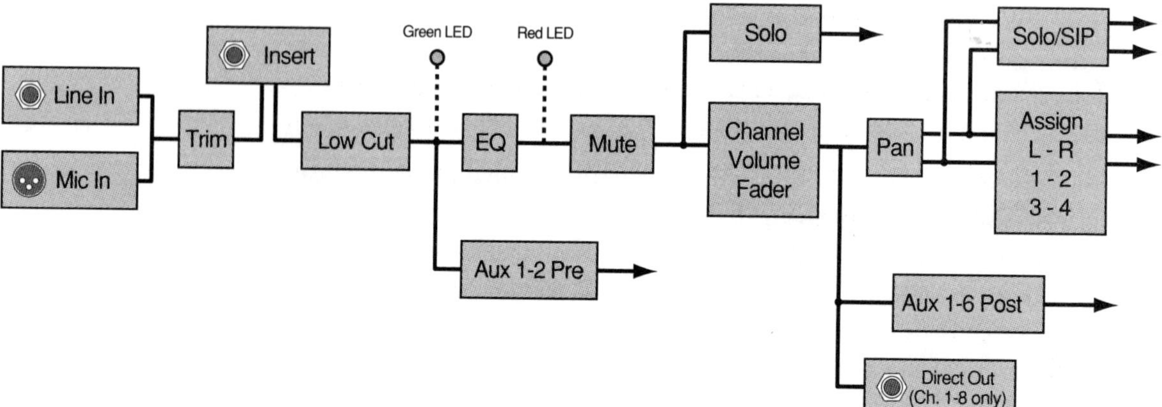

On to the EQ section: This consists of four knobs, low and high shelving and a swept midrange peak/dip EQ (swept-mid EQs, which use two knobs, are described earlier in this chapter). This EQ is essentially the same as that found in SR24•4, 32•4 and CFX series mixers.

Following the EQ is another status LED. This red light comes on when clipping/ distortion occurs. Since the sense point is post-EQ, it will show you if excessive EQ boost is causing distortion. It also illuminates for distortion caused by improper trim levels, or an overly-hot signal coming into the insert connector.

Unlike smaller Mackies, the 1604- and 1642-VLZ have dedicated mute buttons. There is no "ALT 3/4" option on this mixer. Instead, there are four additional output buses with their own assign switches. We'll talk more about these in Chapter 5. Note that the position of the Mute button does not interrupt any signal flowing to the aux 1-2 pre-fader sends. However, any post-fader sends will be muted when the mute button is engaged.

Following Mute is the point monitored by the channel's Solo button, but only when the master "MODE-Normal (AFL) Level Set (PFL)" button is in the down (PFL) position. Since this point is *post-mute*, soloing a muted channel will result in silence.

Next up is the Channel Volume Fader, who's function is pretty obvious. After this point, the signal splits three ways: Pan, Aux 1-6 and Direct outs.

Direct outputs are provided on channels 1-8 only. These connectors are an ideal way to run individual mixer channels to track inputs of a multitrack audio recorder. Since this output follows the Trim, Insert, Low cut, EQ and Fader, any adjustments made to these controls will affect the signal "going to tape" (or whatever storage medium your multitrack uses). Since only a single mic or line input will appear at this channel, you should use direct outputs for signals you want on their own tape tracks, such as individual drum mics, a lead vocal, keyboard output, etc. To route multiple mic or line inputs to a single tape track, you'll use one of the assignable output buses (more on these later).

Also following the Channel Volume Fader are Aux sends 1-6. Sends 3-6 always take their signal from this point in the circuit. Aux 1 and 2 flow from this point when the channel's PRE button is *not* engaged. This *post-EQ*, *post-fader* location of the aux sends is ideal for using with effects.

The last fork in the signal's path leads to the Pan control. The position of the Pan, before the Assign buttons, makes it a critical player in the mixer's signal routing capabilities, as you'll learn shortly.

The second Solo point in the circuit also follows the Pan control. It's also after the EQ, Fader, Mute, Low cut and Insert, so just about anything you do on a channel will be audible while soloing. Since solo also follows mute, you can't hear a solo'd channel that is also muted. This second solo point is active when the Solo button in the master output section is in the "Normal/AFL" position.

Finally, after the Pan control, the (now-stereo) signal reaches the channel Assign buttons. We'll delve deeply into this subject in Chapter 5. Briefly, to use the 1604-VLZ as a regular stereo mixer, be sure that each channel's "L-R" button is pressed. Any channel with this button in the up position won't be audible in the main mix, unless you use an alternate (and more complicated) signal routing scheme, which we'll cover later.

1642-VLZ PRO Inputs 9-16

The 1642-VLZ PRO has a couple of input channel types not found on 1604-VLZ models. First, it includes four stereo input channels. Channels 9/10 and 11/12 include an XLR mic input and trim. If you're not using the stereo line inputs of these channels, you can plug a single microphone into them instead. Note that the 75 Hz high-pass filter on these two channels affects the mic, but not the line inputs (see below).

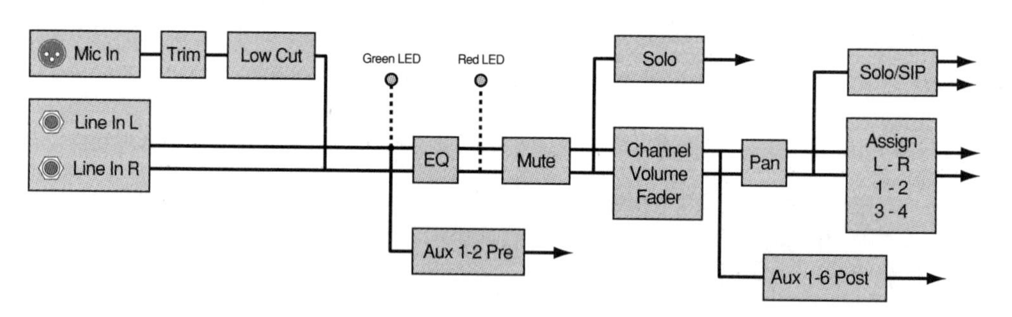

Also, the last pair of input channels, 13/14 and 15/16, have no mic input or high-pass filter, but they add a trim control to the stereo line inputs (see below). These two inputs also have 4 band fixed, rather than 3 band swept-mid EQ.

While the 1642-VLZ PRO's main input channels 1-8 are identical to those of the 1604-VLZ PRO (see previous pages), the last four channel strips are a bit different (see text at right). The unique mono mic/stereo line input from channels 9/10 and 11/12 is shown above.

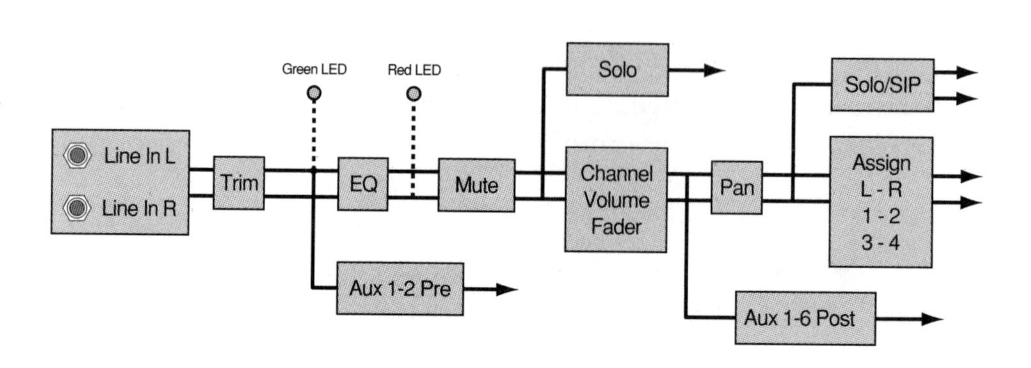

SR24•4/SR32•4

The SR24•4 and SR32•4 mixers are designed primarily for live sound applications, but are suitable for recording as well. As far as their input channel section is concerned, the SRs are very similar to the 1604-VLZ. Of course, the number of input channels, as well as the physical size of this mixer is significantly larger. The SR-series mixers have two slightly different types of input channels. Most are normal mic/line inputs; the last two are stereo line input channels. Let's begin with the mic/line inputs, which are channels 1-20 on the SR24•4, channels 1-28 on the SR32•4.

MIC/LINE INPUTS

The line and mic inputs both run through the same Trim control. Remember that connecting line level signals to the XLR mic input can result in distortion; connect line-level signals to the 1/4" Line connector instead.

Next stop is the Insert point, which we'll cover in Chapter 7. Briefly, this jack allows you to connect, or "insert," an external signal processor into the signal path of a given channel. Compressors and equalizers are common examples. Inserting such devices provides more sophisticated control than the mixer's built-in processing can offer. If nothing is connected to an insert jack, the signal automatically flows around it; in this case, from the trim to the low cut switch.

Following the Insert point is the Low cut switch, which removes the lowest of the low frequencies from the sound passing through a given input channel. Also following the insert connector is each channel's green "signal present" LED indicator. The level at this point in the circuit is sensed, and the LED blinks in response to the level of signal passing through the channel.

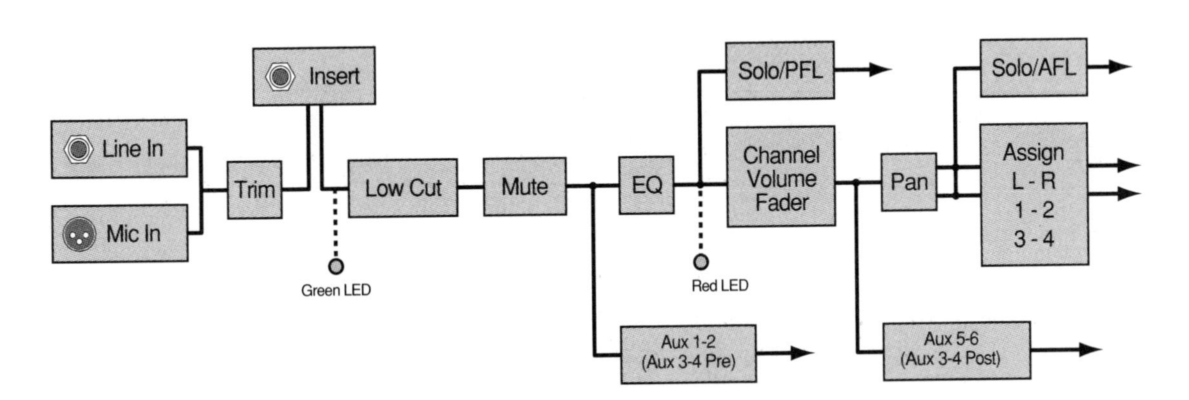

Following the Low cut switch, the signal passes on to the Mute button. Following the mute, we reach the point where Aux 1-2 (and optionally Aux 3-4) sends get their signals. Each channel has a PRE switch. Pressing it makes the Aux 3 and 4 tap their signals here, after the Insert, Low cut and Mute, but before the EQ and fader.

The SR-series input channel (above) and corresponding signal flow diagram (above-left).

SR-series

Since each channel has its own PRE button, you can set some channels aux 3 and 4 to PRE (*pre-fader*) and others to "not-PRE" (*post-fader*). Regardless of the position of the PRE switch, Aux 1&2 are always *pre-fader*; aux 5&6 are always *post-fader*. Note that pressing the Mute button will silence that channel's signal in all aux sends, both *pre-fader* and *post-fader*.

Leaving the Mute button, we go to the EQ section: This consists of four knobs, low and high shelving and a swept midrange peak/dip EQ (swept-mid EQs, which use two knobs, are described in the previous chapter). This EQ is essentially the same as that found in the 1604-VLZ and CFX series mixers.

Following the EQ is another status LED. This red light comes on when clipping/ distortion occurs. Since the sense point is post-EQ, it will show you if excessive EQ boost is causing distortion. It also illuminates for distortion caused by improper trim levels or an overly-hot signal coming into the insert connector.

Nothing new here, just a reprint of the SR-series input channel block diagram from the previous page. Provided for your viewing convenience at no extra charge.

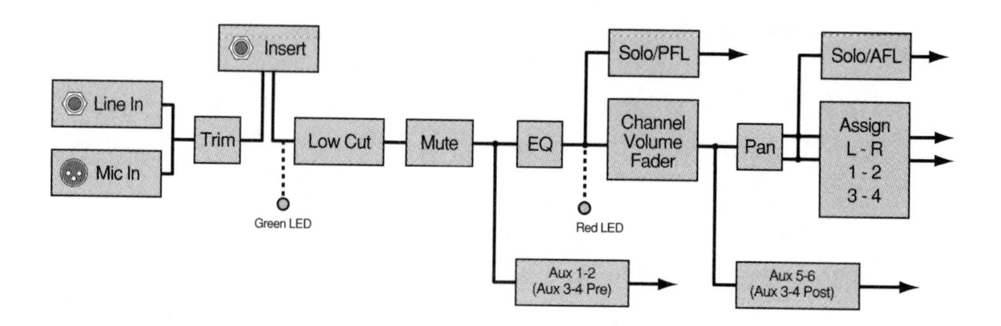

Also following the EQ is the point monitored by the channel's Solo button, but only when the master "Pre Fader/ In-Place AFL" button is in the up position.

Next up is the Channel Volume Fader, who's function is pretty obvious. After this point, the signal splits to both the Pan and Aux 3-6 sends. Let's look at the aux sends first. Sends 5&6 always take their signals from this point in the circuit. Aux 3 and 4 flow from this point when the channel's PRE button is *not* engaged. This *post-EQ, post-fader* location of the aux sends is ideal for using with effects. Aux 1&2 are always *pre-fader*.

The signal leaving the Channel Volume Fader also flows to the Pan control. The location of the Pan, coming prior to the Assign buttons, makes it a critical player in the mixer's signal routing capabilities, as you'll learn shortly.

The second solo mode (AFL) takes its signal following the Pan control. It's also after the EQ, Fader, Mute, Low cut and Insert, so just about anything you do on a channel will be audible while soloing. Since solo also follows mute, you can't hear a solo'd channel that is also muted. When the master "Pre-Fader/In-Place AFL" button is in the down (AFL) position, you'll hear the signal from this point in the channel.

Finally, the signal reaches the channel Assign buttons. We'll be delving deeply into this subject in Chapter 5. Briefly, to use your SR-series model as a regular stereo mixer, be sure that each channel's "L-R" button is pressed. Any channel with this button in the up position won't be audible in the main mix unless you use an alternate (and more complicated) signal routing scheme, which we'll cover later.

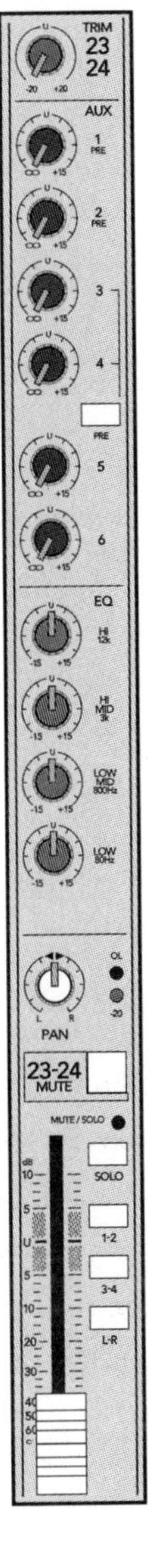

STEREO LINE INPUTS

The last two channels on an SR-series mixer are stereo line-input channels. These are quite similar to the mic/line channels just described. Let's list the differences:

- No mic inputs
- No low cut switches
- No insert points
- 4 band fixed, rather than 3 band swept-mid EQ

Please refer to the previous section describing the mic/line channel type for a narrated walk through the signal flow diagram, ignoring the comments about mics, low cuts and inserts.

One additional item of note: The aux sends, both pre- and post-fader, all tap off a mono summed mix of the left and right inputs. This means that while you can certainly patch a nice stereo signal into these channels, your aux sends will be a mono blend of left and right.

The SR-series stereo input channel (above) and corresponding signal flow diagram (left).

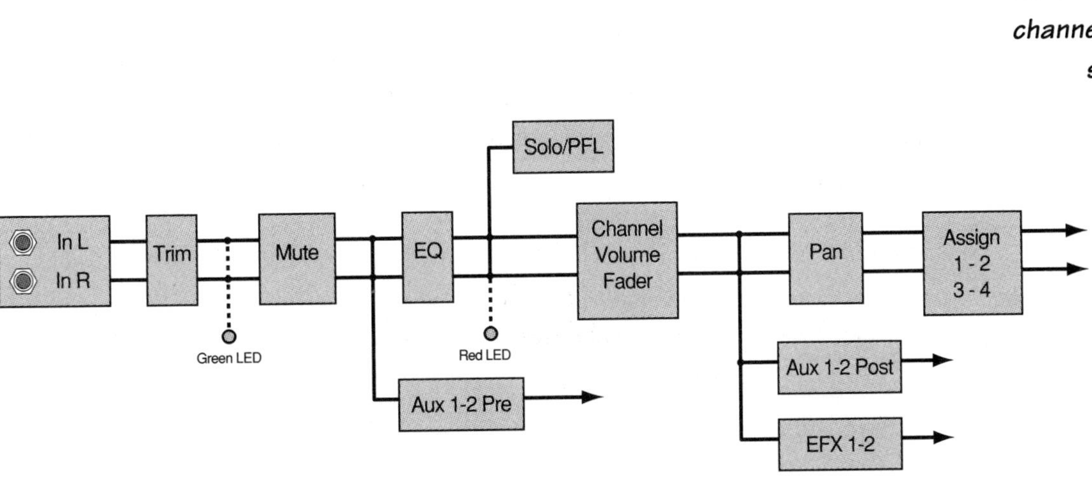

PPM Series

The PPM Series debuted with five models, the 808S and 808M, 408S and 408M and 406M. All share most of the same features, the largest difference being that the "S" models include a stereo mixer while the "M"models are mono. Of course, these mixers include beefy built-in power amplifier(s), graphic EQ(s) and an "emac" digital effects unit making them an ideal all-in-one box for gigging musicians. Still, the PPMs have mixers at their hearts.

This section will cover inputs 1-6 of both the mono and stereo PPM models, as their internal signal flow is nearly identical. Mono models (406M, 408M & 808M) are illustrated on this page, while stereo models (408S & 808S) are shown on the facing page. The two dual line input channels of the "08" models are shown on the page after next.

MIC/LINE INPUTS (CH. 1-6)

Although PPM inputs 1-6 have both a mic and line input, you really should just use one or the other. Plugging a mic and line source into the same channel will be less than satisfactory, because there will be no way to adjust their levels independently from the mixer. This is because each channel's line and mic inputs run through the same Input Level Set knob. As you can probably guess, input level set is another name for trim; it is used in exactly the same way.

Following the Input Level Set is a red LED used in the input level setting proceedure. If at any point during a performance you see this light illuminate on a channel, that signal is beginning to distort. To fix the problem, turn that input level set knob down slightly. Then, turn the channel volume knob up a bit to compensate. The end result should be a sound at the same volume, but without distortion.

Next in line is the Insert point. You can use your PPM for years and never plug anything into this jack, so don't lose any sleep if you're not sure what to do with it. However, you can use the insert to connect external signal processing tools like compressors or fancy equalizers to apply more processing horsepower to a particular channel. We'll talk more about inserts in Chapter 7. Just remember that if you don't plug anything into the Insert, the signal automatically flows right around it to the next part of the input channel.

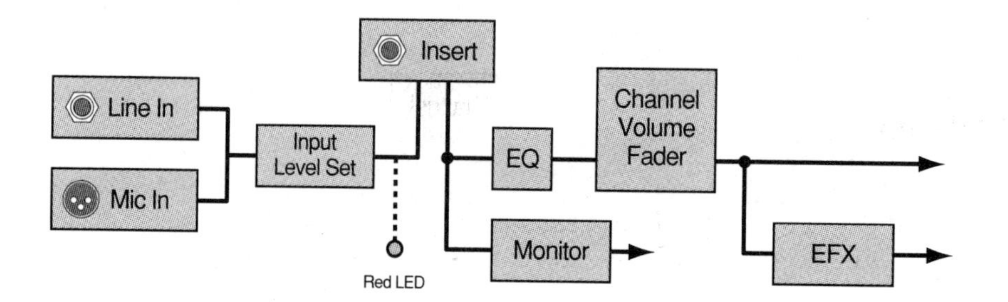

The input channel of a mono "M" series PPM powered mixer (above) and corresponding signal flow diagram (right). Sorry, no Pan control.

Following the Insert point, the signal splits two ways, to the monitor and EQ controls. The Monitor knob lets you tap off a portion of each channel's signal and route it to a monitor loudspeaker that is pointed at the stage, rather than at the audience. This helps you and other musicians hear what's going on during the performance. Note that the position of this control in the signal path comes before the EQ and Channel Volume Fader. This means that changes to EQ or channel volume will have no effect on the signal you are hearing through the monitors!

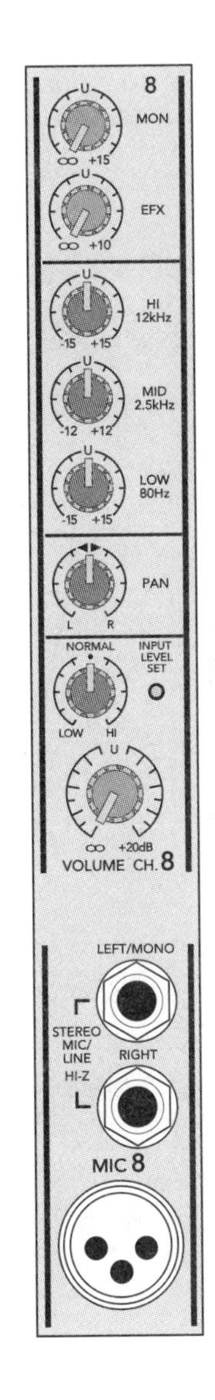

Also following the Insert point is the channel EQ section. This consists of three controls, low and high shelving and a midrange peak/dip EQ. This 3-band EQ is essentially the same as found on the original CR-1604, 1202- and 1402-VLZ mixers; its use is described in the previous chapter.

Following the EQ section comes the Channel Volume Fader. This is the knob you should use to mix the volume levels of all your channels relative to one another. The Input Level Set control should not be used for changing the balance between channels. Instead, adjust the Input Level Set knob *once* before the performance (following the instructions mentioned earlier in this chapter). During the show, make all your mix changes using the larger Volume control.

Following the channel volume knob, the signal splits off to the EFX control. By turning up this knob, you'll be able to send part of each channel's sound to emac, the PPM's built-in effects unit. Note that this knob comes after the channel volume knob in the signal flow. This means that when you turn the channel's knob all the way down, your signal will also be prevented from flowing to the emac. In practice, this is a helpful feature, not a limitation. You'll see why as we talk more about effects and aux sends in Chapter 7.

If you have a stereo model PPM mixer, your signal reaches the Pan control. Whether or not your mixer has a Pan control, this is the end of the line for the input channel. From here, each channel's signal flows on to the master output section of the PPM mixer. We'll look at that in the next chapter.

Finally, one item absent from the PPM channel configuration is a mute button. While most Mackies offer independent mutes for each input channel, the PPM-

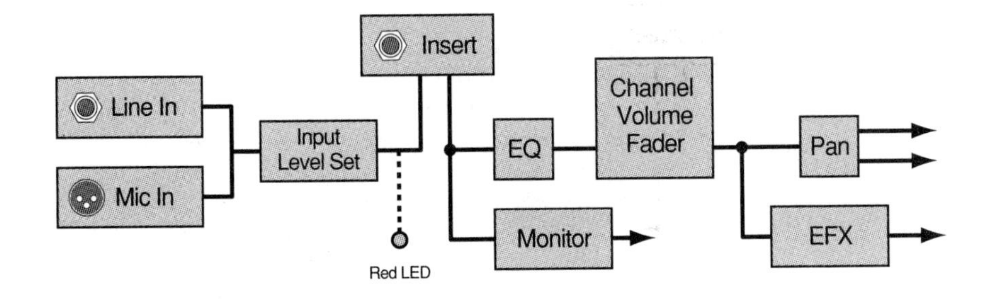

The input channel of a stereo "S" series PPM powered mixer (above) and corresponding signal flow diagram (left). Dig the pan knob!

series mixers have a single button, labeled "Break (mutes channels 1-6)." Pressing this button turns off the first six channels, allowing you to leave each channel volume knob undisturbed. When you get back from your break, simply pop the Break button again and you'll be ready to roll.

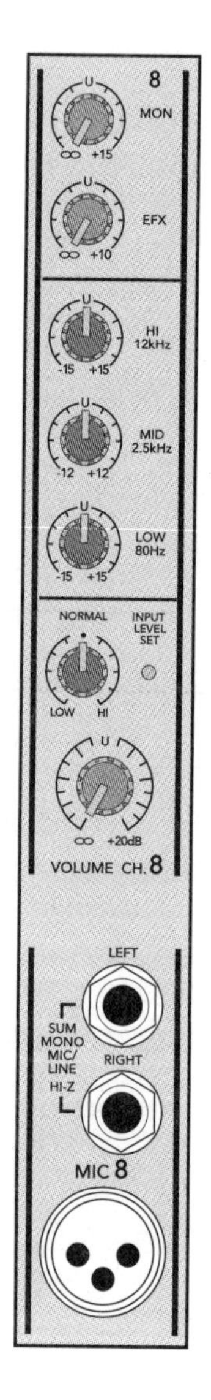

PPM STEREO INPUTS (CH. 7-8)

PPM mixers with an "08" in their name include a pair of dual-input channels. These two input strips include a pair of line inputs (1/4" connectors) and an XLR mic input. On stereo "S" model PPMs (pictured on the facing page), these two channels are true stereo inputs. On the mono "M" model PPMs (shown at left and below) the "stereo" line inputs are there for your convenience only—they immediately get mixed into a mono signal. Let's stroll through the dual input channels, shall we?

If you're short of inputs, this is the only Mackie to date where it is practical to connect mic and line inputs to the same channel. This is because *only the mic input passes through the Input Level Set control.* Since the line inputs don't go through the input level set, you could, in a pinch, plug a mic and a mono or stereo line input into one channel. The Channel Volume Control would turn both mic and line input(s) up and down together, while the input level set would adjust only the mic level. Perhaps a bit inconvenient, but a potential godsend if you're short of inputs—and who isn't?

Note that the Input Level Set's red LED monitors only the signal levels coming from the mic input. If your line inputs are clipping, you may hear the distortion but you won't get any visual indication from this little light.

Following the union of the mic and line inputs, the signal splits two ways; to the Monitor and EQ controls. The Monitor knob lets you tap off a portion of each channel's signal and route it to a monitor loudspeaker that is pointed at the stage,

The dual-line input channel of a mono "M" model PPM powered mixer (above) and corresponding signal flow diagram (right). Note that "left" and "right" line inputs are immediately combined to mono.

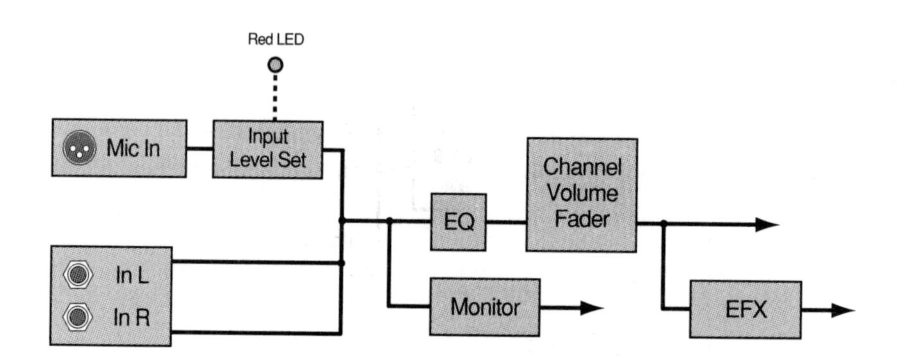

rather than at the audience. This helps you and other musicians hear what's going on during the performance. Note that the position of this control in the signal path comes before the EQ and Channel Volume Control. This means that changes to EQ or channel volume will have no effect on the signal you are hearing through the monitors!

The EQ section consists of three controls, low and high shelving and a midrange peak/dip EQ. This 3-band EQ is essentially the same as found on the original CR-1604, 1202- and 1402-VLZ mixers, and its use is described in the previous chapter.

Following the EQ section comes the Volume knob. This is the knob you should use to mix the volume levels of all your channels relative to one another. The Input Level Set control should not be used for changing the balance between channels. Instead, adjust the Input Level Set knob *before* the performance (following the instructions mentioned in the previous chapter). During the show, make all your mix changes using the larger Volume control.

Following the Channel Volume Fader, the signal splits off to the EFX control. By turning up this knob, you'll be able to send part of each channel's sound to emac, the PPM's built-in effects unit. Note that this knob comes after the channel Volume knob in the signal flow. This means that when you turn the channel's knob all the way down, your signal will also be prevented from flowing to the emac. In practice, this is a helpful feature, not a limitation. You'll see why as we talk more about effects and aux sends in Chapter 7.

If you have a stereo model PPM mixer, your signal next reaches the Pan control. Whether or not your mixer has a Pan control, this is the end of the line for the input channel. From here, each channel's signal flows on to the master output section of the PPM mixer. We'll look at that in an upcoming chapter.

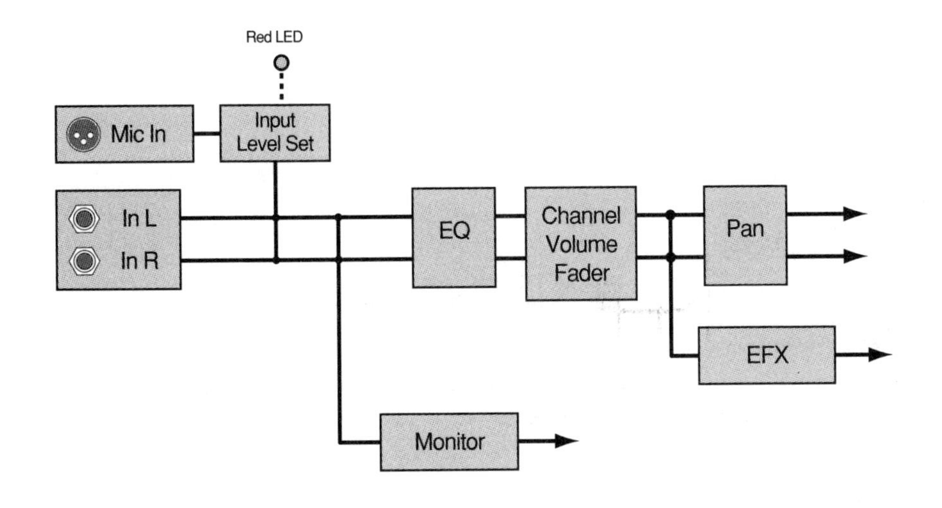

The dual-line input channel of a stereo "S" model PPM powered mixer (above) and corresponding signal flow diagram (left). Left and right line inputs are stereo through the entire input channel and the remainder of the mixer.

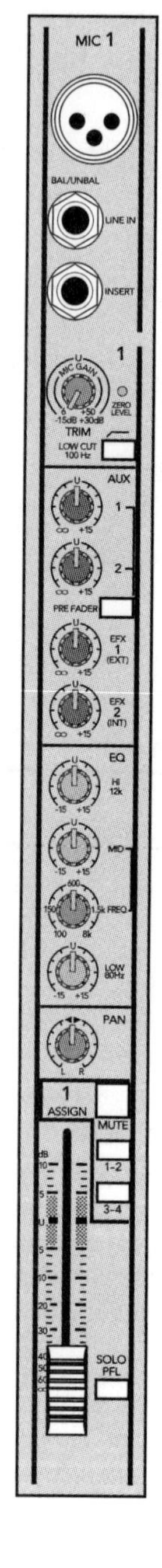

CFX Series

The CFX series debuted with 12, 16 and 20 input model mixers. All are identical except for the number of input channels, so we'll cover all of them at once. The CFX series includes elements of the VLZ and PPM series mixers with a few unique twists of its own. While designed with live performance in mind, the CFX works for recording tasks too. Like the 1202-, 1402- and 1642-VLZ PRO, the CFX mixers include both mono mic/line inputs as well as a few stereo line input channels. Let's cover the mic/line inputs first.

MIC/LINE INPUT CHANNELS

The line and mic inputs both run through the same Trim control. It's not a good idea to use both connectors at once, because you have no way to adjust their levels individually. Also, connecting line-level signals to the XLR mic input can result in distortion; connect line-level signals to the 1/4" Line connector instead.

Next stop is the Low cut switch. While you might think of a Low cut switch as being part of the EQ section, you can see it is a separate animal, both from a front panel layout and signal flow standpoint. By placing the Low cut *before* the Insert (coming up next), you can prevent extreme lows from getting into any device you connect to your mixer's insert points. This can be especially helpful when using a compressor, since low frequency thumps (such as those caused by "stage rumble") can cause compressors to overly-squash things like vocals.

Following the Low cut is the Insert, covered in-depth in Chapter 7. Briefly, this jack allows you to connect, or "insert," an external signal processor into the signal path of a given channel. Compressors and equalizers are common examples. Inserting such devices provides more sophisticated control than the mixer's built-in processing can offer. If nothing is connected to an insert jack, the signal automatically flows around it; in this case, from the low cut switch to the EQ section.

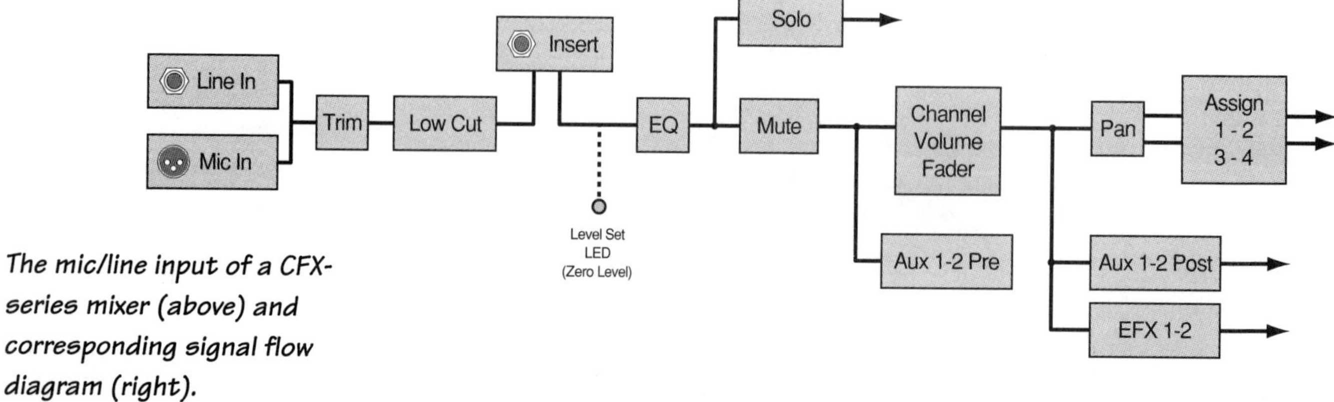

The mic/line input of a CFX-series mixer (above) and corresponding signal flow diagram (right).

Immediately after the Insert, the level of the signal is sensed and if too high, the Zero Level LED blinks in response to the audio signal. If you see this light illuminate, it means the signal in that channel is starting to distort. If this happens, turn the Trim control down slightly. Then, you may want to turn the Channel Volume Fader up a bit to compensate. Lather, rinse and repeat as necessary.

Next comes the EQ section. This consists of four knobs, low and high shelving and a swept midrange peak/dip EQ (swept-mid EQs, which use two knobs, are described in the previous chapter). This EQ is essentially the same as that found in 1604-VLZ and SR-series mixers. Note that you can cause distortion if the EQ is excessively boosted. However, this won't cause the previously-mentioned Level Set LED to illuminate, as that comes before the EQ in the signal path. Use those ears!

Following the EQ, the signal splits two ways. This is the point in the circuit monitored by the Solo control. This means that your solo'd signal will reflect current EQ settings, but won't be affected by Fader and Pan controls.

While the solo tap is a useful feature, the main path for your audio exits the EQ and proceeds to the mute switch. Engaging the Mute button not only silences the main signal coming through a given channel, it also mutes *all* Aux sends, both pre- and post-fader.

After the mute switch, the signal splits off to supply Aux 1-2 when the Pre-Fader switch is engaged. Each channel has a PRE switch. Pressing it makes the aux 1 and 2 tap their signals here, before the Channel Volume Fader. Since each channel has its own PRE button, you can set some channel's aux 1 and 2 to PRE (*pre-fader*) and others to "not-PRE" (*post-fader*).

The signal also flows from the Mute switch to the Channel Volume Fader, who's purpose should be clear. After this point, the signal splits three ways: pan, Aux 1-2 (in *post-fader* mode) and EFX sends 1-2.

Aux 1-2 only take their signals from this point when a given channel's Pre-Fader switch is in the "up" position. This is the smart choice when using Aux 1-2 for effects sends. If you're using them to create a stage monitor mix, you might have better luck with the Pre-Fader switch "down." EFX 1-2 sends always take their signals from this point, regardless of the position of the Pre button. EFX 2 is normally automatically routed to your mixer's internal emac DSP effects unit.

The last fork in the signal's path leads to the Pan control. The position of the pan, before the Bus Assign buttons, makes it a critical player in the mixer's signal routing capabilities, as you'll learn shortly.

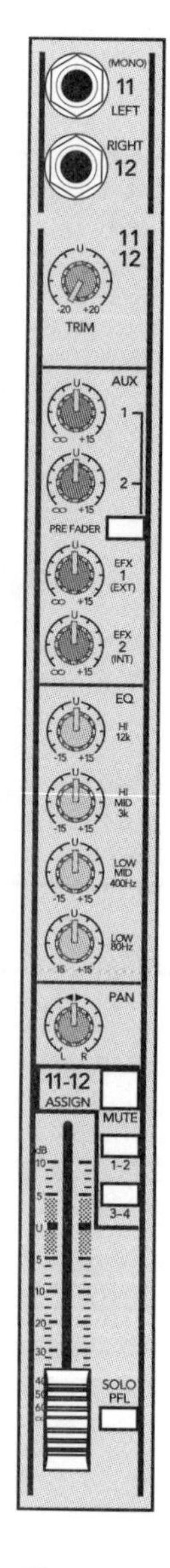

Finally, after the pan contol, the (now-stereo) signal reaches the channel bus Assign buttons. Briefly, the CFX-series provides four output buses. Using the Pan control and the two bus Assign switches, you can send a given channel to one or more buses, labeled Sub 1,2,3 and 4. In the master output section, these Subs are in turn assigned to your mixer's main output. We'll be fully covering this in Chapter 5. For simple stereo operation, be sure each channel's Assign 1-2 button is pressed, and that the blue Sub 1 and Sub 2 faders are turned up. Finally, the Sub 1: Left button and Sub 2: Right buttons must be pressed!

STEREO INPUT CHANNELS

The last two channels on a CFX-series mixer are stereo line-input channels. These are quite similar to the mic/line channels just described. Let's list the differences:

- No mic inputs
- No low cut switches
- No insert points
- No Level Set LEDs
- Four bands of fixed-frequency EQ

Please refer to the previous section describing the mic/line channel type for a narrated walk through the signal flow diagram, ignoring the comments about the above.

One additional item of note: The aux sends, both pre- and post-fader, all tap off a mono summed mix of the left and right inputs. This means that while you can certainly patch a nice stereo signal into these channels, your aux sends will be a mono blend of left and right.

The stereo line input of a CFX-series mixer (above) and corresponding signal flow diagram (right).

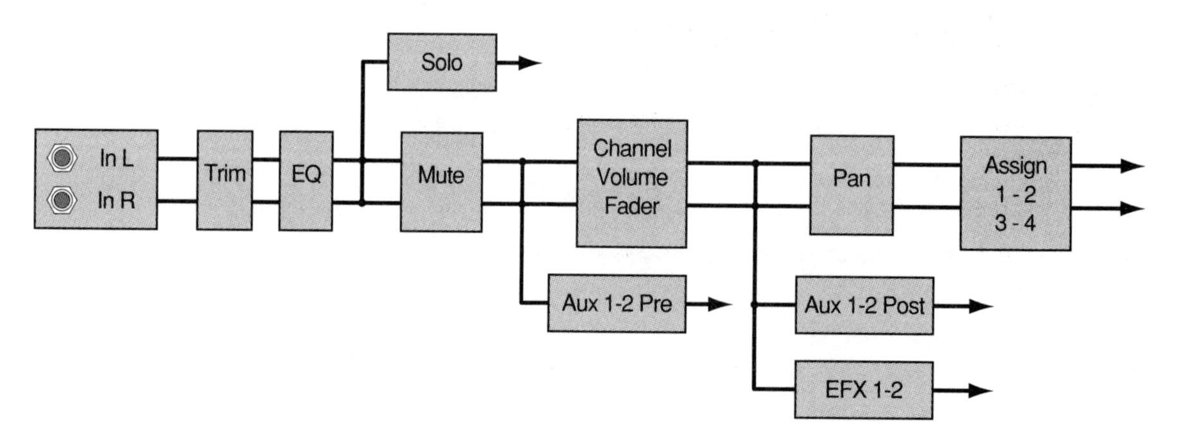

CFX-series

XDR Mic Preamps

In 1999, Mackie finished the design for a new microphone preamplifier. They called it "Extended Dynamic Range," or XDR for short, and began using it in the 1202-, 1402- and 1604-VLZ models. These VLZs with added XDR mic-pres got a "PRO" tacked on to the end of their names and production of the original VLZ models ceased. Then, Mackie released the 1642-VLZ PRO, also with XDRs (there was never a 1642-VLZ "not-PRO" model). Don't be surprised if XDR mic inputs start cropping up in other Mackies as well.

The XDR outperforms earlier Mackie mic preamps in several important areas. First, they have much wider bandwidth, extending well beyond the theoretical limit of human hearing. While this might seem like overkill, it does indeed improve transient response. There are also things you won't hear from an XDR—they have significantly less distortion than previous Mackie designs. Also, they offer much greater immunity to radio-frequency interference.

However, earlier Mackie mic preamp designs were quite good to begin with, so one might well ask what is the audible difference between designs? From an even more pragmatic standpoint, if you already have a VLZ, do you need to upgrade to a VLZ PRO?

In my own listening tests, I found the XDR mic preamps to offer slightly clearer attack transients and a little extra body or warmth to the sound. In short, I prefer them to the original designs, but the improvement is a subtle one. If I was currently using a 1202-VLZ or 1402-VLZ and wanted more channels, I'd certainly opt for a new 1604- or 1642-VLZ PRO. On the other hand, if I wasn't upgrading to a model that offered new functionality (more channels, control room outputs, assignable buses, etc.) then it's a closer call. If you're doing critical recordings and you're very happy with your mics, then you're a candidate to upgrade from VLZ to VLZ-PRO. Otherwise, you'll probably notice a bigger difference if you buy new mics instead of a new mixer.

Input Channel Post Script

You've reached the end of the channel input section, but astute readers will note that we've skipped a few controls, specifically the red-capped Aux knobs. While these surely are part of the input channels, we're going to cover these controls, along with the channel inserts that we mentioned briefly, in an upcoming section on the use of external signal processing and effects units. That information appears in Chapter 7.

But first, we're going to look at the signal routing capabilities of your mixer.

CHAPTER 5

Channel Routing Controls

In the last chapter we covered most, but not all the controls in your mixer's input section. While we touched on the ALT 3/4 and the 1-2, 3-4 and L-R buttons, you might still be wondering exactly what those controls are for. Well, my inquisitive friend, it's time to find out.

Here's the deal: In a simple stereo mixer, all the inputs flow to the same main left and right outputs. However, many Mackies have more than one stereo output. When a mixer has multiple outputs, you will also find controls to route individual inputs to these different outputs.

Meet the signal routing controls! These include Pan, Mute-ALT 3/4 and the Assign buttons (1-2, 3-4 and L-R). They determine which output a given input channel ends up at. Your Mackie won't have each and every one of the controls discussed in this chapter, but I suggest you read about them all. That way, when you encounter a mixer with different types of routing controls than your current model, you'll have a head-start on figuring out how they work.

Like the previous chapters, we'll begin our discussion with the basic concepts behind routing controls—that starts on the next page. The chapter following this one will take the generic concepts you're about to learn and apply them to each model Mackie mixer.

Pan Is a Signal Routing Control!

Each channel has a Pan control, which directs, or routes, that channel's input signal to the left or right main output. Let's look at a very basic example—a little four-channel stereo mixer (of course, your mighty mixer has more than four channels, but the concept is the same). Check out the illustration at left. Assume that an input signal is coming into each of the four channels at the top, although we haven't shown this explicitly in the drawing. That input signal passes through each channel volume fader, then to the pan knob. That's the order of the signal flow in your mixer, which is why I've drawn the pan below, rather than above the fader. Continuing downwards, the signal from each channel flows to the main left and right signal path, or "bus." A bus (shown as a fat black line) is a point in the circuit where multiple signals merge together. The dot shown at the intersection of multiple lines confirms that a connection is made at that point. Finally, the left and right bus flow through the Main Mix fader and on to the main output jacks.

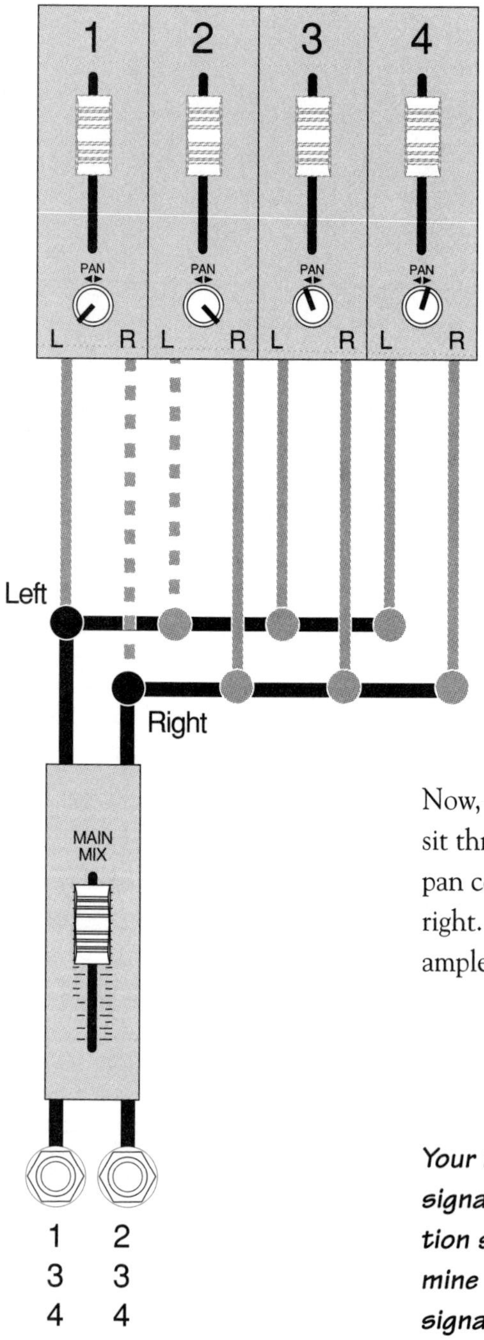

Now, see how channel 1 is panned hard to the left? This means channel 1's signal doesn't go to the right bus, which is why this connection is shown as a dashed line. Similarly, channel 2 is panned hard right, and consequently doesn't get to the left bus. Channels 3 and 4 are panned somewhere in between hard-left or right, so their signals make it to both buses, although not necessarily in equal amounts.

Note the little numbers at the bottom of the drawing. These indicate which input channels are reaching a given output. As you can see, inputs 1, 3 and 4 make it to the left output; 2, 3 and 4 to the right.

Now, I realize that you probably already know what pan does, but I'm forcing you to sit through this for a very important reason: In this example, you can see how the pan control acts as a signal routing device between the two main outputs—left and right. The pan control will continue to serve this important function in the examples ahead, but it won't act alone.

Your mixer's pan knobs are actually signal routing controls. This illustration shows how pan settings determine which output (left or right) the signal from a given channel ends up coming out of.

Master Volume (a quick preview)

Before we get to the next level of routing controls, let's talk about one more volume fader. Our little four-input stereo mixer example includes a main mix fader. As you can see in the signal flow diagram on the opposite page, all input channels flow through this control. The Main Mix fader is also called a *master volume* because it sets the overall (master) level of the combined sound from all the individual input channels. In practice, this means that in order to hear any sound, the master volume must be turned up, at least a little.

Signals leaving the master volume fader flow to two outputs—left and right. You'll probably be connecting the main left and right outputs to an amplifier and speakers or an audio recorder. And again, remember that the pan control of each channel is what routes a given input signal between the main left and right outputs.

Technically speaking, the master volume and main left/right outputs are part of the mixer output section, which is covered in the next pair of chapters. But it seemed unnecessarily rigid of me to withhold this information from you for another dozen pages. After all, what are friends for?

Now, a special note to users of PPM-Series, DFX-Series and classic MS1202 mixers: Your Mackie does not have any of the signal routing controls featured in the rest of this chapter. While I recommend reading this information to prepare for your inevitable encounter with a larger mixer, you may be excused the first time you read this book. But don't let me catch you skipping this chapter the second time around!

ALT 3/4 Is a Routing Control

So, we've seen how a simple four-input, two-output (stereo) mixer works. Now we're going to see what happens when we add a second pair of outputs, in *addition* to the main left right mix. There are a lot of reasons to have more than just a single pair of stereo outputs. For example, extra outputs are great for working with multi-track audio recorders. There are many other applications which we'll get to later. First, let's talk about how multiple outputs work.

Next to each channel fader on the CR-1604, 1202-VLZ and 1402-VLZ, you'll find a button marked "Mute-ALT 3/4." Like the pan knob in the last example, this is a signal routing control. But rather than providing the variable control of a knob, ALT 3/4 is a switch that's either on or off (pan still plays an important role, as you'll see in a moment).

Take a peek at the illustration below. Again, we have a 4 channel stereo mixer, although to avoid confusion between channel "3" and ALT output "3," I've numbered the input channels 11-14. This example adds a Mute-ALT 3/4 switch to each channel, and a corresponding pair of additional outputs—ALT 3 and 4. Channels 11 and 12 feed the main stereo output, because their ALT buttons are not pressed. On the other hand, channels 13 and 14 do have their ALT assign buttons engaged. This "assigns" them to the ALT 3 and 4 output bus, and *disconnects* them from the main stereo output!

Now you might see why this control is labeled "mute," too. If you're only using the main left/right outputs, hitting the ALT button simply removes a channel's sound from your mix. On the other hand, if you connected ALT 3 and/or 4 to the input of an audio recorder, hitting a channel's ALT button routes that signal to your recorder. But the same recorder would *not* record any signals that were going to the main stereo out, only those assigned to ALT 3/4. (We'll talk more about other ways to use this new-found flexibility shortly.)

As promised, the pan control is still part of the signal routing picture, just as it was in the first stereo mixer example. Here, channel 14 is panned hard right, and its ALT button is pressed. The result? Channel 14's sound only ends up in output ALT 4, not in ALT 3, nor will it be present in the main Left or Right outputs. Another thing to notice is the location of the Main Mix control in the signal path—it doesn't affect the levels of the ALT outputs! In other words, turning the Main Mix fader down, or even off, will have absolutely no effect on the volume coming out the ALT 3/4 jacks.

There are shortcomings to doubling up mute and ALT 3/4 on a single button: If you *are* using the ALT outputs, you can't mute a channel—all you can do is change its output assignment from the main to ALT outputs. Furthermore, there is no way to route one channel to *both* the ALT and main outputs—it's an either-or deal.

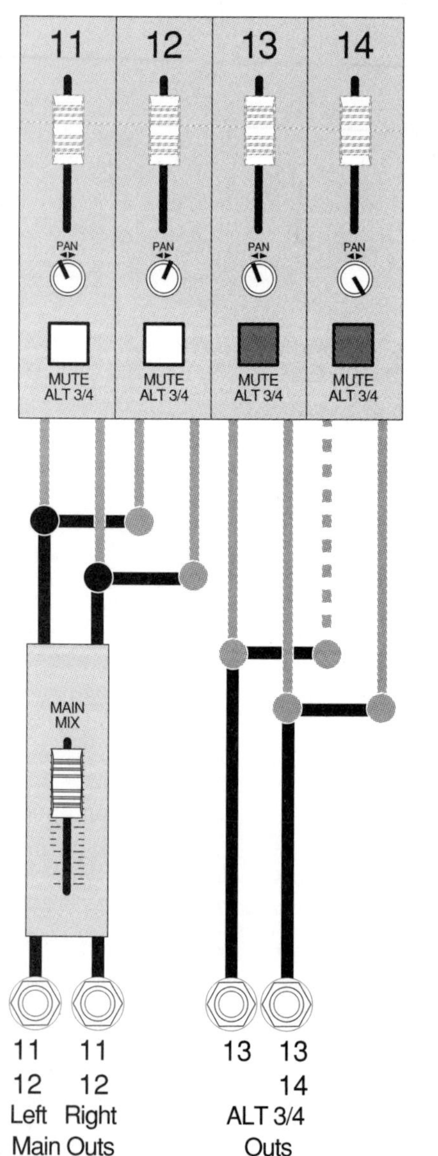

Adding ALT 3/4 controls to each channel makes it possible to route a given input between two pairs of stereo outputs—main left-right or ALT 3/4. Here, input channels 13 and 14 are routed to the ALT outputs.

"Real" Four-Bus Mixers

For many applications, greater flexibility in output assignments is required. It's also nice to have dedicated mute buttons that don't mess with a channel's signal routing. Oh yeah, and how about individual output master volume control of *all* the output buses, instead of just left/right? Let's talk about the Mackies that address all these wants and needs.

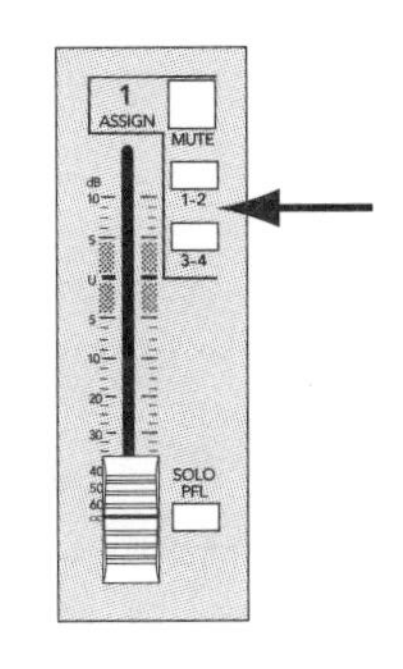

Bus assign switches, like the ones on the CFX-Series mixer shown here, route a given channel to one or more pairs of stereo outputs.

Many Mackies (including the 1604-VLZ, 1642-VLZ, CFX- and SR-series mixers) have "bus assign switches," which are those little buttons down by each channel fader marked 1-2 and 3-4 (see illustration at right). Mackie 8 bus boards and most other large mixers also include this feature, so you'll definitely want to learn about this if you plan on operating these bigger consoles. (Other common names for assignable buses are *groups*, *subgroups*, *sub outs*, and *subgroup outputs*.)

Mixers with "real" assignable buses go beyond the ALT 3/4 scheme we just covered. Instead of providing a signal routing switch (ALT 3/4) that toggles one channel's routing between the main and alternate stereo outputs, these consoles have individual routing buttons on each input channel.

Pressing any one of these buttons assigns that channel's signal to one of the corresponding output buses. And, you can press more than one at a time. Yup, unlike the ALT 3/4 scheme, this allows the same channel to feed one or more pairs of output buses! I hasten to add that in most cases (not all), you'll probably want to assign each channel to only one output bus. Let's look at a simple diagram to illustrate the signal routing possibilities.

In the illustration at right, we're looking at three channels of a four-bus mixer. Let's start with channel 9: Neither of its assign buttons are pressed. This means we'll never hear that channel from any of the four main outputs. The position of its pan knob isn't shown, because it's irrelevant. It doesn't matter where the pan or fader of channel 9 is set. Since it isn't assigned, you won't hear it in the main outputs.

Moving on to channel 10. This is assigned to 1-2 and panned hard-left. This puts its signal in output 1 only. See the little "10" below the output 1 jack? That signifies that channel 10's signal reaches output 1, but not 2, 3 or 4.

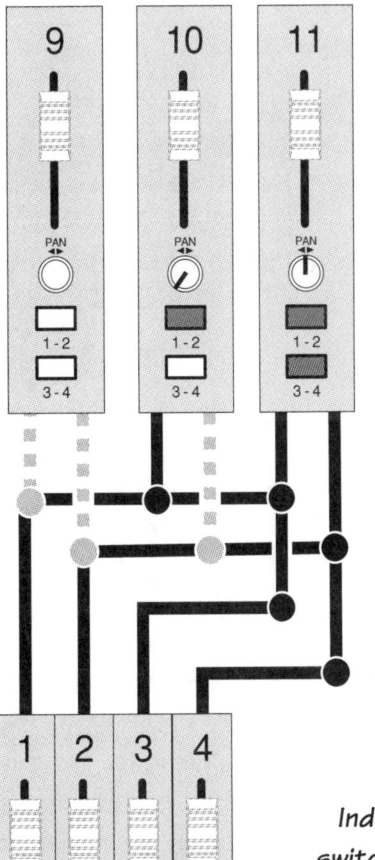

Independent bus assign switches make it possible to route any input to one or more output pairs. Here, Channel 9 is unassigned, while 10 goes to output 1 and 11 is routed to all four outputs.

Check out channel 11 in the illustration on the previous page. It's assigned to *both* 1-2 *and* 3-4. Furthermore, it's panned to the center. This causes the signal from channel 11 to appear in all four outputs! Now, there aren't many cases where you'll want to assign one channel to multiple bus pairs, but it is possible to do so should the need arise. The more likely case is that you'll inadvertently push down two assign buttons when you only meant to press one. Knowing that one channel can be assigned to multiple buses will help you troubleshoot accidental bus assigns. Because channel 11 is assigned to both 1-2 and 3-4, you might look at the illustration and wonder if some signal from channel 10 could flow "upstream" through channel 11's assignment and end up coming out of output 3. This will never happen—the assignment (or lack thereof) of any one channel has no effect on the signal routing of any other channel.

Independent bus assign switches make it possible to route any input to one or more output pairs. Below, channel 9 is assigned to output 2, while 10 and 11 go to output 1. Channels 12-15 are panned in stereo across outputs 3 and 4.

Finally, you can see that each of the four outputs has its own master level fader. This makes the assignable output buses even more useful. Remember, on mixers using the ALT 3/4 scheme, there was no way to adjust the overall level of the ALT outputs. However, with independent bus master faders, you have a fader for the master level of each output. This is extremely handy when using bus outputs to feed a multitrack recorder—you can use the bus master fader to adjust the overall record level going to an individual track on your recorder, without having to adjust each channel's fader individually.

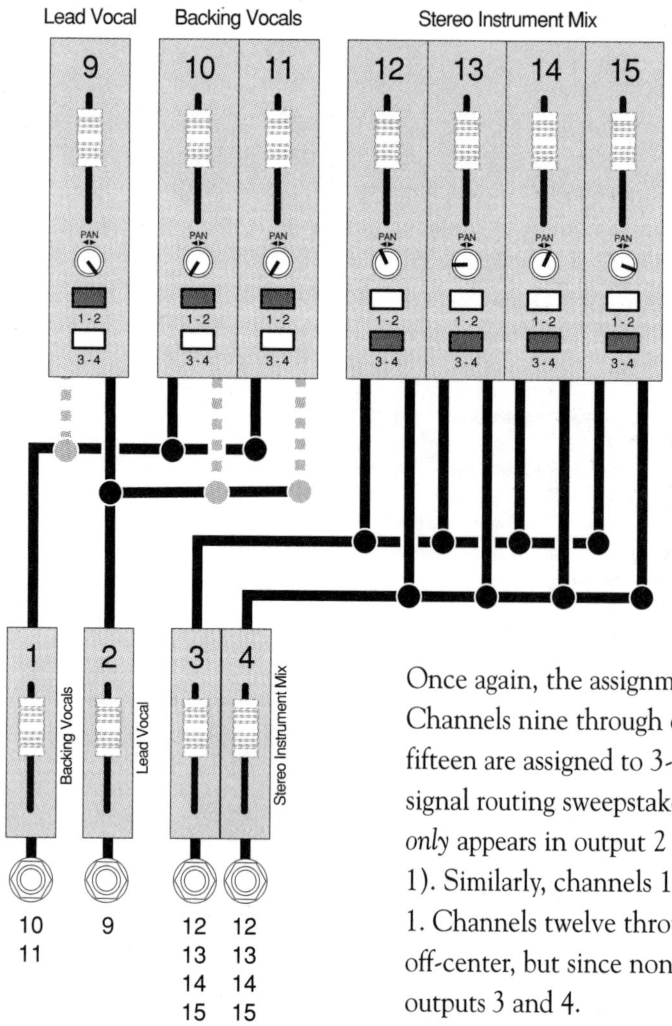

Now, let's look at a slightly bigger (and more typical) assignment example. In the illustration at left, we're looking at seven channels of a larger mixer. Above each group of channels is an example description of what might be connected to these inputs. By "grouping" related signals on the same output buses, you can simplify your mixing chores. Here, for instance, all the instruments (channels 12-15) could be turned up or down simply by adjusting the master faders for outputs 3 and 4.

Once again, the assignment buttons are key to understanding what's going on. Channels nine through eleven are assigned to 1-2, while channels twelve through fifteen are assigned to 3-4. The pan control continues to be a key player in the signal routing sweepstakes. Channel 9 is panned hard right, which means its signal *only* appears in output 2 (this is also indicated by the dashed line flowing to output 1). Similarly, channels 10 & 11 are panned hard left, so they only appear in output 1. Channels twelve through fifteen, assigned to 3-4, are all panned to various places off-center, but since none are panned hard-left or right, all four channels appear in outputs 3 and 4.

Four Buses plus L-R!

"But wait," as they say on TV, "there's more!" As if the prospect of four assignable bus outputs wasn't enough, we can add a stereo mix bus back into the picture for a total of six assignable output buses or perhaps more accurately, three stereo buses! Now how much would you pay? (Your actual mileage may vary: Additional left-right bus not available on CFX-series mixers.)

In the illustration below, channels seven and eight are assigned to L-R. As always, the pan control affects the routing, in this case, directing each of the two channels to the extreme left or right.

Channels nine through eleven are assigned to 1-2. Again note the panning of these three channels, which puts nine in output 2 and ten and eleven in output 1.

Finally, channels twelve through fifteen are assigned to 3-4. They are all panned across the stereo image, so each reaches outputs 3 and 4 in varying levels.

Adding a third pair of assignable stereo outputs (L-R) gives you even more options for signal routing. At left, Channels 7 and 8 are assigned hard-left and right, respectively. Channel 9 is assigned to output 2, while 10 and 11 go to output 1. Channels 12-15 are panned in stereo across outputs 3 and 4.

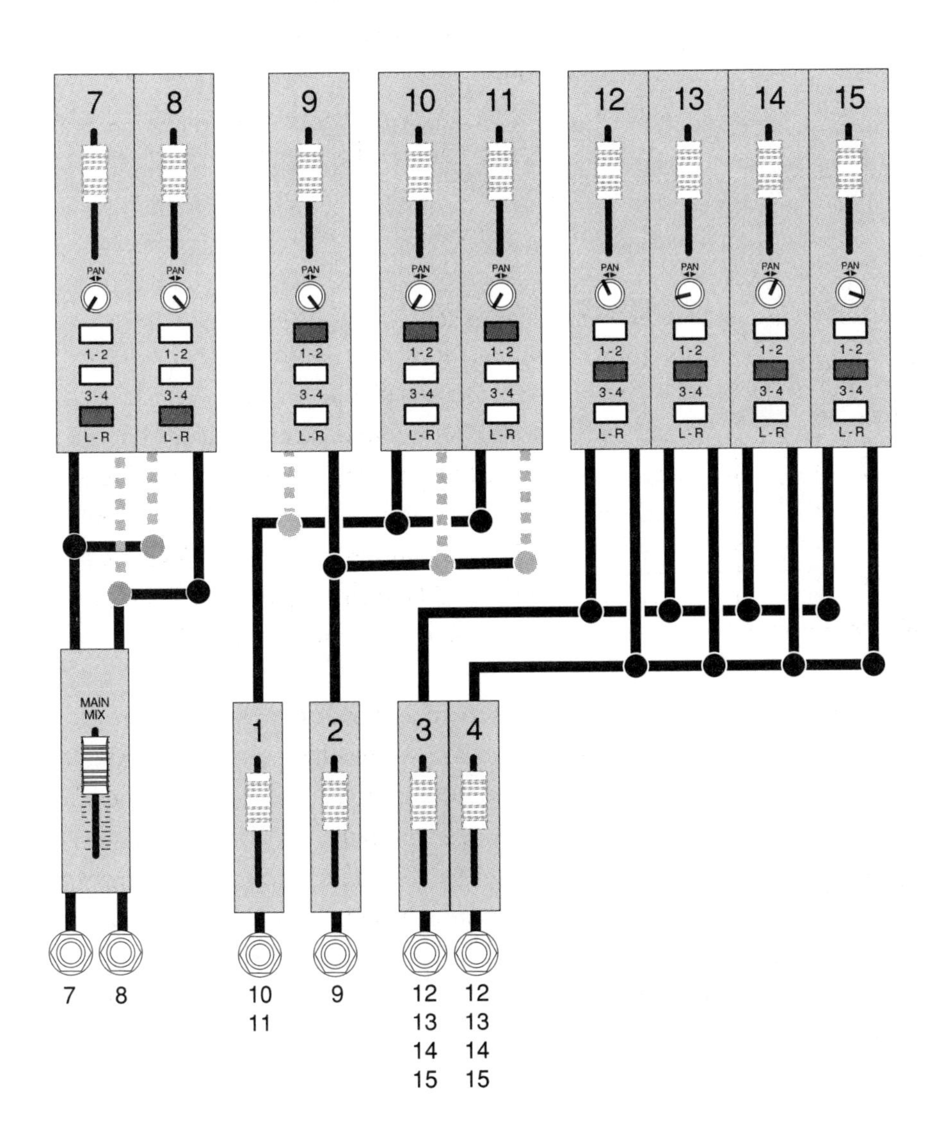

Time to Recap

OK, you've made it through your mixer's signal routing controls. I appreciate your patience, given that we haven't said much about the practical applications for these routing controls. That will come soon, I promise. For the moment, let's reflect on what we have learned: You now know that many mixers include more than a single set of stereo outputs. Mixers that do have multiple output buses, beyond a simple left and right, will also include signal routing controls that let you determine the final output destination of each individual input channel on your mixer.

Mackie's ALT 3/4 scheme is a cost-effective way to provide limited routing functionality. However, the "traditional" (and more flexible) way to handle signal routing is with individual bus assign switches. These little buttons are labeled 1-2, 3-4 and L-R. FYI, larger "8 bus" consoles have two extra assign buttons, labeled 5-6 and 7-8.

Finally, here's a rule-of-thumb for simple stereo operation of your mixer. On models with ALT 3/4 buttons, make sure each channel's ALT button is in the up position. This will insure that all input channels flow to the main left/right output. If your mixer has bus assign switches, press the L-R buttons on all channels (CFX users press 1-2). Also, remember that while it is possible to assign one channel to multiple buses, you must e-mail me for permission first. Just kidding. But seriously, unless you have a very clear idea of why you need to assign one input to multiple buses, don't do it—it gets confusing pretty quickly.

CHAPTER 6

Mackie Signal Routing

The previous chapter explained how generic signal routing controls work. This chapter will take that knowledge and apply it to the specific controls present on your model Mackie. No fair reading this chapter if you didn't read the last one—if you haven't got the generic concepts firmly in mind, the terse descriptions presented for each mixer here probably won't make much sense.

MS1202, PPM: The Mighty Pan Pot

I'm afraid there isn't much to say here, except that these model Mackies have no channel assignment capabilities beyond the pan control (and the mono PPMs don't even have that). Instead, you're working with a basic stereo mixer—no fancy stuff. Your pan control performs the signal routing tasks described in the last chapter, but that's about it. But just so you don't feel short-changed, I'm going to tell you a secret that I'm not going to share with users of those fancy, high-falutin', bigger Mackie mixers. Listen closely—I'm going to whisper so those other guys don't hear.

You may have heard about "pots." Pan pots, volume pots, even EQ pots. "Pot" is short for "potentiometer," the electronic component that is used to adjust the level of a signal. In fact, the pot is actually the thing that your mixer's knobs are attached to. So, a "pan pot" is simply another way of saying "pan knob." If you run into a real old-timer, you might even hear them ask you to "pot up" a particular input. They aren't asking you to boil your mixer, they just want you to turn up a channel's volume knob.

1202-VLZ, 1402 VLZ, CR-1604: ALT 3/4

These three models use the ALT 3/4 scheme that debuted on the original CR-1604. As you can see in the illustration, these mixers have two pairs of stereo outputs: main (L & R) and alternate (3 & 4). Given that mixers in this price range used to be plain old stereo-only models, adding ALT 3/4 gives you extra flexibility for a minimal cost.

As discussed before, the main limitations of this scheme are:

- No master faders for ALT 3/4 levels.
- Channel mutes are effectively disabled when using both pairs of outputs, since hitting mute simply switches that channel between outputs; it doesn't actually block that signal from reaching an output.
- You can't assign a single channel to multiple stereo outputs (although there are only a few cases where you'd ever want to do this...).

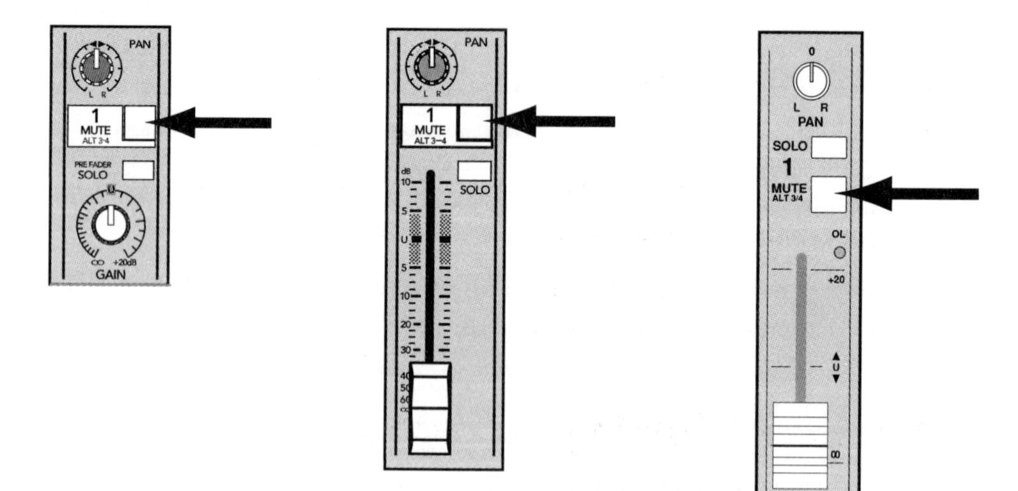

Routing controls for the 1202-VLZ, 1402-VLZ and CR-1604 (left to right).

CFX-Series: Four Bus

Like the previous models mentioned, the CFX series mixers have two pairs of stereo outputs. The difference is that rather than the single Mute-ALT 3/4 button, the CFX features individual assign 1-2, 3-4 and mute controls. In addition, each of the four buses has its own sub master fader.

This change overcomes all three limitations of the ALT 3/4 scheme, so that:

- All four bus outputs have master level controls.
- Channel mutes operate independently from bus assigns.
- You may assign a single channel to multiple stereo outputs (although there are only a few cases where you'll want to do this...).

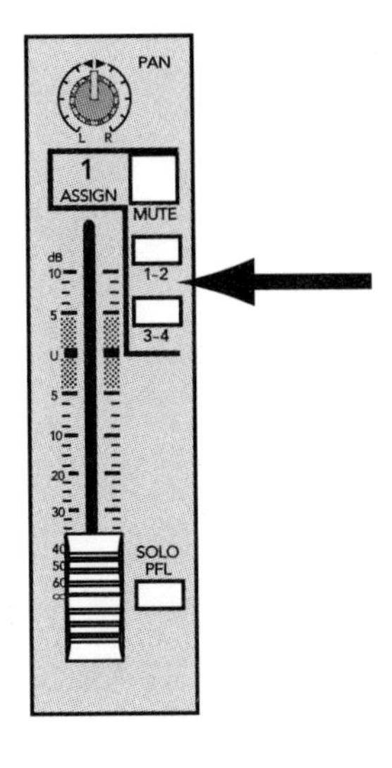

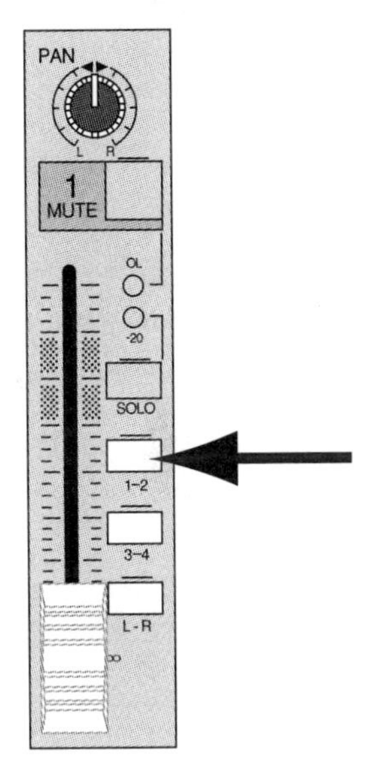

CFX (above) and six-bus mixer routing controls.

1604-VLZ, 1642-VLZ and SR-Series: Six Bus

At the top of the heap, assignment wise, are the 1604-VLZ, 1642-VLZ PRO, SR•24 and SR•32 mixers. Not only do these mixers have four independently assignable buses and individual channel mutes, they have an additional assignable stereo bus. This gives these mixers a total of six bus outputs, or three stereo output bus pairs.

Like the CFX series, these controls overcome all three limitations of the ALT 3/4 scheme, so that:

- All four bus outputs and the main output have master level controls.
- Channel mutes operate independently from bus assigns.
- You may assign a single channel to multiple stereo outputs (although there are only a few cases where you'll want to do this...).

A Preview of Coming Attractions

Now, I realize that you might still be a little confused about how these assignable buses relate to the master output section of your mixer. You probably also have unanswered questions about how you'll use the routing and signal assignment functions of your mixer. Fear not, these questions (and more) will be answered shortly. Hang in there baby!

CHAPTER 7

Auxes, Inserts and Effects

This chapter explains how to connect effects and other external "signal-processing" devices to your mixer. Before we jump into that discussion, you may be excused from reading this chapter if you have an important gig or session coming up. Perhaps you have just 30 minutes before the wedding guests arrive, and you need to make sure the PA is working when your Uncle Albert makes a toast to the happy couple. Reading this chapter will teach you to add echo, compression and flanging to Al's voice, but that isn't going to do you much good if you aren't sure how to hook up your mixer's outputs to your amp and speakers! In other words, the information provided in these two chapters is extremely valuable, but not essential in your quest for basic mixer mastery.

So, if this is your first time reading through this book, feel free to skip these two chapters and jump ahead to Chapter 9, Output Section Concepts. You can always come back and read these two chapters later. Right—off you go!

OK, glad to see you made it back. Hope the wedding was a roaring success. Let's talk about effects and signal processing.

Signal Processing Road Map

Up to this point, we've talked about connecting equipment to the inputs or outputs of your mixer. While we haven't said so directly, all these examples assume that a single piece of equipment is connected to either the input *or* output of your board, *but not both.*

As it turns out, there is a family of audio equipment designed to be simultaneously connected to both inputs *and* outputs of your mixer. Your Mackie mixer has a special set of connections and controls designed for these devices. (Note that we're going to ignore audio recorders for the moment, although they too are often connected to both mixer inputs and outputs.)

Some devices (like mics and amplifiers) are only connected to your mixer's inputs or outputs. Effects and signal processors (like the reverb and equalizer illustrated here) are connected to and from your mixer by using aux sends/returns or inserts. Note that signals travel in both directions on an insert connection (see the gray arrows), thanks to the special insert jack and cable.

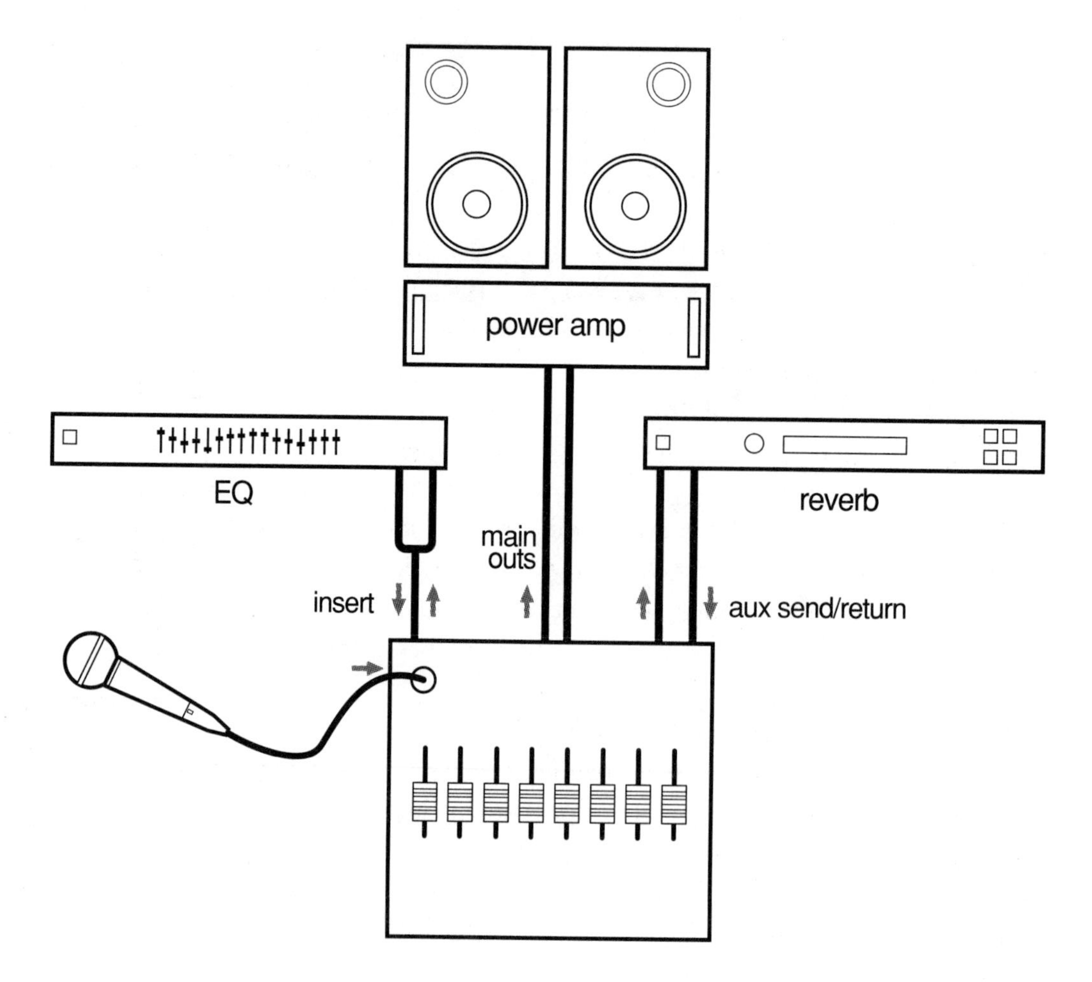

At first glance, it might seem strange that one would connect both the inputs and outputs of the same piece of equipment to your mixer. Isn't that just going around in circles?

Here's why: Mixing involves an almost endless number of variables. Since mixing board designers can't anticipate every possible circumstance, boards are made to work in conjunction with special-purpose audio signal processors. These external devices have capabilities that would be too costly or impractical to include as part of every mixer input channel. Since they are not physically part of your mixer, these units are often referred to as "outboard" gear.

You've almost certainly heard of devices like echo and reverb units, and perhaps graphic or parametric equalizers and compressors. All these devices are available as external pieces of equipment that can increase your sound-shaping capabilities.

The reason these units are connected to both your mixer's inputs *and* outputs is that they are used to modify individual sounds, but then these modified sounds *must then be added back into the final stereo mix.*

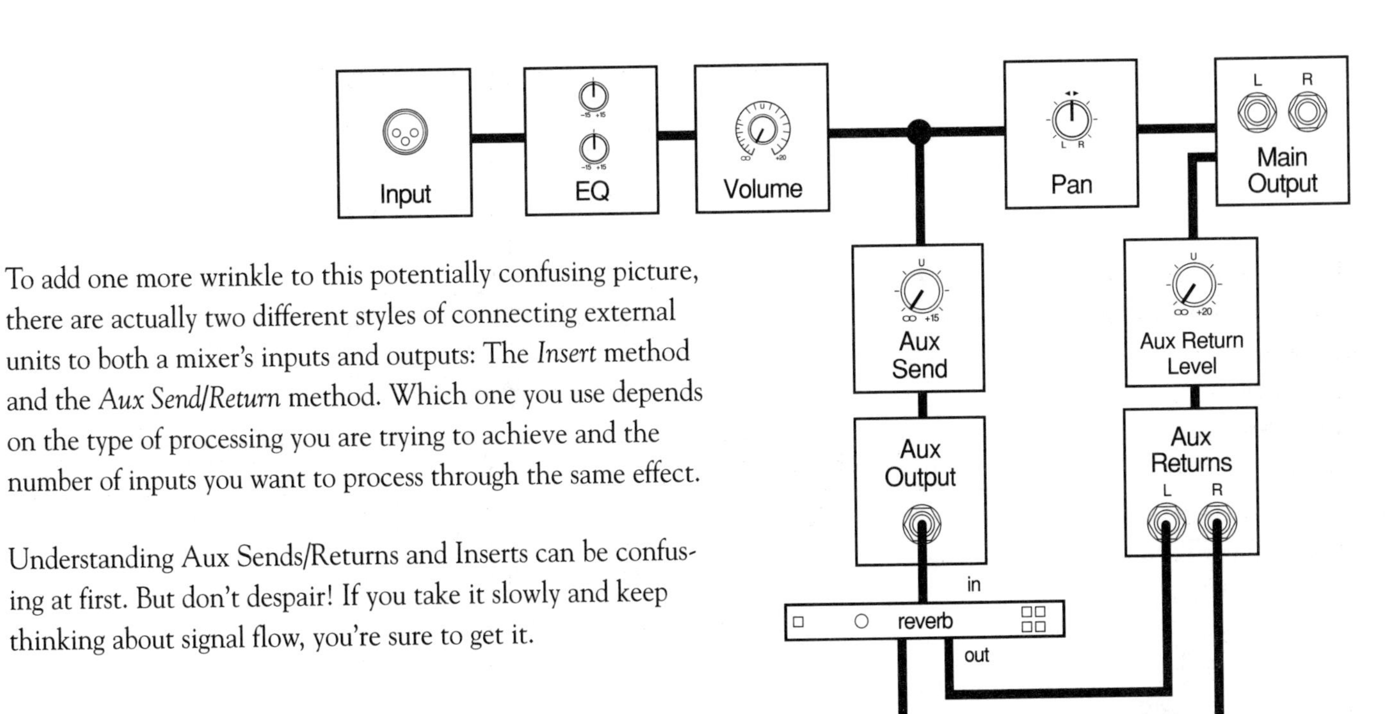

To add one more wrinkle to this potentially confusing picture, there are actually two different styles of connecting external units to both a mixer's inputs and outputs: The *Insert* method and the *Aux Send/Return* method. Which one you use depends on the type of processing you are trying to achieve and the number of inputs you want to process through the same effect.

Understanding Aux Sends/Returns and Inserts can be confusing at first. But don't despair! If you take it slowly and keep thinking about signal flow, you're sure to get it.

A Bit About Aux Sends and Returns

To help you understand aux sends and returns from a signal flow perspective, take a gander at the illustration above. As you can see, the effect send/return configuration taps off a portion of one or more signals without interrupting their original paths. The tapped-off signal is "sent" to an external device, processed and then "returned" into the main stereo mix. In addition, the original signal still makes it to the main output unaffected.

The effect send taps off part of a channel's signal and routes it to an external device, like the reverb shown above. The output of the reverb is connected to an aux return which is mixed back into the mixer's main output. Note that the original signal still passes through the channel, in this case, from volume to pan control.

A signal processor connected through Aux (short for "auxiliary") sends and returns is available to process *all* channels, although in practice you probably won't effect all your channels, just some of them.

Aux send/return configurations are used with devices like echo and reverb units, where you still want to hear the original unaffected sound mixed with the output of the external effects processing unit. Sometimes these devices are referred to as "effects processors," a definition that often is meant to exclude the equalizers, compressors and other "insert-oriented" devices mentioned above.

Some Mackies feature a built-in effects processor, called the "emac." In this case, the emac is accessed through an aux send and return that is internally wired within your mixer. While you won't need to patch any cables in to use emac effects, the same aux send and return concepts apply to its operation.

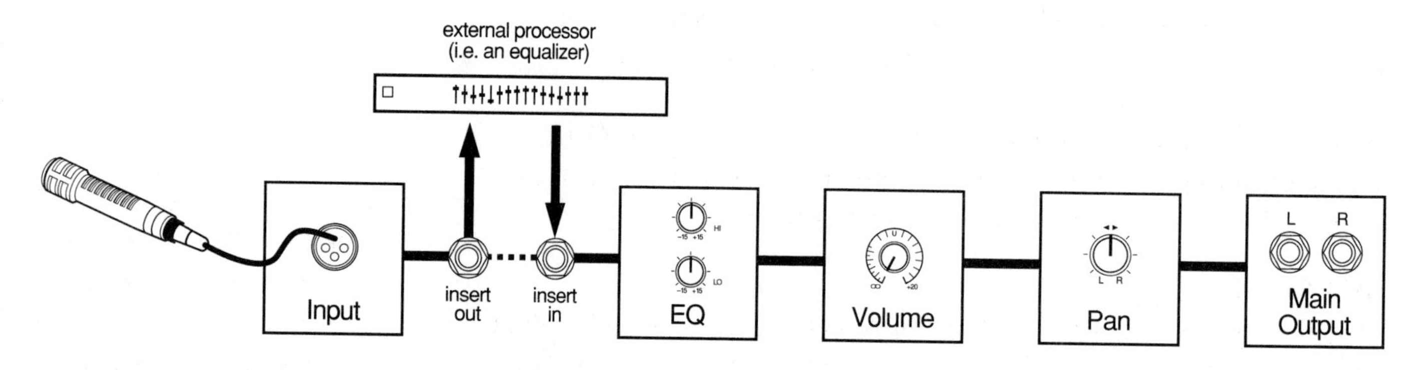

An insert connection interrupts the signal path within an input channel and routes the sound to an external device. The output of the device then returns the processed signal back into the channel, where it continues its normal signal path. If nothing is connected to the insert connector, the signal automatically flows through the channel. This "normal" path is shown by the dashed line.

Introducing Inserts

Understanding the difference between inserts and aux send/returns is easy when you see how they interact with signal flow.

As you can see in the illustration above, the insert signal path simply places the external processor right "in-line" or "in series" with one audio signal. This means the entire channel's signal will go through the device, get processed and then continue through the mixer's signal path to the output. Unlike signal processing connected to an aux send (shown on the previous page), devices inserted in an individual channel can process *only* that one channel.

Insert signal flow configurations are appropriate for equalizers, compressors, limiters and other devices that are applied to the entire signal, not just a part of it. Sometimes this category of devices are referred to as "signal processors," but this is a pretty broad, generic term.

Note that some Mackies include inserts on their main stereo bus. In this case, inserting an effect on the main bus would apply the signal processing to all sounds passing through the mixer's main output. We'll talk more about inserts later in this chapter. But first, let's take a deeper look at aux sends and returns.

Aux Sends and Returns In-Depth

You may be eager to experiment with effects units like digital reverbs. If so, sends and returns are the right controls for the job. With a properly connected effects unit, aux sends can be used to add varying amounts of effects like reverb or echo to any or all input channels.

Hooking up an effects unit is a two-part process. The signal must be sent to the external device, and then the resulting "effected" signal must be returned to the mixer to be recombined with the rest of your mix.

These two connections are called *sends* and *returns*. The aux knobs and aux output jacks make up the sends, and the stereo aux return jacks and knobs are the returns.

In practice, this means a cable from an aux output will go from your Mackie to the effect unit's input, and the effect unit's output—or outputs in case of stereo effect units—will be connected to your mixer's aux return jack(s).

WHY YOU NEED 'EM

Once this connection is made, you'll be able to add varying amounts of an effect to one or more channels. A typical effect used in this way is *reverb*, which makes a signal sound like it's playing in a spacious-sounding room. With a reverb connected to your mixer's send and return, you'll be able to take any of your input signals, and give them different amounts of reverb (or none at all), simply by turning their aux controls up or down while you mix.

If you have two separate effect units, you can hook them up to different aux sends and returns; then while mixing, you can send any input to either or both effects simultaneously, or neither.

In practice, if you are using two effect units, you'll probably want to have different, complementary effects dialed into each unit, perhaps a big-room reverb in one, and a single short echo in the other. Then you can use whichever effect suits a particular input during mixdown.

However, just because you have effects *doesn't mean you have to use them*! It's a common mistake to over-use effects, which leads to a mushy, un-focused mix (more about this in Chapter 13). Don't become another tragic statistic! Use effects wisely!

Here, the MS1202 Aux Send 1 and Aux Return 1 are used to connect an external effects unit. The Aux 1 knobs on each channel and the Stereo Aux 1 Return To Master would be used to adjust individual channel and overall effect levels, respectively. In this example, channel 1 is set so the delay effect will be added to its signal.

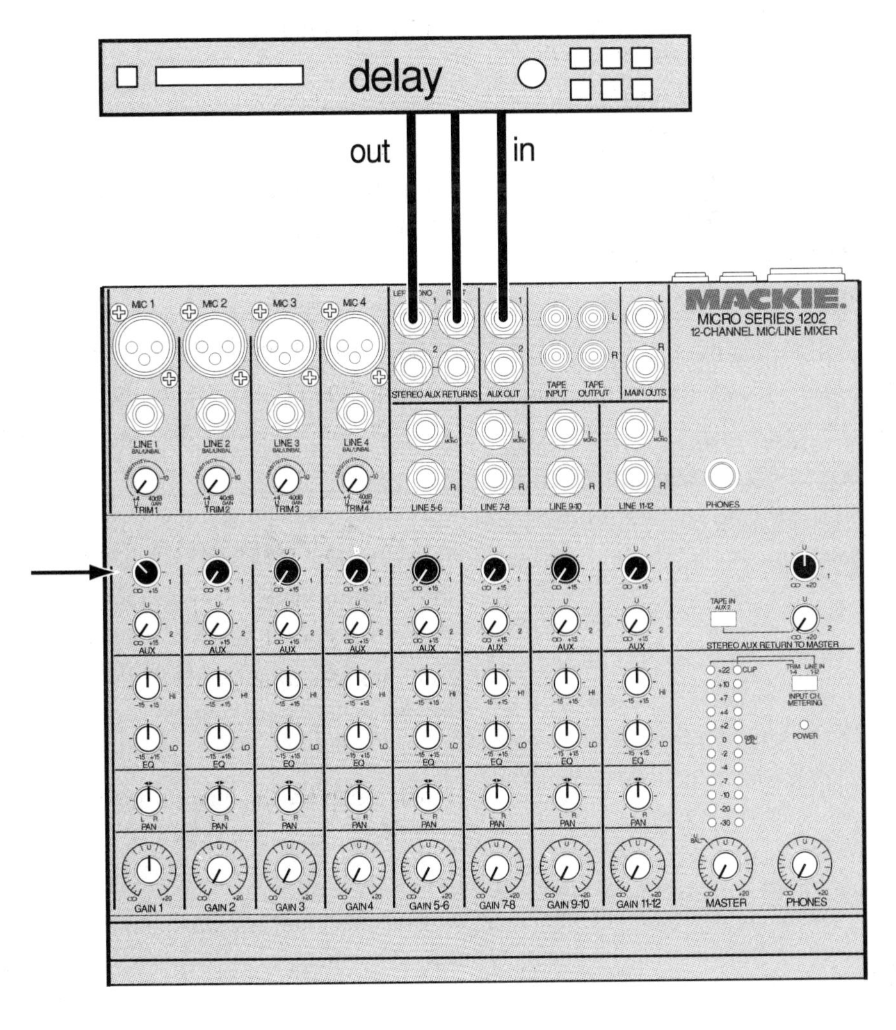

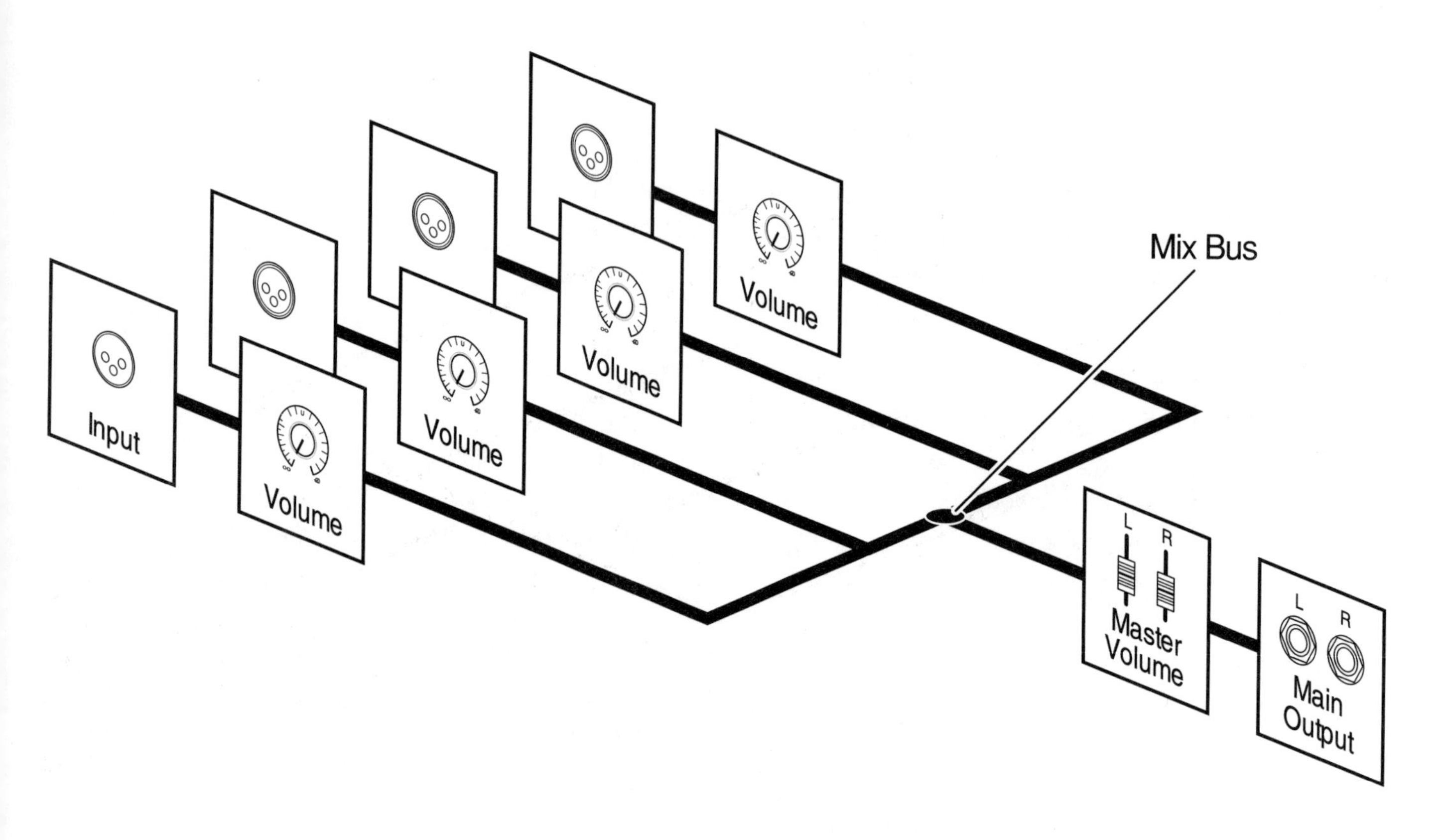

Here's an illustration of signal flow for a simple four-input stereo mixer, shown in a 3D perspective. For clarity, pan controls and individual left-right buses are omitted.

HOW SENDS WORK

An Aux Send (often called an Effects Send) is actually very similar to the gain control found on each input channel. Let's pause a moment to review the signal flow of multiple input channels, their gain controls and their combined output.

As you can see, each input channel can be sent to the main output simply by bringing up its gain control. The signals from all the input channels are combined before they reach the single main out. This combining circuit (shown in the block diagram simply as the intersection of lines) is known as a *summing* or *mix bus*. It's the same idea we discussed in Chapter 5. An aux bus is fundamentally identical to an assignable output bus such as ALT 3/4 output or the four buses of, for instance, a 1604-VLZ. The only difference is that instead of a push-button assign switch, you have a variable knob to control the signal routing.

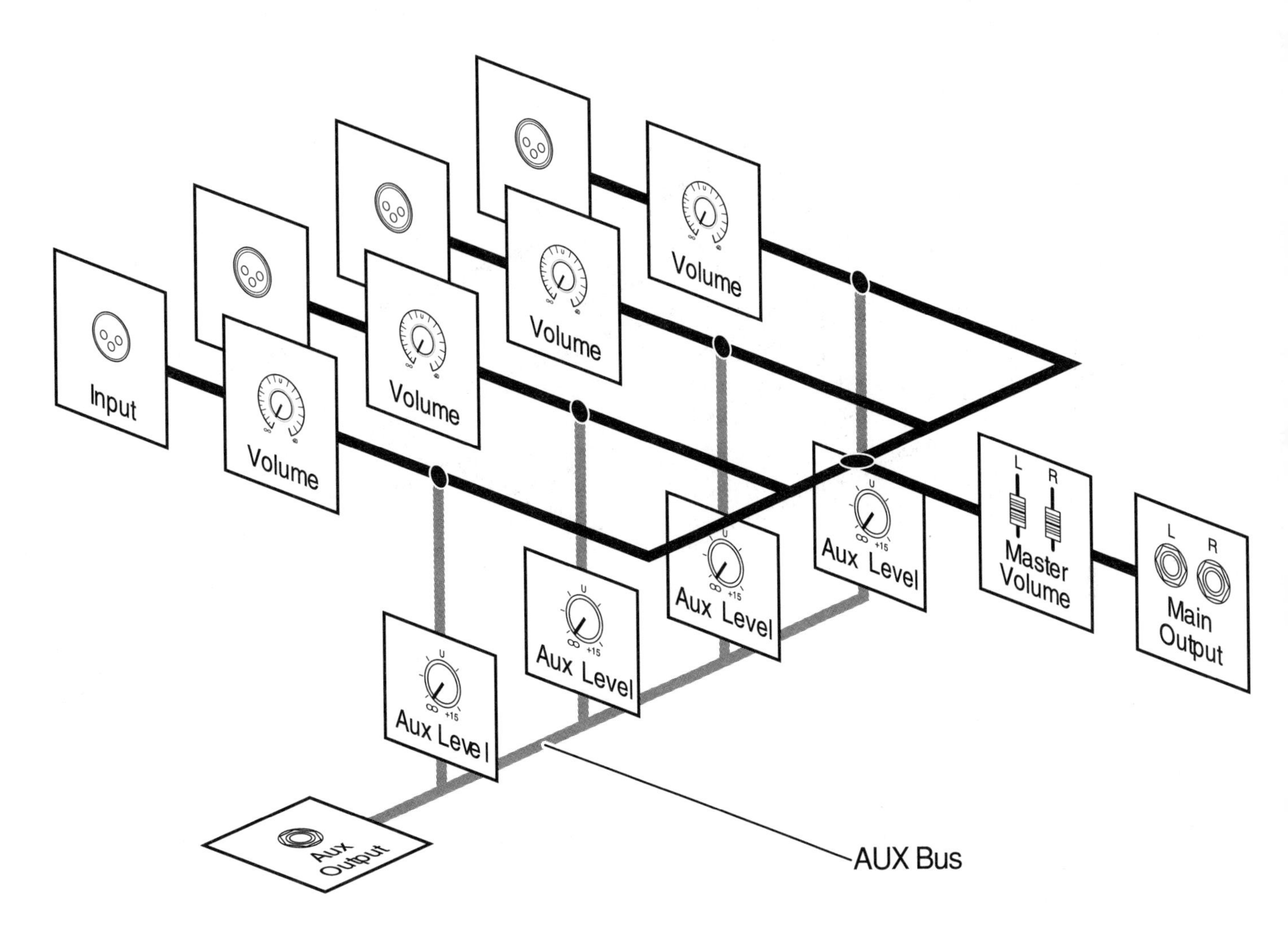

Here's the same mixer from the opposite page, with the addition of an aux send bus. The 3D perspective should help give you an idea of the independent signal routing option an aux bus creates.

This bus-approach makes a single external effects unit accessible to one or all input channels, simply by turning up their individual aux knobs! This is especially useful, as you may want several different sounds to get varying amounts of a particular effect. In those cases where you want to dedicate a particular effect to one channel exclusively, using an individual channel insert may be a better choice.

PRE-FADER AND POST-FADER SENDS

All important questions about mixer behavior come back to signal flow, and aux sends are no exception. In this case, one might pose the question, "what happens to a channel's aux send when its main level control is turned up or down?"

At first, one might not even think to ask this question—isn't this kind of obscure, dude? Well...no! Here's why: Let's say you are mixing, and have a reverb unit attached to Aux Send 1. You've got a working balance of each input signal, as well as pleasing effect levels on several different channels.

Then, as you are mixing, you decide that one of the backing parts (which happens to have a lot of reverb on it) should be turned way down, to improve the musical arrangement. Fine so far. But when you turn that channel down, what happens to the Aux Send 1 level?

If turning down the channel fader doesn't also turn down the send, a funny thing will happen—the "dry" sound will get softer, but the *reverb sound will be just as loud as before*. This can drastically change the mix balance you created and will make mixing much more difficult, since you'll constantly be compensating with send level changes to follow your main channel volume moves.

It is much more convenient if changing the channel volume automatically makes the send get softer or louder too. Then, the balance between the dry and "wet" or "effected" parts of a sound stays the same, even when you turn that channel up or down relative to the rest of the mix. But how would the send level be able to adjust itself automatically when the channel level moved? Invisible monkeys sitting on your mixer, following your every fader move? Not exactly.

The location of the aux control in the input channel's signal path can vary. In the illustrations on this and the facing page, note the location of the aux control in relation to the channel volume fader. A post-fader aux send (this page) is best for working with effects, while a pre-fader send (facing page) works well for stage monitor and headphone mixes.

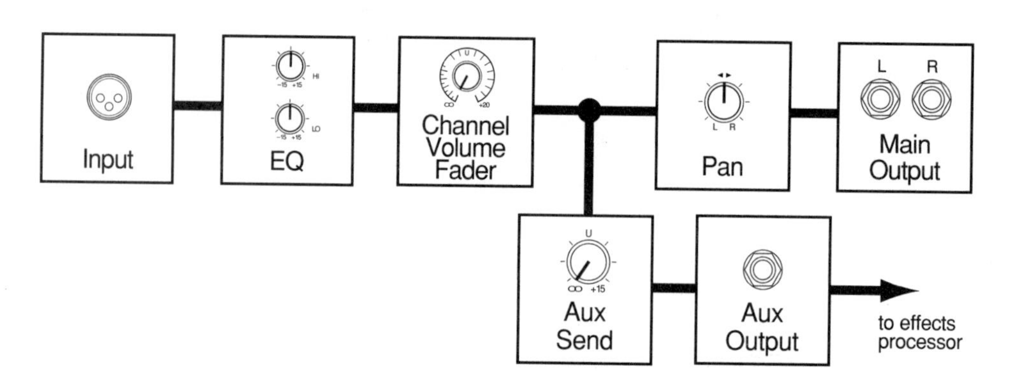

Once again, we return to signal flow for our answers. Have a peek at the illustration above. As you can see, if the send taps off its signal at a point in the signal path that comes *after* the Volume Control or Fader (post-fader), then any changes made to the channel volume fader will change the amount of signal available to the Aux Send. This results in the automatic adjustments we are seeking. (Note that the word *fader* here is meant to be generic—mixers with knobs instead of sliders, such as the 1202 and 1402 still have Channel Volume Faders.)

It's just another example of our "water in the river" analogy. If someone upstream holds back some water, less water will be available to everyone downstream. What is true for level is also true for EQ. If the Aux Send taps its signal from a point after the channel EQ, then any EQ changes will affect the sound of the send's signal.

On the other hand, if the Aux Send gets its signal from a point ahead of the main Channel Volume Fader (pre-fader), it becomes independent. Level changes to the Channel Fader have no effect on the Aux Send's level. Also note that the Channel Fader is not getting its signal from a point after the Aux Send control, so the channel level is also independent from the Aux Send setting. This is shown below:

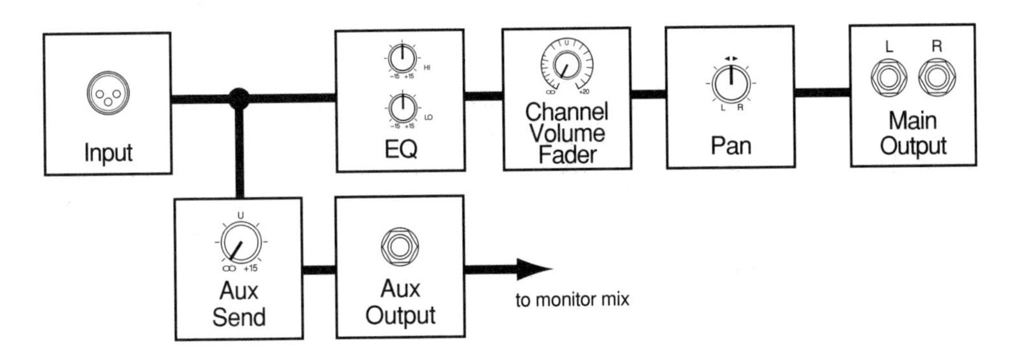

To simplify mixing, post-fader sends are the best choice for working with effects sends and returns. Pre-fader sends are used when you want the send to maintain its own mix, unique and independent from the mix set by the main faders. The most common reason to do this is when you are making a stage monitor mix for musicians during a performance or recording session. They may need a different mix than the main mix, for instance, one with louder vocals to compete with the louder stage volume of other instruments (an example of a monitor mix hook-up appears in Chapter 12).

Using a pre-fader send for a monitor mix means that if you turn down the backing vocalists, they'll still hear themselves just as loud in their stage monitor speaker or headphone mix. Most Mackies include both pre-fader and post-fader aux sends. On some models, you'll see pre-fader sends also labeled as "Mon" (short for monitor) and post-fader sends marked EFX. You'll read all about your particular model's implementation in the next chapter.

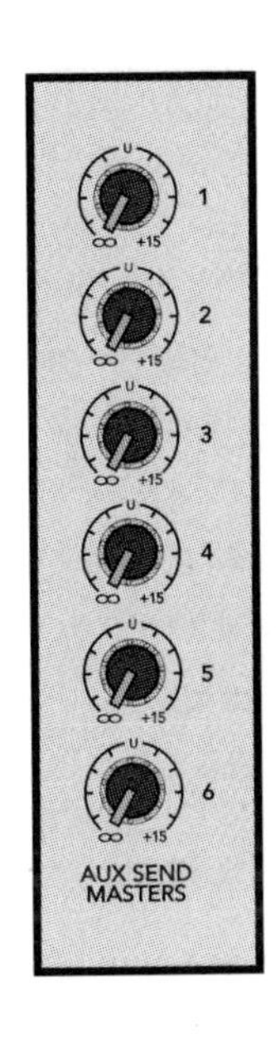

Aux Master level controls provide a way to vary the overall level of an aux send. You'll still use individual channel's sends to set their effect level, while the Aux Master is one knob that controls the overall effects or monitor send level. The example above shows the Aux Masters of an SR-series console. Other Mackies may provide masters for some, but not all Aux Sends, while still other mixers have no Aux Masters at all.

OVERALL AUX OUTPUT LEVELS

Always remember that the input and output levels of every part of an audio system will need to be set appropriately. This is certainly true for aux send output levels. When connecting an external effects unit, see if it has input level meters and an input trim control of its own. Then use your individual channel aux send levels and the effect unit's master input controls (if present) to make sure the levels to and from the effect are set appropriately.

Note that some Mackies include an Aux Send Master control. This is a nice feature, since it lets you change the overall level being sent to the external device without you having to make individual adjustments to each channel's Send knob. Aux masters are also handy when creating headphone or stage monitor mixes, since an aux master gives you an easy way to turn everything up or down at once.

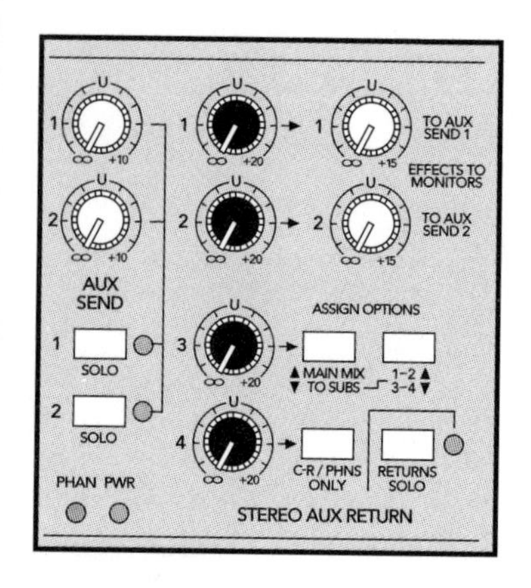

Aux Return controls let you set the overall level of the signal coming out of your effects unit. Here, the return section of a 1604-VLZ is shown. The four Aux Return knobs are highlighted in black.

About Aux Returns

Using your mixer's aux sends, you can feed a signal to an external effect unit. However, you'll need to get the signal out of the effect, and back to the board, so you can add the effect to your overall mix. The Aux Returns are designed for this purpose. Simply connect the output from your effects unit to an Aux Return and you'll be able to control the overall level of that effect in your mix by varying that aux return's volume knob.

To help me keep track of things, if an effect is already connected to Aux Send 1, I often choose to connect its outputs to Aux Return 1, just to keep the numbering consistent. Note that this is not required—an effect who's input is connected to Aux Send 1 can have its outputs connected to Aux Return 4 if you like.

From a signal flow perspective, signals attached to any aux return typically end up in the main stereo mix, just like signals connected to your mixer's main input channels. In fact, it is accurate to think of the returns as simply additional line inputs, because that is exactly what they are.

Of course, there are differences between the returns and full input channels: Aux returns have only a level and in some cases, a Pan control, while a "real" input channel has EQ, Sends, Trim controls and so on. The reason that returns are simpler, stripped-down inputs is that it's assumed that all you need is a line input with just a level control—any fine-tuning will be taken care of within the effect unit itself or at the individual channels that are sending to that aux bus.

This is often true. However, if you need the flexibility a regular input channel offers, you can skip the Aux Return inputs and connect your effect unit's outputs to normal mixer input channels. In fact, when I have spare input channels available, I often prefer this. It makes it easy to adjust the overall effect level with a couple of faders and add EQ, if necessary.

Returning an effect's outputs to a channel also makes it possible to send that effect's signal *to another effect*, such as sending a delayed sound into a reverb. Similar results can be achieved by patching the output of one effect directly into the input of another. Then your aux send would connect to the input of the first effect, and the output of the second unit could feed your mixer's aux returns.

Finally, while all Mackies can route their aux inputs to the main stereo bus, some models allow returns to be routed to alternate buses or even to other sends (this is helpful when trying to put reverb in a stage monitor or headphone mix). We'll discuss these fancier schemes when we cover the individual mixer models in the next chapter.

Effect Unit's "Wet/Dry" Settings

IMPORTANT: When hooking up an effects unit with sends and returns (into the dedicated aux returns or regular inputs) there is one very important setting to make on the effects unit itself. The effects unit will have level controls for both the "effected" version of the sound, as well as the level of the unadulterated "dry" sound. This setting is called by various names: *wet/dry mix, direct/effect level, balance,* and so on. ("Wet" refers to the "effected" sound, "dry" means the original unchanged signal.)

The settings may be controlled by front panel knobs, or, on many digital effect units, may require a lot of button pushing to find that parameter in the unit's tiny little display.

What's important is that you set the level of the dry, or "un-effected" signal *off* or to *zero*. The output level of the "effected" signal can usually be set at the default level. In units where a balance-style control is used, both wet and dry levels are controlled at once. In this case, make the signal 100% wet.

Failure to make this setting to external effects units, connected via Aux Sends and returns, will make mixing difficult—when you turn up a channel's Aux Send, not only will you be making that sound have more of the effect, you'll also be making its direct sound louder, throwing off your mix balance with other sounds.

The emac Effects Unit

The CFX, DFX and PPM-series mixers are Mackie's first analog mixers to feature a built-in digital signal processor, the emac. It offers delays, reverbs, chorus, flanging and phasing effects, a total of 16 different settings each with two knobs for fine-tuning effect parameters. (The DFX version of emac lacks the two fine-tuning knobs.)

Although the emac is built into your mixer, it is accessed through internally wired effects sends and returns, just like an external effects unit would be. So, in order to use the emac successfully, you must understand the general discussion of sends and returns that appeared earlier in this chapter.

The emac effects unit is built-in to the CFX, DFX and PPM series mixers. It offers many commonly used effects.

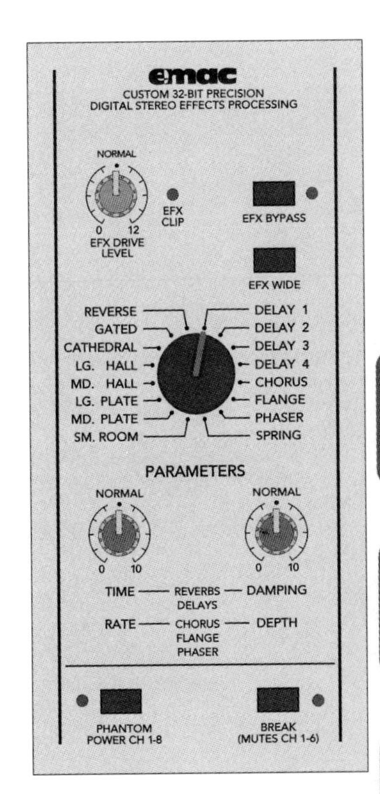

Using emac

Operationally, the emac is quite straightforward. Here's a two-minute guide to adding effects to your mix.

- Turn your mixer on!
- Connect a mic to an input channel and turn it up.
- Set its Trim (or input level set) knob appropriately.
- Turn that channel's EFX control up half way.
- Set the EFX Drive level (or EFX 2 Send) knob to "normal" (12-o'clock on DFX mixers)
- Using the main 16-position dial, choose the desired effect type.
- Make sure the EFX Bypass button (or footswitch) is NOT turned on!
- Turn up the EFX To Main knob to its half-way point (EFX Return fader on DFX-series mixers).
- Speak into the mic. You should hear the effect applied to your vocal.
- Use the individual channel EFX send knobs to add effects to other input channels as needed.
- Turning each channel's EFX send up or down controls the amount of effect on each channel, but does not change the level of that channel's "direct," or unaffected signal.
- Turning the EFX to Main control up or down adjusts the overall master level of the effect signal, but does not change the level of the unaffected parts of your mix.
- Don't over-use effects!

Mackie's manuals do a fine job of explaining the particulars of each effect, so I won't repeat that here. Instead, I'll describe what each of the effect types sounds like and define some effects-related terminology you might not be familiar with.

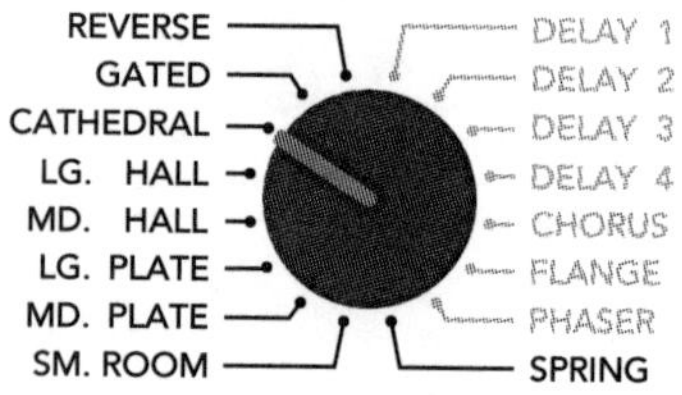

REVERSE
GATED
CATHEDRAL
LG. HALL
MD. HALL
LG. PLATE
MD. PLATE
SM. ROOM
DELAY 1
DELAY 2
DELAY 3
DELAY 4
CHORUS
FLANGE
PHASER
SPRING

Reverbs

Nine of emac's sixteen effect presets provide "reverberation" (reverb for short). Reverb is an electronic simulation of how sound bounces around within a room, hall, cathedral, football stadium, etc. Reverb is a popular effect, but in my opinion, one that is frequently over-used. In most live performance spaces, the natural reverb of the room is already causing problems with clarity and definition of your overall mix. Adding lots of reverb will not improve intelligibility or definition, especially if your music is up-tempo with strong percussive accents.

On the other hand, in smaller or "deader-sounding" venues (i.e. rooms with lots of carpet, curtains and other sound-absorbing materials) adding reverb to featured sounds, such as a lead vocal or melodic instrument gives a nice sense of space that would otherwise be lacking.

A pair of knobs (not available on DFX mixers) lets you fine tune the character of each emac reverb effect. The Time knob determines the duration of the reverb. A reverb's duration is the length of time the reverb "tail" takes to fade away after the original sound stops. Setting the Time control to minimum makes for very short reverbs, while increasing this control can create reverb tails that stretch out for many seconds.

The exact range of reverb times varies with each of the nine reverb settings and is documented in your Mackie manual. What's worth noting is that several different reverb settings are essentially identical except for their Time setting ranges (Long and Medium Plate differ only in their durations).

The second control, Damping, doesn't change the duration of the reverb tails, but it does affect their character. At minimum settings, the damping knob gives the reverb a very dark, warm character. Imagine clapping your hands in a theater with drape-lined walls and upholstered seats. Now imagine how it would sound if the building was gutted and all that remained were concrete walls and floors. Low damping settings approximate the heavy carpet, while high damping values provide the bright, sizzling reverb characteristic of hard wall surfaces. At first listen, many people will be drawn to brighter-sounding reverbs, but I tend to prefer the darker ones, especially when working with instruments with strong transients—percussion, acoustic guitar, etc. I find that a bright reverb adds an unnatural "zing" to attacks that can tire my ears after a while.

Besides the duration and character of the reverb tails, electronic reverbs vary in the quantity and spacing of their so-called "early reflections." Briefly, early reflections are like a bunch of individual little echoes that cluster right around the beginning of the reverberation. They mimic the effect of sound bouncing off the nearest wall, floor or ceiling and reflecting back to the position of the listener. As the early reflections are dying out, the main body of the reverb is building up, and continues well past the time that the early reflections have faded to silence. While emac doesn't give you direct control over early reflections individually, each family of reverb algorithms has its own early reflection character.

Another parameter that helps define different sized acoustical spaces is "pre-delay." This is a short delay that postpones the onset of early reflections and the reverb tail. By adding pre-delay to a reverb effect, the sense of a large hall is created.

It is as though the nearest large wall is some distance away, and the sound has to travel that distance (and back) before you hear any reflections. Pre-delay is not directly adjustable, nor does it vary with different settings of the Time control.

Note that "Spring" is the ninth reverb effect, even though it appears next to the flanging, chorus and phaser settings. Spring is so named because it attempts to recreate the sound of a mechanical spring reverb common on guitar amplifiers and other older sound equipment. Unfortunately, I have yet to hear a digital effects unit that can convincingly recreate this effect, because when you drive a real spring reverb with a loud enough signal, you get an amazing "boinging" sound as the spiral windings of the spring crash against each other. If you really want a spring reverb sound, get a spring reverb! I did.

Delay Effects

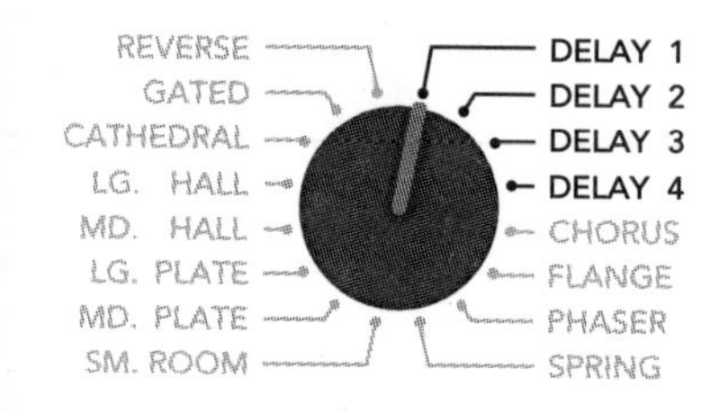

Delay is a simpler effect than reverb. Instead of a burst of early reflections followed by a reverberant cloud trailing off into silence, delay effects simply capture the incoming sound and spit it back out after a short pause. This simplicity is often an advantage in live performances—a delay can be less likely to clutter the mix than a dense reverb fog.

While the emac includes four delay settings (Delay 1-4), all are minor variations on a theme. To understand how they differ, one must be familiar with the features common to all delay units.

A basic delay offers three parameters: The first is delay time. This is the duration between the moment you shout "Hello" and the moment you hear yourself echo "Hello" back. As you might expect, the Time knob varies the delay time (between about 30 milliseconds and half-a-second).

The second most common delay parameter is "feedback." Not to be confused with the nasty squealing sound your PA makes when you turn up a mic too far, this sort of feedback takes part of the echo sound and feeds it back into the effect, so that HELLO...HELLO becomes HELLO...HELLO...HELLO...HELLO. On the emac, feedback is controlled by switching between the four different delay presets. Delay 1 features a single repeat, Delay 2 has two, Delay 3 has three distinct repeats of the original sound, etc.

Finally, the addition of some EQ or "filtering" in the delay's feedback path creates echoes that get darker in tone for each subsequent repeat. The Damping knob performs this function on the emac. Turning it up to full means that each echo has exactly the same tonal character as the original sound. Reducing the damping makes the echoes warmer and darker-sounding. This mimics the behavior of older tape-based echo units, and also is somewhat close to how sound behaves in the real world. In many cases, I'd start with Damping settings at the half-way mark or below, because it helps the listener discern the original sound from the subsequent repeats.

Flanger, Chorus and Phaser

The three remaining emac settings are part of the family known as "modulation effects." Flangers, Chorus and Phaser effects all work by using a very short delay mixed back in with the dry signal. Furthermore, the duration of that delay time is varied slightly, back and forth. This delay time modulation creates their characteristic swooshing or rippling sound.

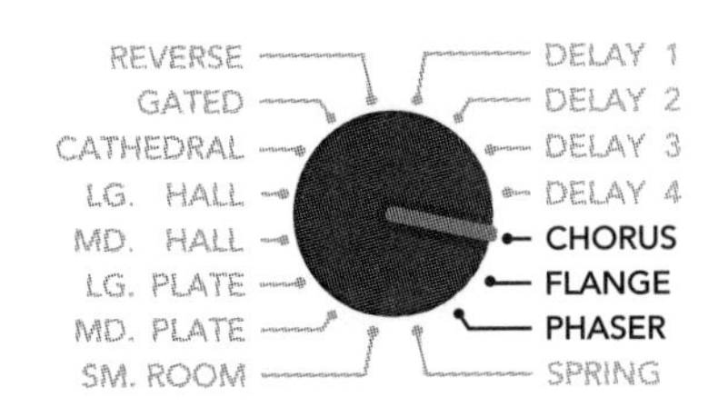

To my ears, flanging is the most vivid, imparting an almost metallic or robotic character to the sound. Chorus provides a somewhat more subtle effect, which could be described as adding an airy swirl or liquidity to the sound. Phasing is also a pretty fluid sound, but thicker and a bit darker than chorusing. The emac phaser mimics the sound of vintage phase shifter effects from the '70s. While one might apply reverb to many channels of a mix, modulation effects are often used on just one or two instruments for effect.

As with the reverbs and delays, the two knobs can fine-tune each modulation effect. However, rather than assigning the knobs to control delay and damping, they control Rate and Depth when the Chorus, Flange or Phaser effects are selected.

Rate and Depth both refer to the modulating signal that varies the delay time at the heart of these effects. This delay time is sped up and slowed down by a very low frequency signal. It is the frequency and amplitude of this modulating signal that are modified by the Rate and Depth knobs. As you might guess, the Rate knob changes the speed of the modulation signal. At its lowest setting, you'll hear a couple of sweeps per second. At higher settings, the sweeping effect happens so rapidly that you won't be able to follow individual swooshes.

The Depth knob controls the amplitude of the modulating signal, which in turn increases the variation in delay time of the effect. In English, this simply means that turning up the Depth makes the effect more pronounced.

Super-Cool EFX Volume Pedal!

Emac provides a Bypass button that lets you turn the effect on and off. I like this, because I always find it silly when a performer speaks to the audience between songs with a ton of reverb or echo on their voice.

If you want to turn effects on and off without having to walk over to your mixer, you can hook up an EFX bypass foot switch. However, I'd like to suggest an even cooler alternative: a volume pedal-controlled aux send level. To pull this off, you'll need a regular volume pedal and an "insert cable," which has a three-conductor stereo 1/4" connector on one end and two regular "guitar cord" cables at the other.

Simply plug the stereo end of the cable into the emac Effects Send jack. Make sure it's pushed in all the way. Now, connect the other two ends of the insert "Y" cable into the input and output of the volume pedal. At this point, turning the pedal up and down with your foot will make the effect louder and softer. If you don't hear any change in the effect level, swap the cables between the foot pedal's input and output (FYI, the cable from the "tip" connector should go to the volume pedal's input jack).

Now, here's the coolest part of the trick: Set the effect to a very long reverb or delay time, and turn up the effect levels so that your signal is just *drenched* in the effect. Now, turn the foot pedal down or off completely and begin playing. As you reach a particularly dramatic phrase or lyric, push the volume pedal up for a moment, then quickly bring it back down as you continue playing or singing. What you'll hear is the key phrase with a huge reverb or echo tail which dies away very slowly and naturally as you continue to play (without adding yet more reverb or echo to the sound).

From a signal flow standpoint, the foot pedal comes before the effects processing. This is why you can turn the pedal *up* to "load new sounds" into the reverb, but then turn *down* the pedal so that additional material is not being fed to the effect. What makes this so cool is that the sounds loaded into the effect will continue to decay naturally, regardless of the position of the volume pedal—they won't be abruptly cut off, as is the case with the EFX Bypass foot switch.

Try this setup with an unaccompanied guitar or other solo instrument. You can use a long reverb to sustain individual notes or chords while you play melodic figures over the top of the reverb tail—it's a gas!

Inserts, In-Depth

Most of a mixer's signal path is hidden away inside its case. You can see inputs and outputs, but the mixer's internal signal path can be modified only through its knobs and switches.

However, a channel's signal path can reach the outside world between input and output, almost as if it were coming up for air. At this *insert* or *channel access* point, you can extract the channel's signal and process it in ways that can't be done within the mixer itself. Then, you have the option of returning the processed signal back into the mixer to rejoin the rest of the mix.

If we look at inserts in signal flow terms, I think you'll understand why they can be such a powerful tool. In contrast to the aux send/return routing scheme, the Insert signal path places the external processor directly "in-line" or "in series" with one audio signal.

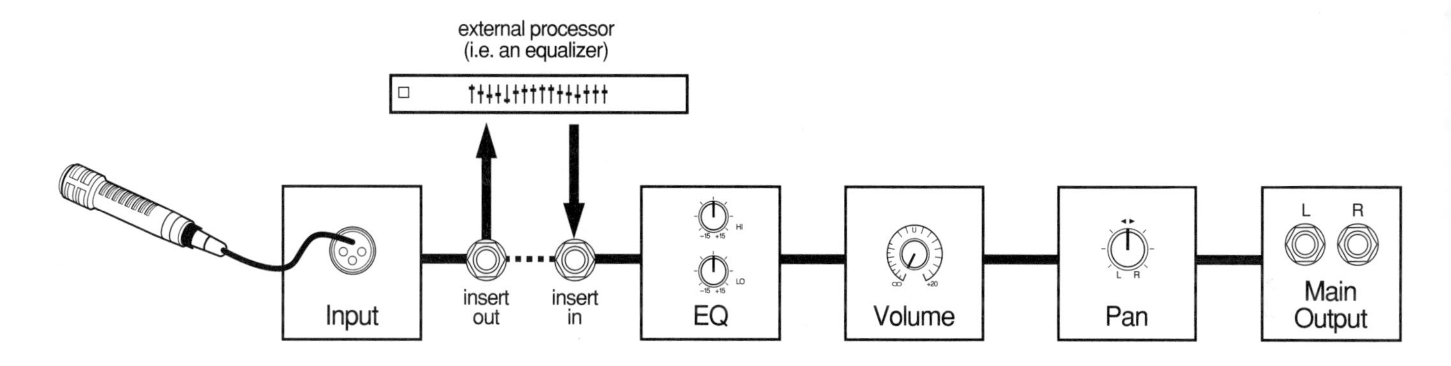

Therefore, the channel insert allows you to add extra processing right in the middle of an individual input channel! For example, the original MS1202's two-band EQ, while useful, is pretty basic as EQ goes. The channel insert allows you to take a sophisticated external equalizer and graft it right into your 1202's signal path, allowing much greater tonal shaping power for an important input signal (like a lead vocal, for instance).

An insert connection interrupts the signal path within an input channel and routes the sound to an external device. The output of the device then returns the processed signal back into the channel where it continues its normal signal path. If nothing is connected to an insert connector, the signal automatically flows through the channel. This "normal" path is shown by the dashed line in the illustration above.

Inserts create a signal path significantly different from aux sends and returns. While sends and returns allow some *portion* of multiple input channels to share a single effects unit like a reverb, the insert signal flow configuration applies processing to one *entire* audio signal. This makes it better suited to equalizers, compressors and other devices that are meant to shape a complete signal, rather than add an extra effect to it.

ONE JACK DOES IT ALL

How exactly are inserts implemented on Mackie mixers? As you can see in the illustration below, the insert is a single connector. But an insert needs two connections—input and output. How can a single jack be both? Good question.

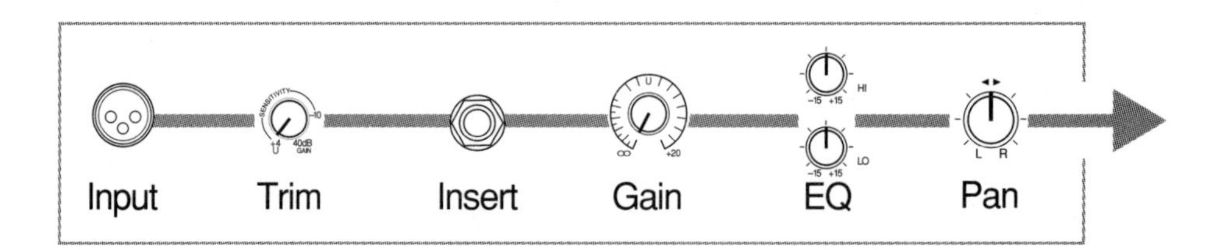

While an insert uses only one jack (as shown above) that connector serves as both input and output, when used with a proper "insert cable." When connected, the signal flow through the insert jack is as shown below.

As you may know, 1/4" phone plugs and jacks come in two varieties—two-conductor "tip-sleeve" and three-conductor "stereo" or "tip-ring-sleeve" (TRS) connectors. If all this talk of tips and sleeves is new to you, a "tip-sleeve" plug is like a guitar cord, while a "tip-ring-sleeve" connector is what you'll find at the end of a pair of stereo headphones (see Chapter 11 for more connector info).

Insert jacks use TRS connectors. The tip connection carries the output signal to the external device's input, and the ring connection goes from the output of the external processor back into the mixer.

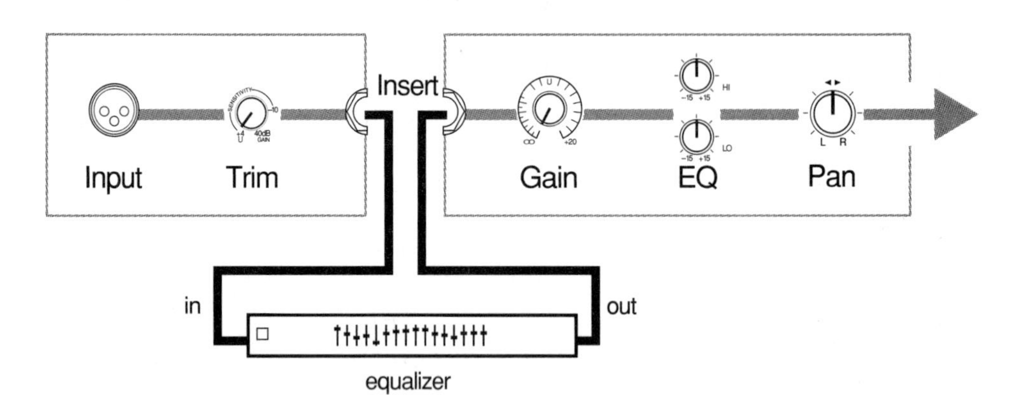

Hooking up to an insert jack requires a special Y cable. This type of Y cable—not coincidentally called an insert cable—has a TRS plug on one end that Y's out to two mono 1/4" plugs (see below). The stereo end goes into the mixer's insert jack. Then the two mono ends go to the input and output of the external processor, as shown on the following page:

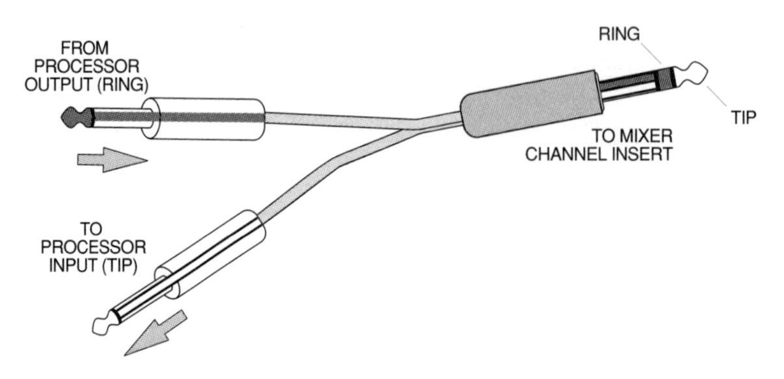

Another important point to notice is that the insert jack automatically passes the signal through unless a plug is inserted all the way. This lets you ignore an insert jack when you don't need it. A link that's made automatically when *no* cable is connected is called a "normalling" connection. The name refers to the "normal" signal path automatically made when no cables are plugged into the connector.

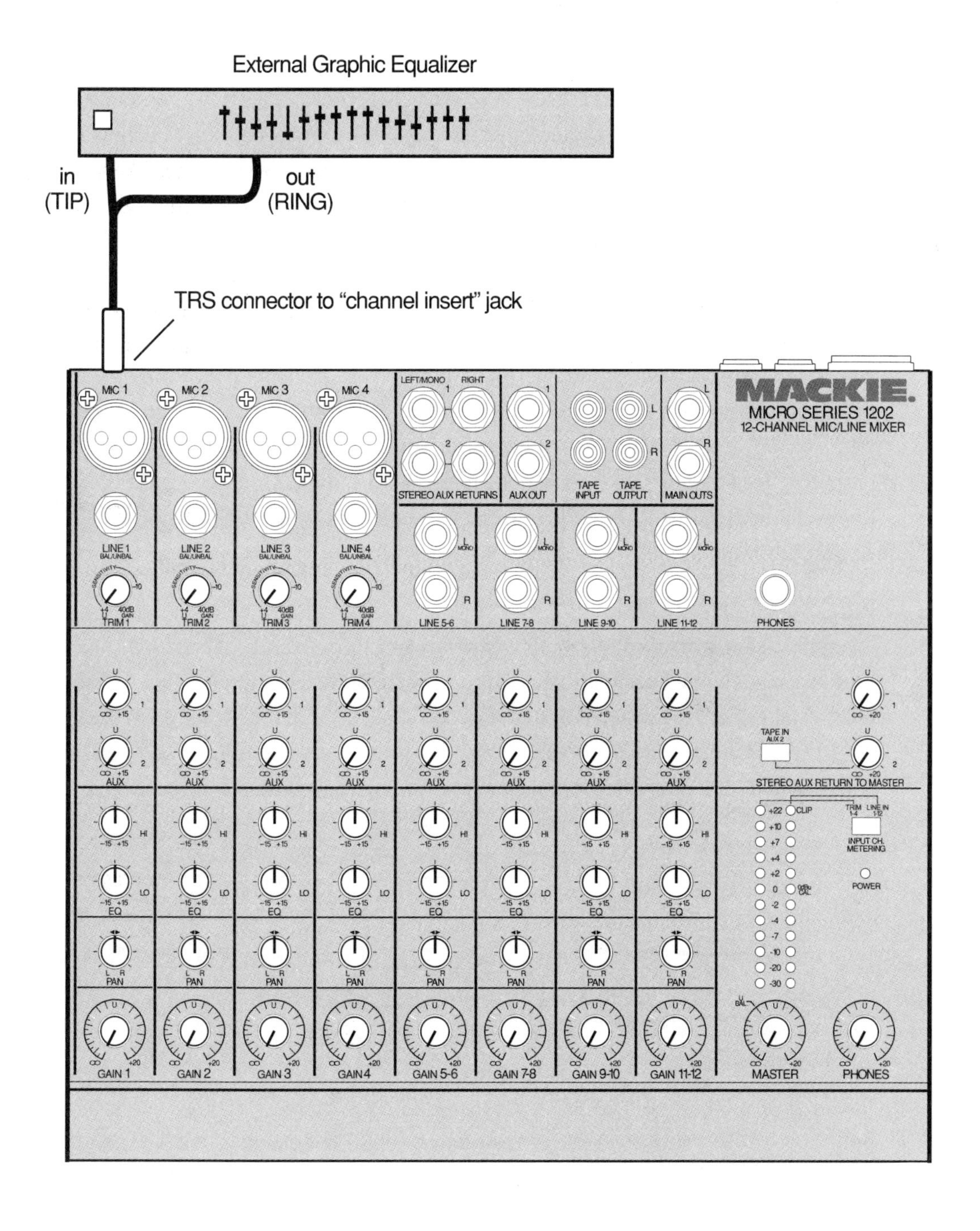

INSERTS AS DIRECT OUTS

There is another useful purpose for the channel inserts. You can use them as *direct outputs*. A direct output is an output from a single channel that is extracted from your mixer at an insert point, but *doesn't come back to the mixer*. (Note that some Mackies include separate direct outputs in addition to inserts.)

To use a Mackie insert as a direct out, connect a 1/4" mono cable pushed in all the way to the second clock.

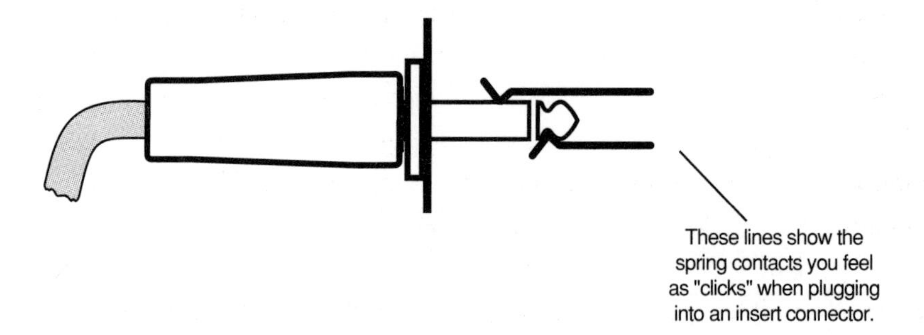

You can use an insert as a direct output when you want to take a single input signal and route it to an external audio device *by itself*, without tying up one of the mixer's main outputs.

The most common practical use for direct outputs is in multitrack recording. Imagine working with an 8-track recorder and a Mackie with ALT 3/4 outputs, such as a CR-1604, 1202-VLZ or 1402-VLZ. You could connect your ALT 3/4 outputs to feed two inputs of your multitrack recorder. Then, to record additional tracks simultaneously, you could use inserts as direct outputs, connected to additional multitrack inputs.

INSERTS AS SPLIT OUTS

It's also possible to tap a direct output from a Mackie Insert *without* breaking the normal input channel's signal path. This means that channel's input signal will still reach your mixer's main stereo output, but you'll have a *split* of that signal to use elsewhere. The term *split* refers to a single signal that is separated into two or more copies, which can be connected to the inputs of different devices.

To use a Mackie insert as a split out, use a 1/4" mono cable pushed in only to the first click.

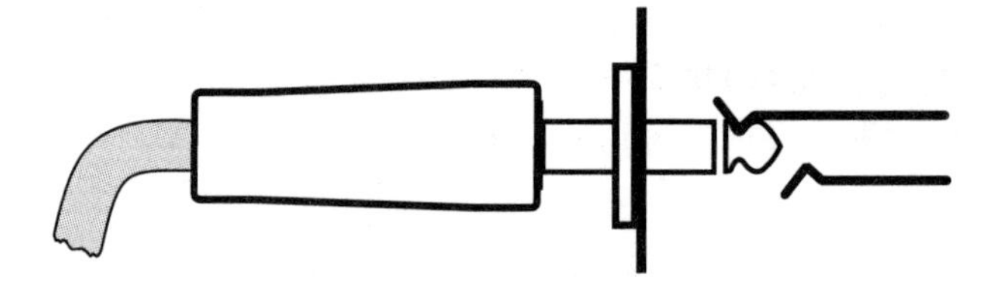

Split outs can be handy if you've run out of aux sends. For example, the 1202 and 1402-VLZ have just two aux sends. If you wanted to use a third effect unit on a solo instrument, you could take a split out from that channel and connect it to the input of the effect unit. The effect output would still have to be connected back into a mixer input.

Some Mackies also have insert jacks on the main stereo output bus. You can use a split connection to get an extra pair of main stereo outputs, tapped off a point before the main volume faders. Note that since these bus inserts are pre-fader, you won't be able to create fade-ins or -outs on the device receiving the split.

Other ideas? Splits are also common in live sound systems. A flexible sound reinforcement system will often have a dedicated mixing board for setting levels in the stage monitors, while another mixer handles the audience mix. Alternatively, a second mixing board could be used to create a mix for a live recording.

In either case, rather than use two microphones for each singer (like you'll see taped together in some '60s concert documentaries), a split can be used to provide both consoles independent access to a single microphone.

Remember to look at the signal flow diagram for your mixer when using inserts, splits and direct outputs. The position of the insert jack, relative to the Trim, Low Cut, EQ and Channel Volume Fader will affect the signal present at the insert connector. For example, on the original MS1202, the channel insert location causes the split to happen right after the microphone's output, while the classic CR-1604's split is post-EQ and post-fader. With an MS1202, the equipment receiving the split from your Mackie will be subject to your trim settings. CR-1604 splits will follow EQ and fader changes, as well as trim settings. Keep these details in mind when using splits and direct outs.

Also note that sharing a split signal creates an electrical connection between consoles (unless transformers are used to do the splitting, which won't be the case here). It is possible that connecting the two boards can lead to "ground loop" problems, created when more than one path back to ground exists in a sound system. This can lead to annoying buzzes and hums.

In the worst possible case, severe faults in a building's electrical wiring can lead to smoke, flames and serious shock hazards when two consoles on different electrical circuits are interconnected. I don't mean to frighten you off the idea, just be aware that successfully splitting a signal between two consoles is not necessarily a trivial task.

Inserts Are Not Balanced!

Please remember that Mackie inserts are not balanced, even though they do use TRS connectors. This means you should never connect a balanced TRS signal to an insert jack!

For example, if you are using a video tape-based modular digital multitrack machine, you may have a multi-pin ELCO or D-sub connector which breaks out to multiple 1/4" TRS plugs. These should be connected to your mixer's inputs or outputs (as appropriate) but not inserts.

Along the same lines, although an insert connector uses the same type of three-conductor 1/4" jack as a stereo headphone does, inserts are not stereo! They are an input and an output that for convenience (and space-saving) sake, share a single connector.

To make full use of an insert connector, use an insert Y cable, with a 1/4" tip-ring-sleeve connector that splits off into two 1/4" guitar-type cables. If all you want to do is use an insert connector as a direct or split output, you may use a simple guitar cord.

CHAPTER 8

Mackie Auxes and Inserts

MS1202 Sends and Returns

The classic MS1202 has a pair of Aux Sends and two stereo Aux Returns. The sends are post-fader, post-EQ, which makes them best suited for using with effects, as opposed to creating headphone or stage monitor cue mixes.

The returns (Stereo Aux Return To Master 1 and 2) are equally straight-forward. They are stereo line-level inputs that are always routed internally to the main stereo bus. Each return has a single stereo level control. Either of the 1202's Aux returns can be made mono by connecting a 1/4" cable from your effect unit to only the Left/Mono jack of the Stereo Aux Return. This is convenient if you have an effect with a single output, as the effect signal will then appear panned to the center of the stereo field.

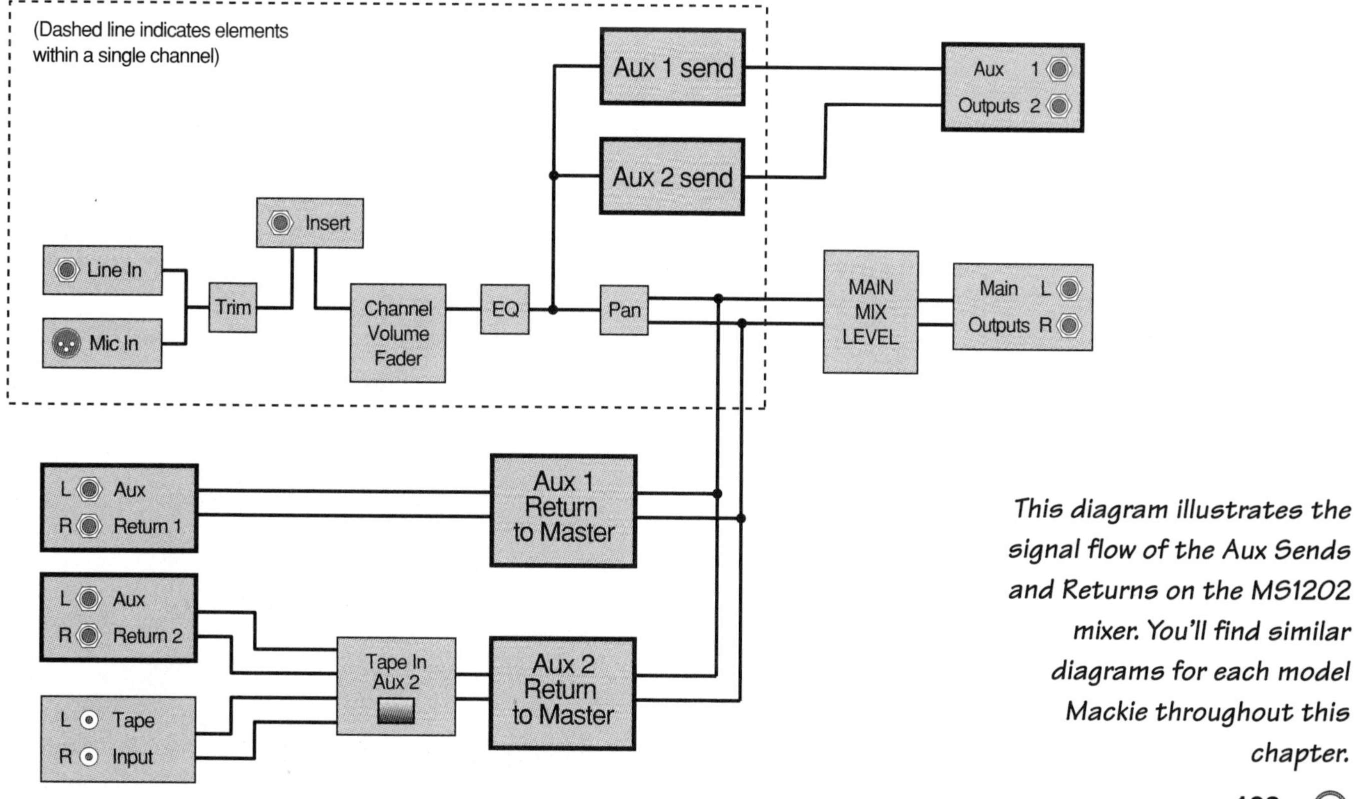

This diagram illustrates the signal flow of the Aux Sends and Returns on the MS1202 mixer. You'll find similar diagrams for each model Mackie throughout this chapter.

The only wrinkle in this otherwise simple picture is that the tape input shares the aux 2 input level control. Pressing the Tape In switch disconnects any signal feeding the Aux 2 return jacks and instead passes the signal connected to the Tape Input's RCA jacks. The Aux 2 return knob controls the tape input level. There's no way that you can hear signals from both the Tape In and Aux 2 returns at once. Although both may be connected, only one can be heard at a time.

Inserts

If you lean over the back of your 1202, you'll see four connectors marked *Channel Inserts*. Only the first four channels (those with XLR mic inputs) have inserts.

From a signal flow perspective, these jacks come immediately after the input preamp of 1202 channels 1-4. This type of insert is called *pre-fader* and *pre-EQ*. This means that twiddling these controls won't affect the signal going out the 1202's insert, regardless of whether you are using it in the *insert*, *direct out*, or *split* hook-up configuration.

The MS1202 inserts are pre-fader, pre-EQ.

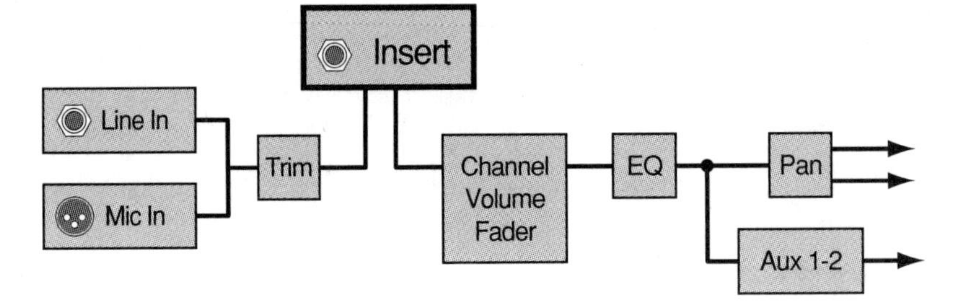

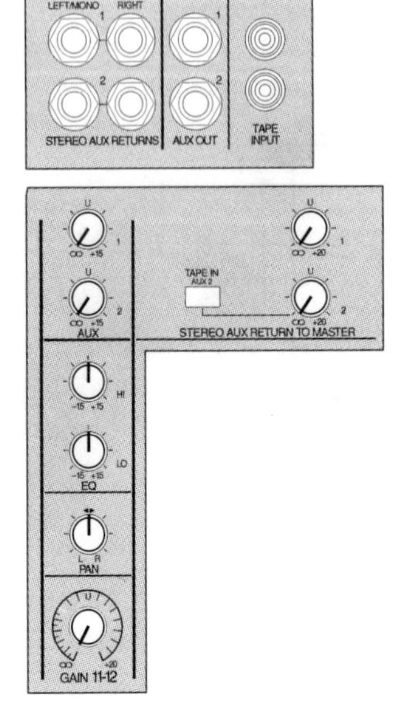

Here are the front panel knobs, switches and jacks on the MS1202 that relate to the Aux Sends and Returns. The internal signal path relationships between these elements are shown in the diagram on the previous page.

1202- and 1402-VLZ

The 1202- and 1402-VLZ have identical send/return and insert configurations, so both will be covered together in this section. The signal flow diagram for these models appears on the following page.

SENDS

The 1202- and 1402-VLZ each have two Aux Sends and two stereo Aux Returns. Aux 1 is labeled Mon/EFX, because it can be switched between pre-and post-fader operation. Aux 2 is best used for effects, hence it is marked EFX on your mixer's front panel.

AUX 1 SELECT (PRE/POST-FADER)

Aux 2 is always a post-fader, post-EQ send. While Aux 1 is also post-EQ, it can be set to pre- or post-fader operation. This is accomplished by a switch in the master section labeled Aux 1 Select. In the down position, Aux 1 takes its signal from immediately after the channel volume fader. With the Aux 1 Select in the up position, Aux 1 taps off the signal from the point after the EQ and immediately before the channel volume fader. Operating the switch affects Aux 1 on all channels, there is no way to change the pre- or post-fader configuration on individual channels.

MUTE-ALT 3/4 AND SENDS

Engaging a channel's Mute-ALT 3/4 button silences its post-fader aux sends, even though that control appears after the aux sends in your signal flow diagram. Aux 1 will be unaffected by the Mute-ALT 3/4 button if the Aux 1 Select switch is in its up (pre-fader) position.

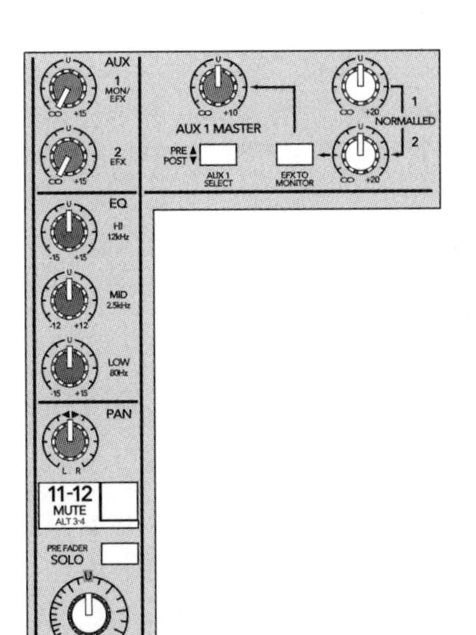

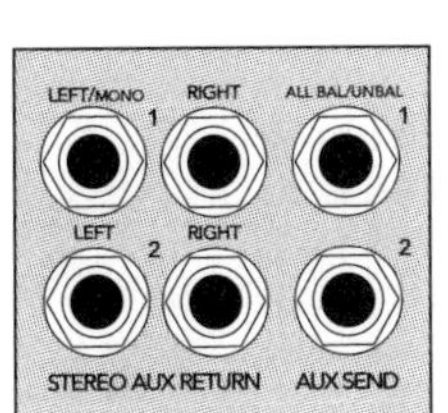

Here are the front panel knobs, switches and jacks on the 1202-VLZ that relate to the Aux Sends and Returns. The internal signal path relationships between these elements are shown in the diagram on the following page.

Returns

The dual stereo Aux Returns on the 1202- and 1402-VLZ are relatively straightforward to use, but share a couple of unique features.

Here's the whole shebang—the 1202- and 1402-VLZ effects send and return story. The section within the dashed line indicates a single input channel, so imagine that part duplicated for each of your mixer's main inputs.

WHO'S TO SAY WHAT'S NORMALLED?

When you plug in something to Aux Return 1, but leave Aux Return 2 un-connected, an interesting thing happens. The signal connected to Aux Return 1 automatically shows up in Aux Return 2, too. But why? The signal from both aux return level controls go to the same place—the main stereo mix. Therefore, having two level knobs that control the same signal going to the same destination seems like a useless feature. But hang on, those folks at Mackie aren't stupid, there's a reason behind this madness...

EFX TO MONITOR

When the EFX To Monitor button is pressed, the signal connected to the Aux Return 2 is disconnected from the main stereo mix and is re-routed to the Aux 1 Send. Take a breath and read that again, because this is a bit unusual.

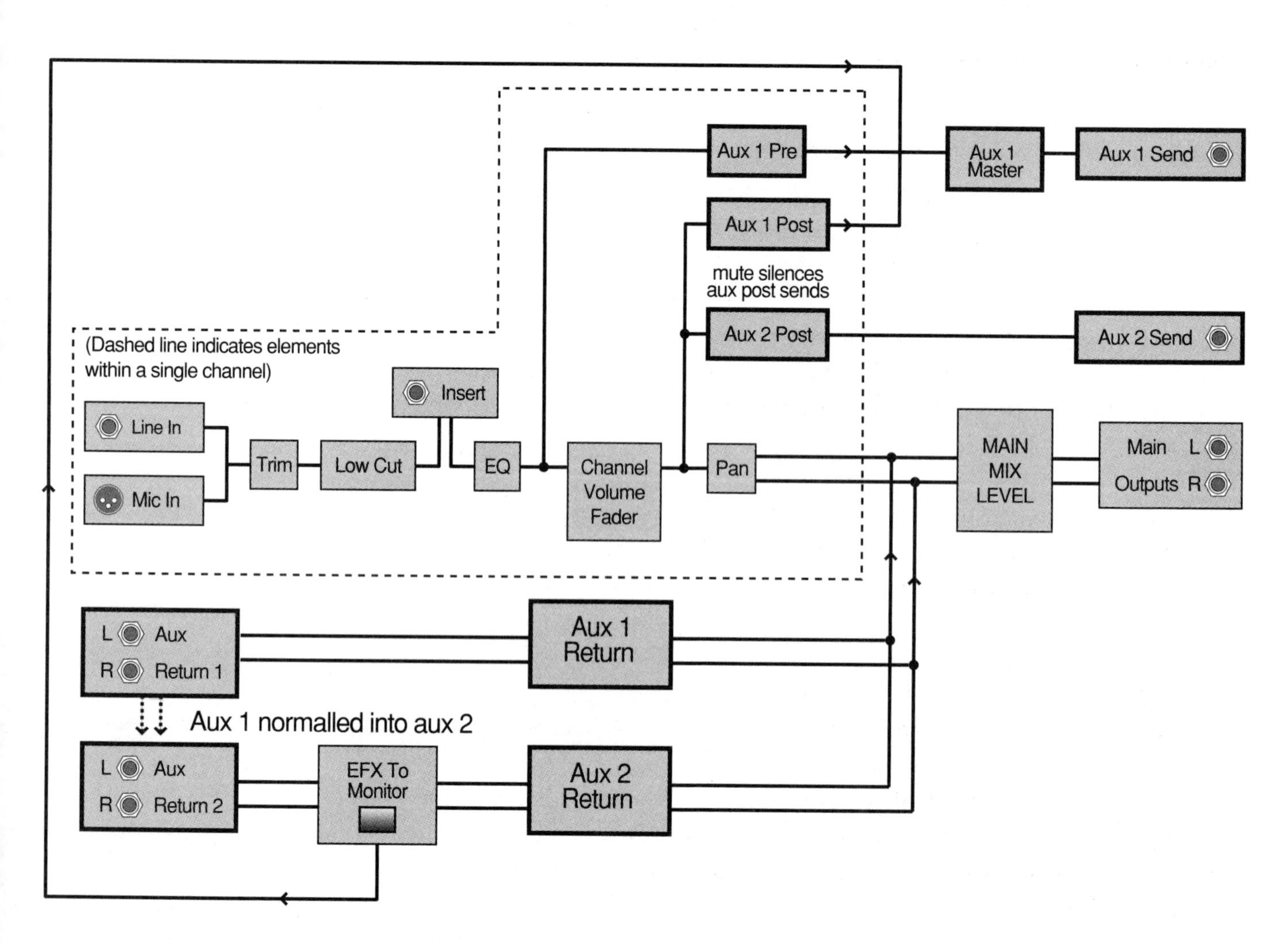

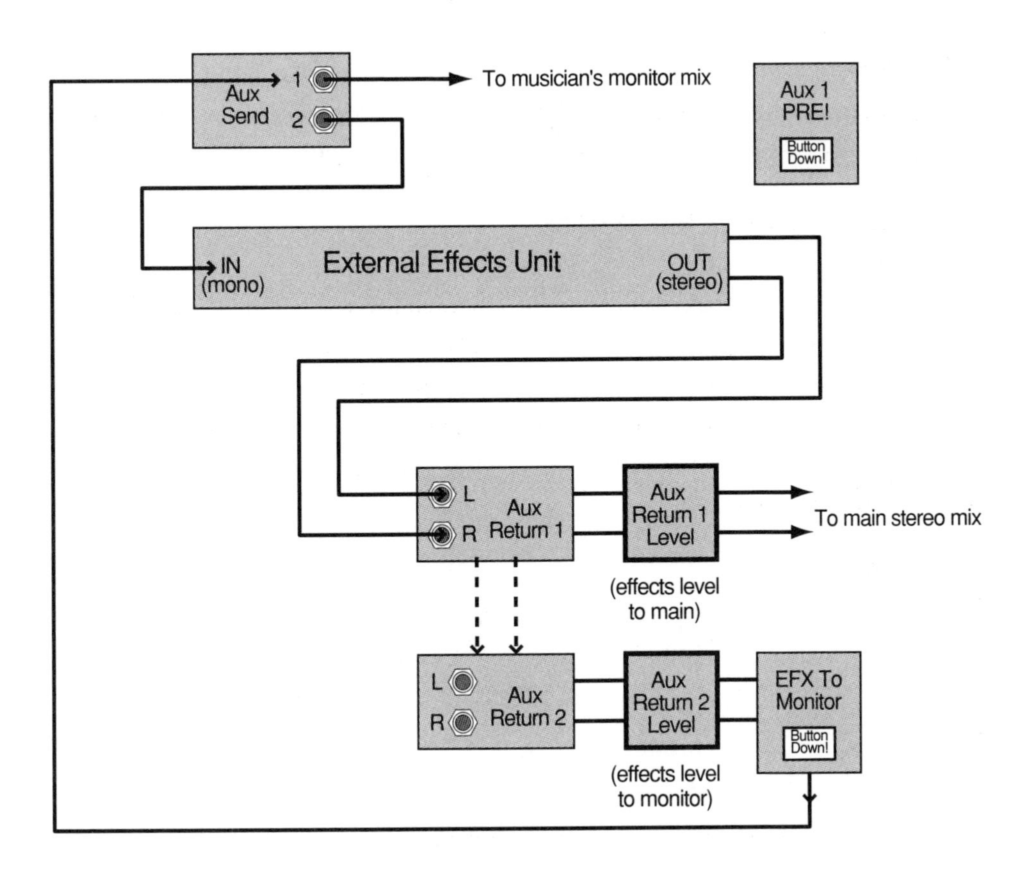

Want reverb in your stage monitors? I thought you would. Here's how to get it, assuming you've got a Mackie 1202- or 1402-VLZ mixer.

This makes it easy to get effects in your headphone or stage monitor mix. The illustration on the opposite page shows you the overall signal flow, while the drawing above shows the specific hook-up to get effects in the monitors.

Start by pressing both the Aux 1 Pre and EFX To Monitor buttons down. Next, connect the external effects unit (presumably a reverb or delay, but it could be anything) to the output of Aux Send 2, but bring it back into Aux Return 1. I realize that the send and return numbers don't match, but this is how it must be done for this particular application.

Aux 2 will determine which channels get reverb on them. Your headphone or stage monitor mix will be running off Aux Send 1, the output of which must be connected to a headphone amp or power amplifier and stage monitors (not shown in illustration).

Since *nothing is connected to Aux return 2*, the signal from Aux Return 1 is normalled into Return 2 (see "Who's to Say What's Normalled," on the opposite page). In our illustration, this is shown by the dashed lines from return 1 to return 2. Next, because the EFX To Monitor button is pressed, the signal from Aux Return 2's level control is routed to Aux Send 1, the monitor mix. Here, it is combined with the rest of the dry signals already feeding the monitors. Also note that this removes Aux Return 2 from the main stereo mix!

Aux Return 1's level control determines the overall reverb level going to the main mix (the audience, in a live situation). Aux return 2's level control sets the overall amount of effect in the monitors. Now you can turn the overall reverb up or down in the monitors without disturbing the effects balance in the audience mix. Pretty cool, huh?

OK, so this might seem a little complicated, but believe me, it's much more efficient than the hook-up required to get effects in the monitors of a classic CR-1604 or MS1202.

Inserts

If you lean over the back of your mixer, you'll see a number of connectors marked *Channel Inserts*. Only those channels with XLR mic inputs have inserts. For the 1202-VLZ, there are four inserts; the 1402-VLZ has six.

From a signal flow perspective, these inserts come after the pre-amp and low cut switch of channels 1-4. This type of insert is called *pre-fader* and *pre-EQ*. This means that twiddling these controls won't affect the signal going out your mixer's insert, regardless of whether you are using it in the *insert, direct out,* or *split* hook-up configuration.

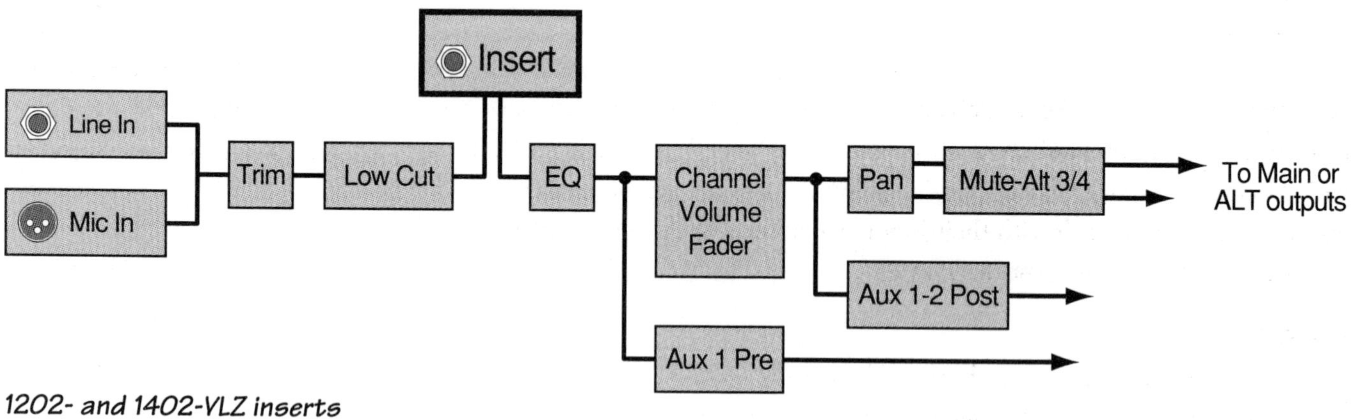

1202- and 1402-VLZ inserts come after the low cut switch, but before just about everything else...

CR-1604

Sends

The CR-1604's six aux sends are also post-fader and post-EQ, just like the 1202. However, the aux section of each CR-1604 input channel has a couple of extra switches that deserve our attention.

The CR-1604 Mon button (shown below) re-routes the signal from the Aux Send 1 knob to the Monitor output jack.

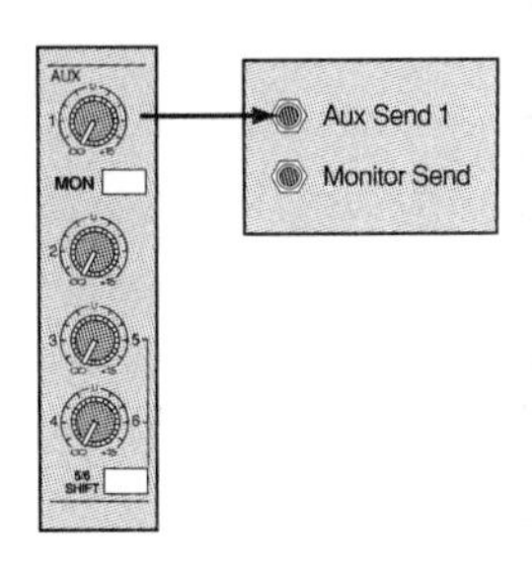

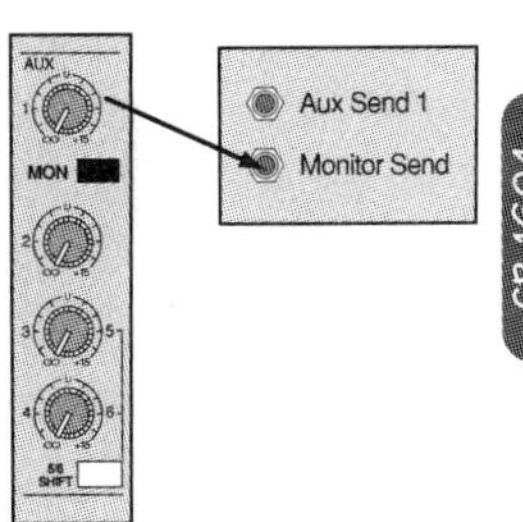

THE MONITOR SWITCH

Below each channel's Aux 1 knob, you'll find a switch labeled Monitor. While all the 1604's Aux sends are post-fader and EQ by default, pushing the Monitor switch on a channel makes that Aux 1 pre-fader and pre-EQ. (See the previous chapter for an explanation about this important distinction.)

Not only does the Monitor switch change Aux 1's position in the signal path, it also disconnects that channel's signal from the Aux Output 1 jack, and instead routes it to the Monitor jack, located next to the other Aux output jacks on your 1604's connector panel.

This feature makes Aux 1 a prime candidate for running a stage monitor system for performing musicians, or a headphone mix for players in a recording session. By doing so, you'll be able to provide the musicians their own mix to help them play at their best, without having to compromise your main stereo mix to accommodate their needs.

Note that using the Monitor switch only affects that particular input channel. You can engage the Monitor switch on any combination of input channels. Channels with Monitor pressed will have a pre-fader Aux 1, with that signal feeding the Monitor output. Channels with their Monitor switches *not* engaged will behave like the other aux sends—post-fader and EQ.

The CR-1604 5/6 Shift button (see below) re-routes the signal from the Aux Send 3 and 4 knob to the Aux Send 5 and 6 jacks.

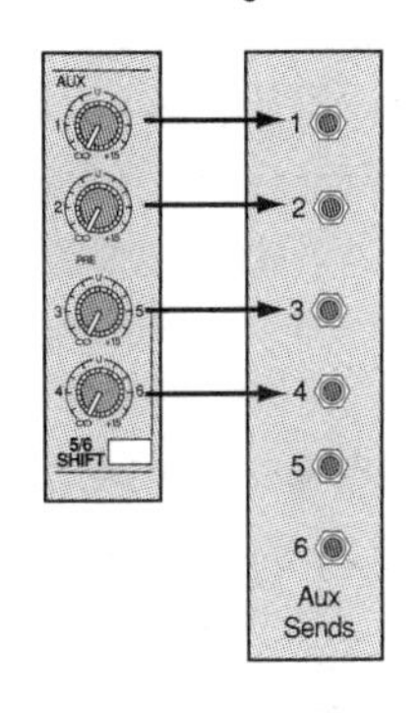

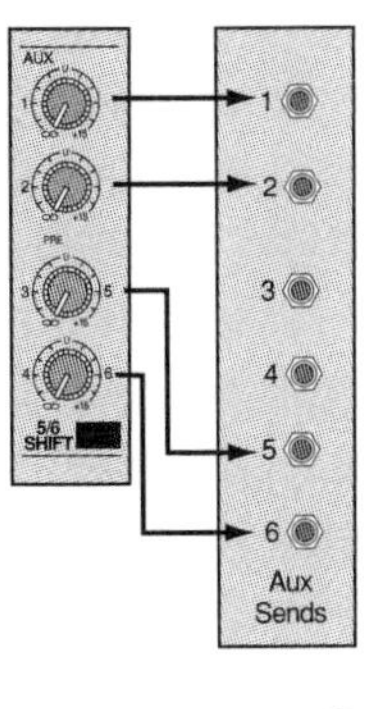

5/6 SHIFT

At the bottom of the Aux controls, there is a button marked 5/6 Shift. While the CR-1604 has six individual Aux sends (or seven, counting the Mon send), there are only four Aux knobs. The bottom pair (Aux 3/4) do double-duty. When the 5/6 Shift button is depressed, these two knobs become level controls for Aux 5 and 6. Turning up either one of these controls with the Shift button pressed means that channel's signal will appear at the Aux Output 5 or 6 jack, instead of Aux Output 3 or 4. Aux 5 and 6 are post-fader, post-EQ, just like Aux 3 and 4.

Just like the monitor button, you can use 5/6 Shift on any combination of channels you like. Channels where it is not pressed will continue to feed Aux 3 and 4 when the bottom pair of Aux knobs are turned up. (The position of the Monitor button has no bearing on the operation of 5/6 Shift.)

Here's the big picture of the classic CR-1604's sends and returns. The section within the dashed line shows just one of your mixer's 16 input channels. Note that the front panel controls illustrated in this diagram are shown on the opposite page.

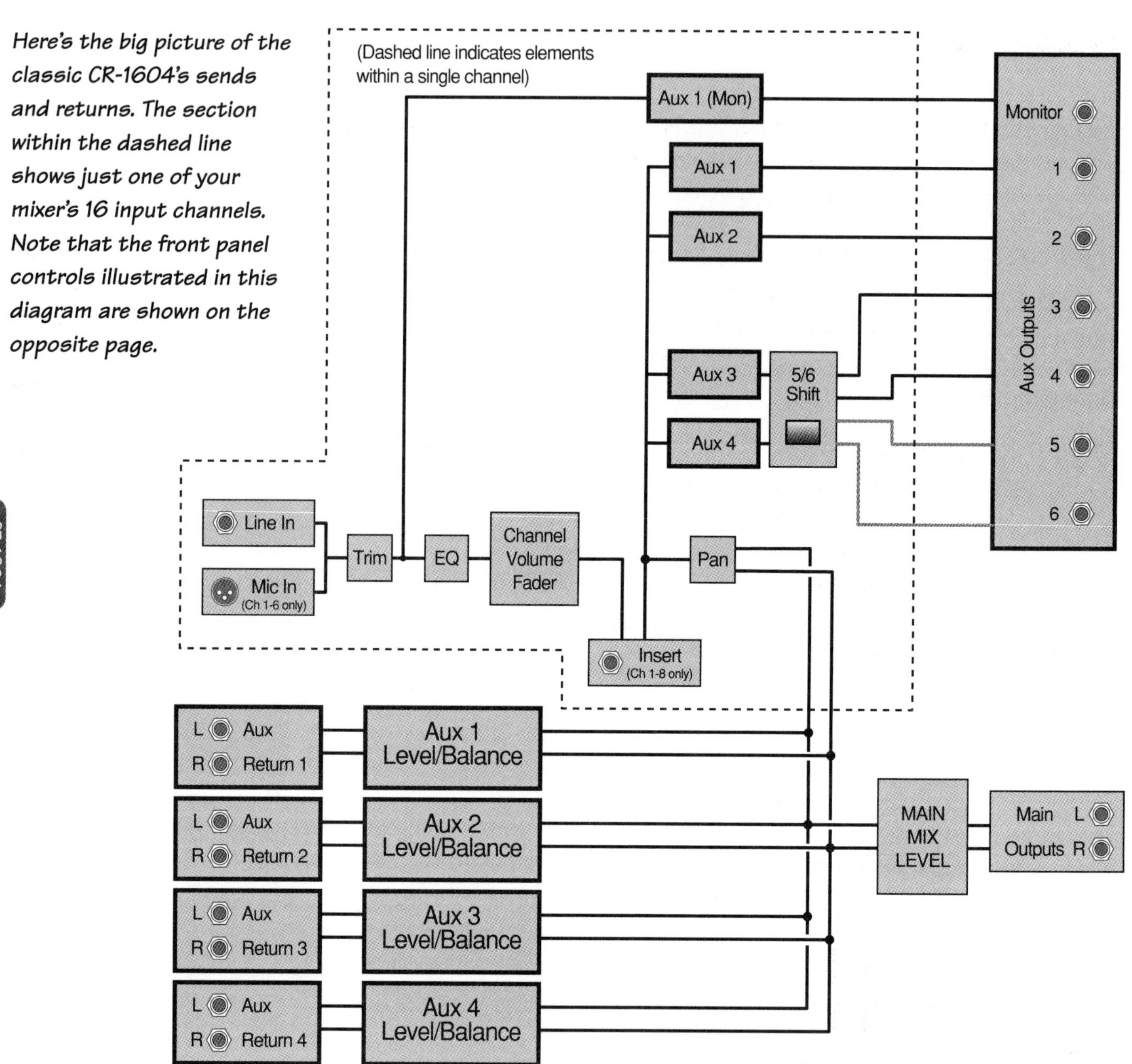

In theory, this means you could feed six different effects units *and* one monitor feed at once! However, knobs that do double-duty can only do one task at a time. Even though you can create these seven aux mixes, not all channels will be able to be sent to each mix.

In practice, this probably won't be much of a limitation, since the main reason to have all these aux sends is the ability to create different, independent mixes. If all seven mixes had the same things in them, you probably wouldn't need seven different mixes!

Returns

AUX SOLO AND MONO

You can solo all four Aux returns at once, by depressing the Aux Solo button. This is a convenient way to hear only your effects, perhaps while you are making adjustments to the parameters on an effect unit in use.

Any of the CR-1604's individual Aux returns can be made mono by depressing the corresponding switch. This is convenient if you have a unit with a single output. Simply connect it to either Aux return jack and press Mono. Then, the Balance control will act like a Pan control, and determine the mono effect signal's position in the stereo field.

Below are the CR-1604 controls that are part of the aux send and return system. The signal path relationship between these are shown in the diagram on the opposite page.

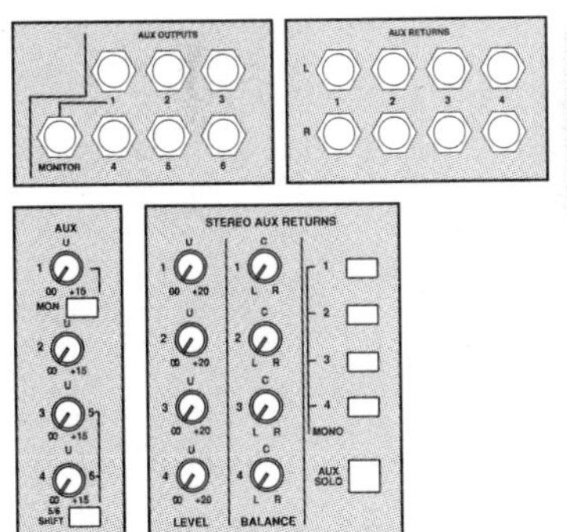

EFFECTS IN YOUR MONITORS

Adding effects to a stage monitor or headphone mix is a two-part process on the CR-1604. First, you must connect your effect unit's outputs to normal input channels, not to an aux returns. Then, you can add effects to the monitor mix by bringing up the effect return channel's Aux send that feeds the monitor mix. Don't turn up the return channel's effect send or feedback will result. The illustration below shows the connections and active controls.

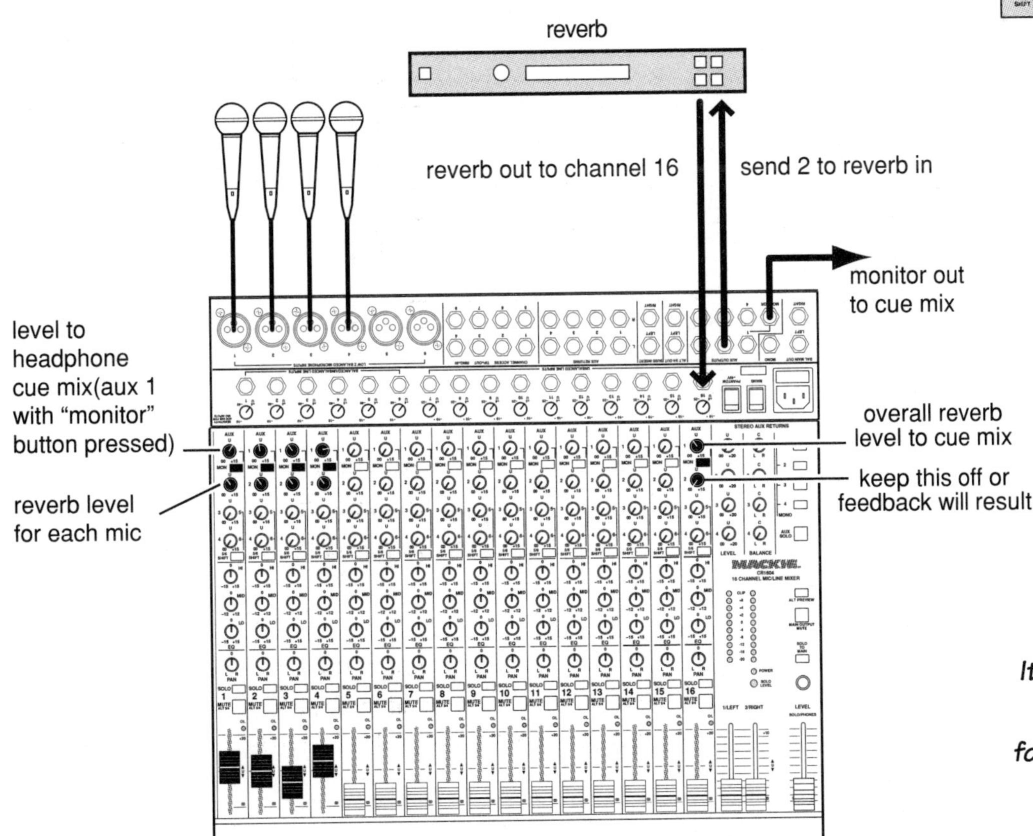

It is possible to get effects in your monitor mix. Just follow the hook-up shown at left. You'll be glad you did. Or maybe not.

Inserts

If you look at the connector panel of your 1604, you'll see eight connectors marked Channel Access. From a signal flow perspective, these jacks come *after* the pre-amp input, Fader and EQ of the 1604's first 8 channels.

It's important to understand the significance of inserts that follow the Channel Volume Fader and EQ. This means that any twiddling you do to these controls changes the signal going out the 1604's insert, regardless of whether you are using it in the *insert, direct out* or *split* hook-up configuration. This type of insert is called *post-fader* and *post-EQ*.

When using a 1604 and a multitrack recorder, *post-fader* inserts mean you have level adjustment and EQ available when recording from an insert/direct out to a tape track input. This is the main benefit to a post-fader insert, the other main repercussion is negative.

Specifically, when inserting a compressor into a 1604 input channel, the *post-fader* inserts will make things more complicated than just about any other model Mackie. This is because a compressor's behavior depends on the input level it's getting, and if the fader comes *before* the insert (as it does in the 1604), then the compressor will be "fighting" you as you raise or lower that channel's fader. In other words, when you turn up the fader, more signal will go to the compressor, which will try to turn the signal *down* while you are trying to turn it up. The result will be an over-compressed signal that may be too squashed-sounding for your needs.

As a work around, you could take an insert out of one channel and feed the compressor, then take the compressor's output and bring it into a second mixer input. Then the first channel's trim and fader would set the compressor's input level, and the second channel could be used to balance that signal's level in the overall mix. The drawback, of course, is that this solution ties up *two*, rather than one input channels. See the opposite page for an example.

The CR-1604's post-fader inserts work great as direct outputs, but create some extra work when using a compressor (see facing page)

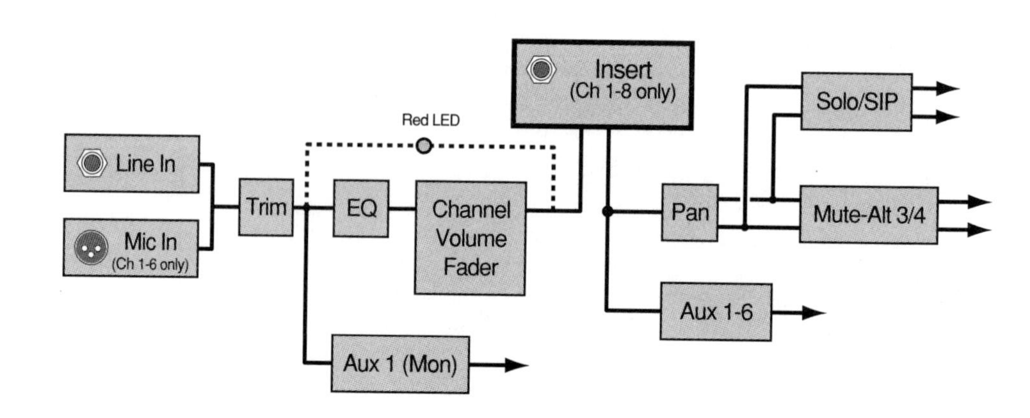

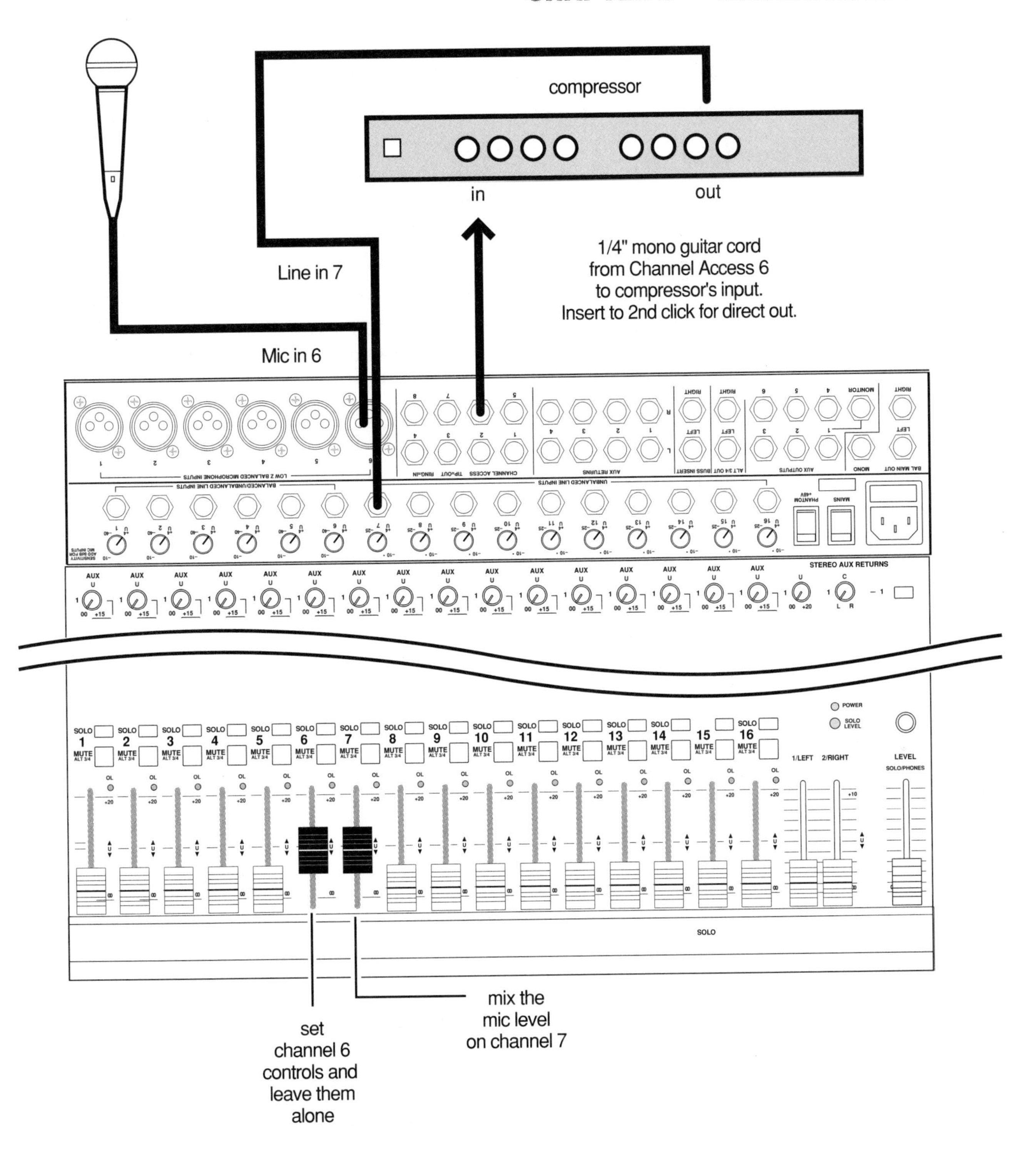

There is a factory-authorized modification that can change the location of the 1604's insert points to pre-EQ, pre-fader or post-EQ, pre-fader. If you need to use compressors often, and don't need direct outs with EQ and level control, this mod may be for you. I made this modification to the first four channels of my CR-1604 and it works great. Check your Mackie manual for tech support's phone number and call for more information.

You can insert a compressor on a CR-1604 (above), but if you must do so often, consider the factory mod described in the paragraph at left.

The 1604-VLZ send and return signal path is diagrammed on the opposite page. The section within the dashed line indicates a single input channel, so imagine that part duplicated for each of your mixer's main inputs.

1604-VLZ & 1642-VLZ PRO

Sends

The 1604-VLZ, and 1604-VLZ PRO have a total of six Aux Sends, controlled via four knobs, while the 1642 has four sends with four knobs. The first two sends may be set to pre- or post-fader, the others are post-fader. Each aux section has a couple of switches that modify the function of the four aux send knobs on each channel. The signal flow of the entire send and return system is shown on the facing page. Turn the page for a front-panel illustration of the knobs, switches and jacks diagramed at right.

PRE

Below each channel's Aux 2 knob, you'll find a button marked Pre, short for pre-fader send. Pressing this button on one or more channels changes the signal flow of those channels so that aux 1 and 2 tap their signals from a point immediately following the insert point and low cut switch, which is before the EQ, mute and Channel Volume Fader. This pre-fader configuration is best suited for monitor mixes (headphone or stage wedge), since it allows you to make EQ and fader adjustments on behalf of the main mix without affecting the mix the musicians are playing along with. Both pre- and post-fader signal paths are shown at letter A in the illustration on the opposite page.

The 1604-VLZ 5/6 Shift button re-routes the signal from the Aux Send 3 and 4 knob (above) to the Aux Send 5 and 6 jacks when pressed (below). This means that a single channel may send to auxes 1-4, or 1,2,5 and 6, but never to all six at once.

5/6 SHIFT

At the bottom of the Aux controls, there is a button marked 5/6 Shift (not available on the 1642-VLZ PRO). While the 1604-VLZ has six individual Aux sends, there are only four Aux knobs. The bottom pair (Aux 3/4) do double-duty. When the 5/6 Shift button is depressed, these two knobs become controls for Aux 5 and 6 (see letter B in the illustration at right). Turning up either one of these controls with the Shift button pressed means that channel's signal will appear at the Aux Output 5 or 6 jack, instead of Aux Output 3 or 4. Aux 5 and 6 are post-fader, post-EQ, just like Aux 3 and 4.

Just like the Pre button, you can use 5/6 Shift on any combination of channels you like. Channels where it is not pressed will continue to feed Aux 3 and 4 when the bottom pair of Aux knobs are turned up. (The position of the Pre button has no bearing on the operation of 5/6 Shift.)

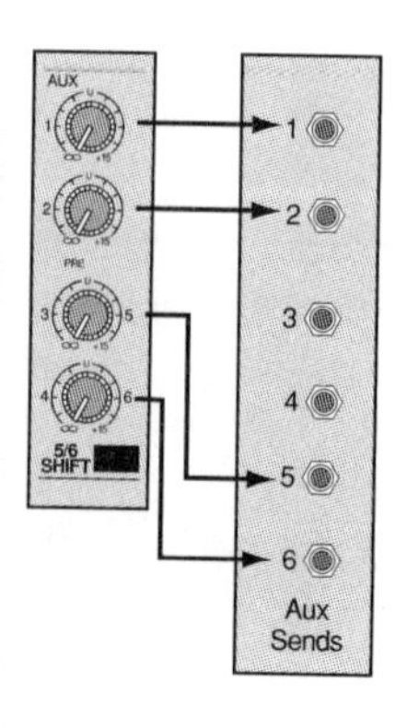

AUX 1 AND 2 MASTER AND SOLO

Aux 1 and 2 each have two extra useful features: an Aux Master control and a Solo button. The aux masters (see letter C at right) let you bring the overall level of Aux 1 and 2 up or down, without having to adjust each of the individual Aux Sends on the input channels.

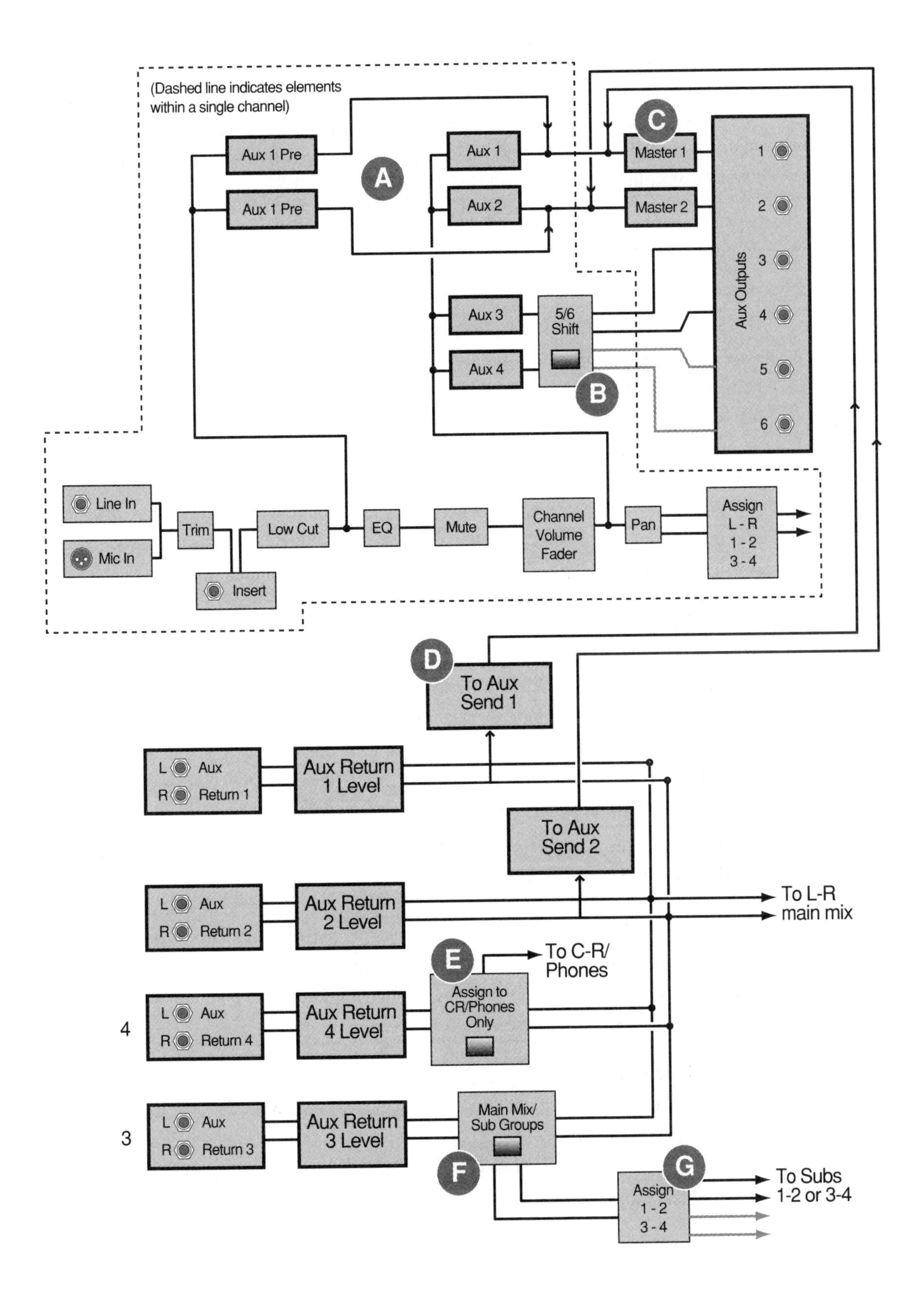
(Dashed line indicates elements within a single channel)
Aux 1 Pre
Aux 1 Pre
A
Aux 1
Aux 2
C
Master 1
Master 2
Aux Outputs
1
2
3
4
5
6
Aux 3
Aux 4
5/6 Shift
B
Line In
Mic In
Trim
Insert
Low Cut
EQ
Mute
Channel Volume Fader
Pan
Assign L - R 1 - 2 3 - 4
D
To Aux Send 1
L Aux
R Return 1
Aux Return 1 Level
To Aux Send 2
L Aux
R Return 2
Aux Return 2 Level
To L-R main mix
E
To C-R/ Phones
Assign to CR/Phones Only
4
L Aux
R Return 4
Aux Return 4 Level
3
L Aux
R Return 3
Aux Return 3 Level
Main Mix/ Sub Groups
F
G
Assign 1 - 2 3 - 4
To Subs 1-2 or 3-4

Along similar lines, the Aux 1 and 2 Solo buttons let you hear what's going to the Aux buses. This is most helpful when Aux 1 and 2 are being used as monitor sends, since this gives you, the mixing board operator, a convenient way to hear exactly what the musicians are hearing at any given moment.

Returns

The 1604-VLZ and 1642-VLZ both have four stereo aux returns. While all four can return their signals to the main stereo mix, there are a few interesting options.

TO AUX SEND 1/2

To the right of the level controls for Aux Return 1 and 2, there are a pair of knobs labeled To Aux Send 1 & Send 2. The "To Aux" knobs route a variable amount of the return to Aux Send 1 and Aux Send 2, respectively. This is most helpful when Aux 1 and 2 are being used as monitor sends, since this gives you a convenient way to add effects to the monitor mix(es). See letter D on the illustration on the previous page for the signal path.

The illustration below shows the hook-up. Connect an external effects unit (presumably a reverb or delay, but it could be any effect) to the output of Aux Send 3, but bring it back into Aux Return 1. I realize that the send and return numbers don't match, but this is how it must be done for this particular application.

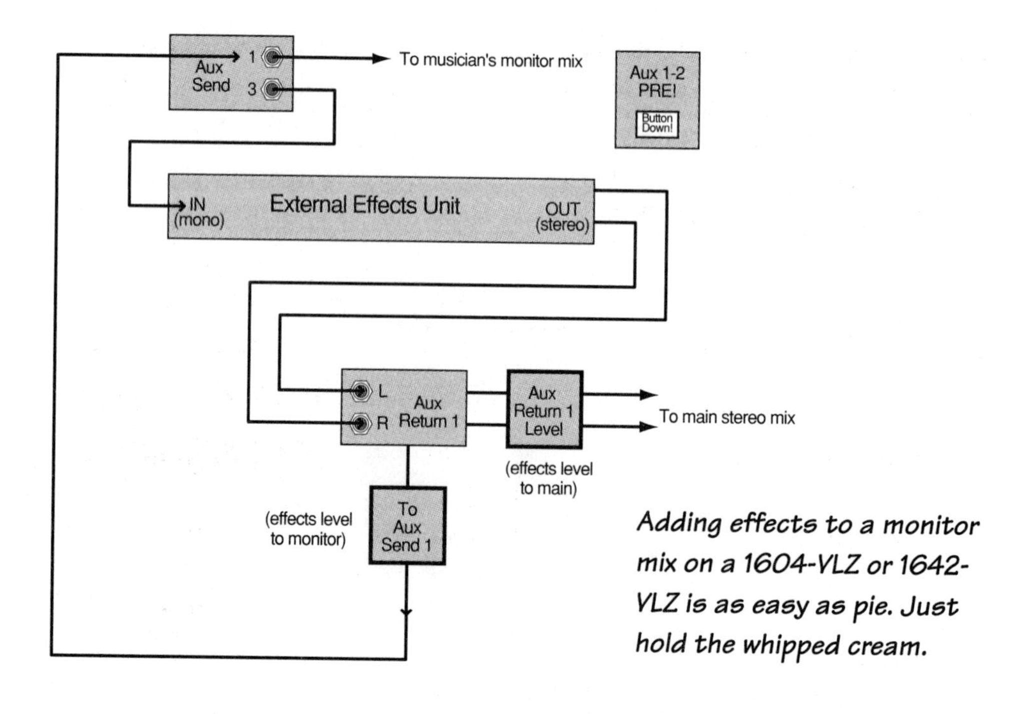

Adding effects to a monitor mix on a 1604-VLZ or 1642-VLZ is as easy as pie. Just hold the whipped cream.

Your headphone or stage monitor mix will be running off Aux Send 1, the output of which must be connected to a headphone amp or power amplifier and stage monitors (not shown in illustration). Aux Send 3 will determine which channels get reverb on them.

Now, you can bring up the main Aux 1 Return control to add effects to the main stereo mix, and use the EFX To Mon 1 knob to control the level of the effects added to the Aux Send 1 monitor mix.

AUX RETURN 3 ASSIGN OPTIONS

While all of the 1604-VLZ and 1642-VLZ's main input channels can be routed freely to the stereo bus, or buses 1-2 and 3-4, the aux returns route only to the main stereo bus. That is, except for Aux Return 3, thanks to its Assign Options buttons (see letter F in the illustration on the previous page). Pressing the first of these, Main Mix/To Sub takes Aux Return 3 out of the stereo mix and puts it into bus 1-2 or 3-4, depending on the position of the second button. As you can see, this makes it possible to route Aux Return 3 to any one of three possible stereo buses, but only one at a time (letter G on the illustration on pg. 115).

It turns out that one at a time is all you'll need. The most likely way to use this option is when creating a submix. Say all your vocals are assigned to bus 1-2, and you have a reverb unit dedicated to only vocals. That vocal reverb is coming back into Aux Return 3. By assigning that return to bus 1-2, your bus master faders 1-2 will control the level of the dry vocals *and* the reverb return levels. This is very cool, because if the reverb return was connected to the stereo bus instead, when you faded out the vocals using the bus master faders, the reverb wouldn't fade, but instead would still be blaring ambiently away after the vocals had faded to silence.

Note that you could have the same capability by returning the output of your effects unit into a regular input channel. Then that channel could be assigned to any bus or aux send, but it's cool that Aux Return 3 gives you this capability without tying up a pair of precious input channels.

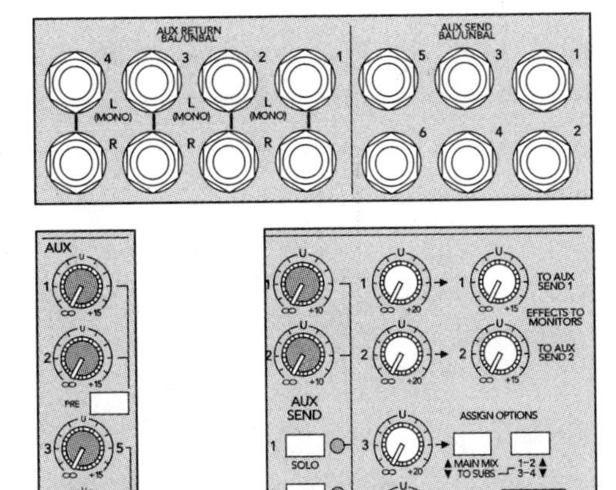

At left are all the knobs, switches and jacks that make up the 1604-VLZ and 1642-VLZ's send and return system. The signal path relationships between these controls are illustrated in the diagram on the previous page.

1604-VLZ
1642-VLZ

CR/PHONES ONLY

Aux Return 4 has a bonus routing option similar to Aux Return 3. The CR/Phones Only button takes Aux Return 4 out of the main stereo bus and routes it to the Control Room/Phones output. This lets you route a signal just to the control room, but not the main mix. Finding a use for this function is an exercise left to the reader. See letter E in the illustration on pg. 115.

RETURNS SOLO

You can solo all four Aux returns at once by depressing the Returns Solo button. This is a convenient way to hear only your effects, perhaps while you are making adjustments to the parameters on an effect unit in use.

Inserts

Each of the 1604-VLZ (and many of the 1642-VLZ's) input channels has its own insert. From a signal flow perspective, these inserts come after the pre-amp and low cut switch. This type of insert is called *pre-fader* and *pre-EQ*. Therefore, twiddling these controls won't affect the signal going out your mixer's insert, regardless of whether you are using it in the *insert, direct out*, or *split* hook-up configuration.

Your mixer also includes eight direct outs on the first eight channels. While inserts, as you recall, contain both an input and an output using a single three-conductor jack, the 1604-VLZ's Direct Outputs are out only! Further, they appear in the signal path after the Channel Volume Fader, just before the Pan control. This means that all EQ, insert processing and level adjustments on a channel will be reflected in the signal that appears at that direct out.

This makes a direct output an ideal way to send a single input channel to one track of a multitrack recorder. That way, you can EQ, level-adjust and even process that channel through an insert device before it gets recorded.

Both mixers includes a pair of inserts on the main stereo bus. These come immediately before the Master Volume Fader, so if you want to insert a compressor on the stereo bus, you won't be changing the input level of the compressor as you move your mixer's master volume around.

Here's how the 1604-VLZ insert fits into the overall channel signal flow. While not part of the insert, the Direct Out has been highlighted, as it offers another way to extract the audio from an individual channel.

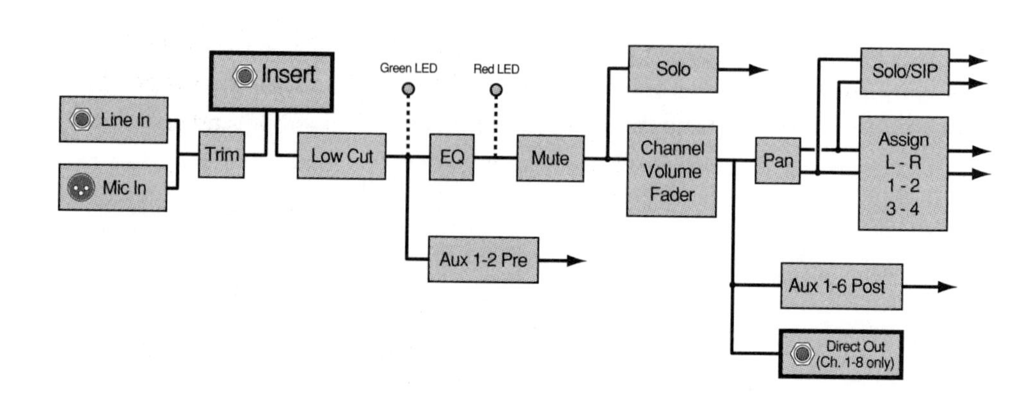

SR24•4, SR 32•4

The SR24•4 and SR32•4 mixers have six aux sends with dedicated knobs, individual solos per send, send masters and four returns with effect sends to monitors. In short, everything you need to create multiple monitor mixes for the band while simultaneously feeding a great mix to the audience or a recorder.

Below are the knobs, switches and jacks associated with the SR's sends and return. Their signal path relationships are illustrated on the following page.

Sends

Aux Send 1 and 2 are always pre-fader, pre-EQ, which is ideal for creating monitor mixes (see letter A in illustration on following page). Aux Send 5 and 6 are always post-fader sends, best for using with effects. And what of Aux Send 3 and 4? These are switchable, pre- or post-fader, using the Pre button on each channel positioned between the knobs for Aux Send 4 and 5.

This gives you a choice. Set Aux Send 3 and 4 to post-fader if you just need two monitor mixes and four effects sends, or the reverse to get four monitor mixes and a couple effects. It is possible to mix and match, setting some channels with Aux Send 3 and 4 to pre- and others to post-fader, although all the signals still end up at the same aux send output jacks, regardless of the setting of a given channels Pre button (for the signal path, see letter B on next page).

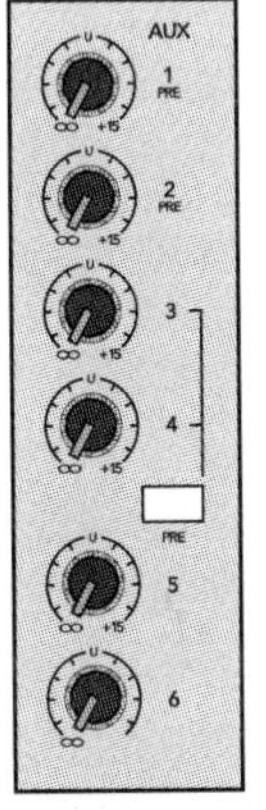

AUX SEND MASTERS AND SOLO

Each of the six Aux Sends has an Aux Send Master control and individual Solo control (see letter C in the illustration on the next page).

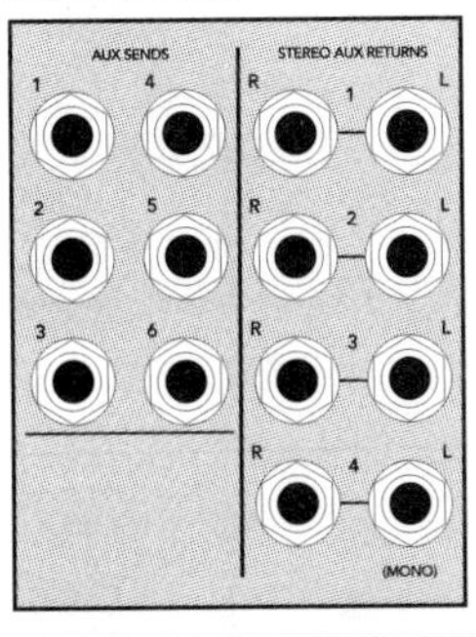

Returns

The SR-series mixers have four stereo Aux Returns. While all four can return their signals to the main stereo mix, there are a few interesting options.

TO AUX SEND 1/2 (EFFECTS TO MONITORS)

To the right of the level controls for Aux Return 1 and 2, there are a pair of knobs labeled To Aux Send 1 & Send 2 (letter D on the next page). The "To Aux" knobs route a variable amount of returns 1 and/or 2 to Aux Send 1 and Aux Send 2, respectively. This gives you a convenient way to add effects to the monitor mix(es).

To do so, connect an external effects unit (presumably a reverb or delay, but it could be anything) to the output of any post-fader send (for example, Aux Send 3). Bring the effect's output back into Aux Return 1. I realize that the send and return numbers don't match, but this is how it must be done for this particular application.

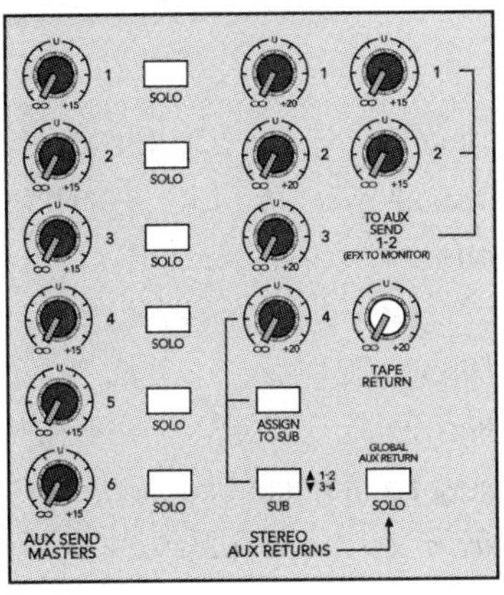

Here's the signal flow diagram for the SR-series' Aux Sends and Returns. The knobs, switches and jacks diagrammed here can be seen in the front panel illustration on the previous page. The section within the dashed line indicates a single input channel, so imagine that part duplicated for each of your mixer's main inputs.

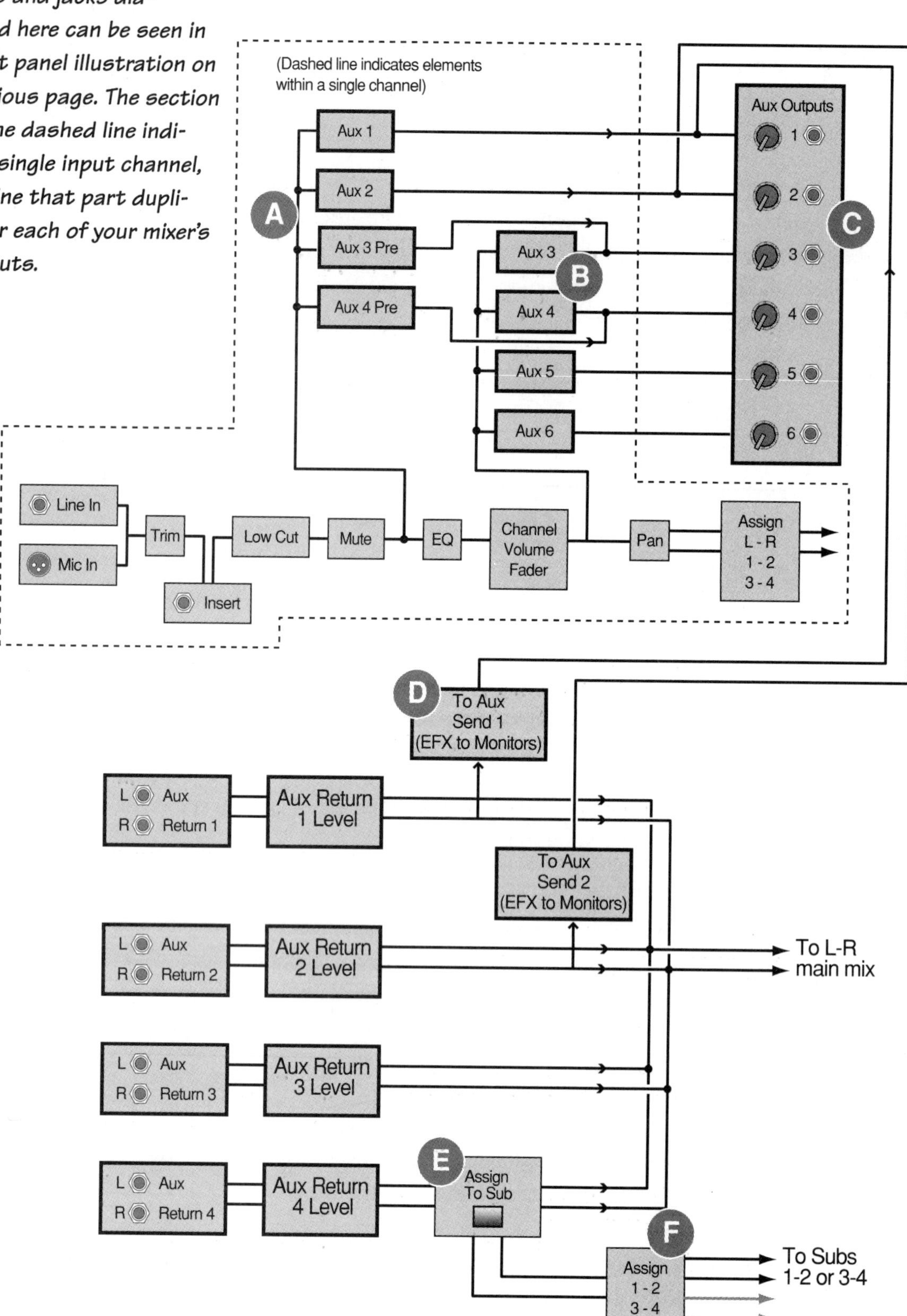

Your stage monitor mix will be running off Aux Send 1, the output of which must be connected to a headphone amp or power amplifier and stage monitors (not shown in illustration). Aux Send 3 will determine which channels get reverb on them.

Now, you can bring up the main Aux 1 Return control to add effects to the main stereo mix, and use the EFX To Mon 1 knob to control the level of the effects added to the Aux Send 1 monitor mix.

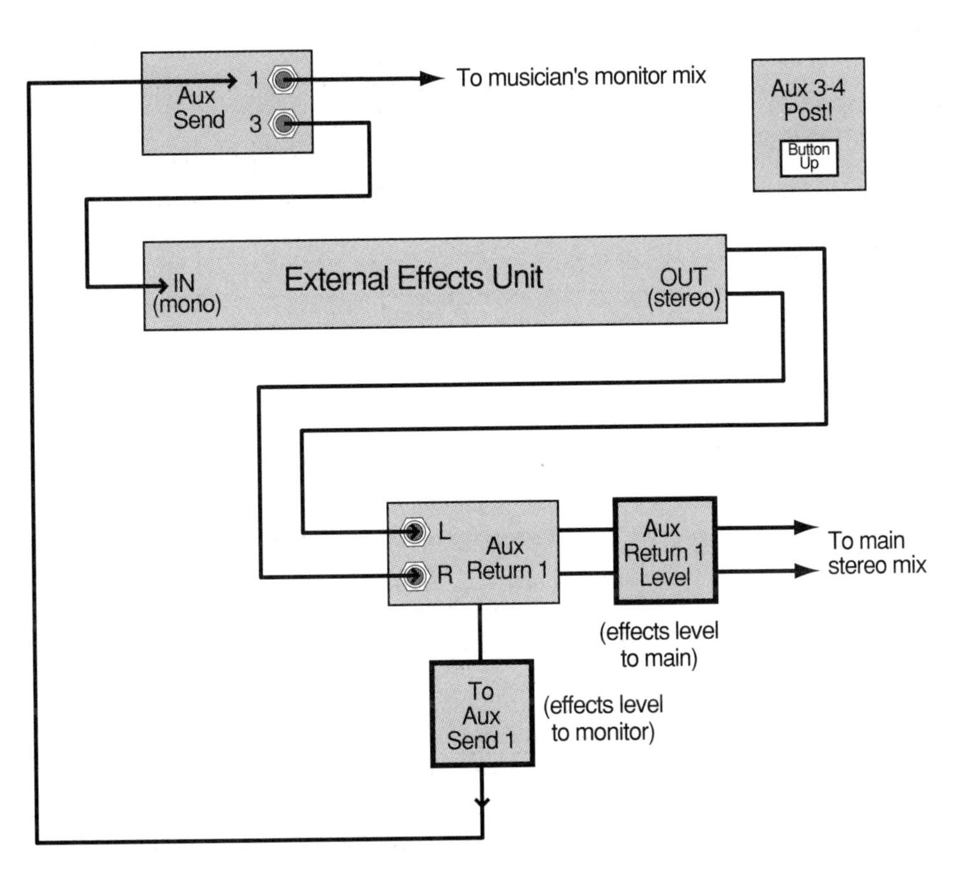

The SR-series' "To Aux Send" knobs make it easy to put effects in the monitor mix. Just follow the example at left. Note that while the effect is shown connected to Aux Send 3, any post-fader send could be used. However, the effect must be returned to Aux 1 or Aux 2 to put effects into the first and second monitor mix.

AUX RETURN 4 ASSIGN OPTIONS

Your SR-series mixer's main input channels can be routed freely to the stereo bus, buses 1-4 and the aux sends. However, Aux returns route only to the main stereo bus. That is, except for Aux Return 4, thanks to its Assign Options buttons. Pressing the first of these, Main Mix/To Sub, takes Aux Return 4 out of the stereo mix and puts it into bus 1-2 or 3-4, depending on the position of the second button. As you can see, this makes it possible to route Aux Return 4 to any one of three possible stereo buses, but only one at a time.

It turns out that one at a time is all you'll need. The most likely way to use this option is when creating a submix. Say all your vocals are assigned to bus 1-2, and you have a reverb unit dedicated to only vocals. That vocal reverb is coming back into Aux Return 3. By assigning that return to bus 1-2, your bus master faders 1-2

SR-series

will control the level of the dry vocals *and* the reverb return levels. This is very cool, because if the reverb return was connected to the stereo bus instead, when you faded out the vocals using the bus master faders, the reverb wouldn't fade, but instead would still be blaring ambiently away after the vocals had faded to silence.

Note that you could have the same capability by returning your effects unit's output into a regular input channel. Then that channel could be assigned to any bus or aux send, but it's cool that Aux Return 4 gives you this capability without tying up a pair of precious input channels.

GLOBAL AUX RETURN SOLO

Sure enough, pressing this button lets you solo all four aux returns at once. Dig that Rude Solo Light.

Inserts

The SR-series provides channel inserts on each and every input channel. These come immediately after the mic pre-amp/line input, before the low-cut switch.

But there are even more inserts lurking on that back panel...You'll also find a pair of inserts on the main stereo bus. These come immediately before the master volume fader, so if you want to insert a compressor on the stereo bus, you won't be changing the input level of the compressor as you move your mixer's master volume around.

Finally, each of the four bus masters has its own insert. This is especially handy. Say you want to add compression to five vocal mics, but you don't have five compressors. Simply assign all the vocals to a subgroup, then insert a single compressor channel on that subgroup insert. While this doesn't quite have the same result as five individual compressors on each individual vocal mics, it's a lot cheaper. It's also much more effective than running a compressor on the entire stereo mix, because when compressing the main mix bus, all your instruments will be pushing the compressor around too. This can squash your vocals in slightly unpredictable ways.

SR-series inserts come right after the trim control and before low cut, mute and EQ.

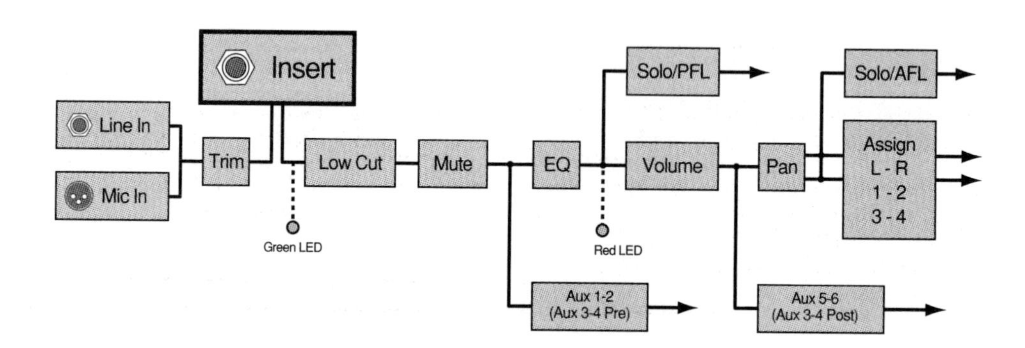

SR-series

PPM Series

The PPM-series mixers' sends and returns are pretty simple, despite the addition of the built-in emac effects unit and its internal signal routing.

Here's the send and return signal path for PPM-series mixers. The section within the dashed line indicates a single input channel, so imagine that part duplicated for each of your mixer's main inputs.

The prototype emac processor (pictured below) was the size of a small motor home. However, recent advances in semiconductor microlithography have reduced its size to that of a postage stamp.

Sends, Mon

PPM's have a pair of Aux Sends for each input channel, Mon and EFX. The Mon send, as you might guess is set as a pre-fader, pre-EQ send ideal for monitors (see letter A in the illustration below). The combined signal of all channels' Mon send goes through the Monitor Master control and then exits the mixer at the Monitor Line Out jack. If the Power Amp Routing button is in the down position (left=main, right=monitor) then the Mon Aux Send is also connected to one channel of the PPM's built-in power amplifier (the amplifier connection is not illustrated). This in turn ends up at the rear panel Monitor Speaker output jacks. Although these are labeled Left and Right, the same signal shows up in both—the Mon send is 100% mono.

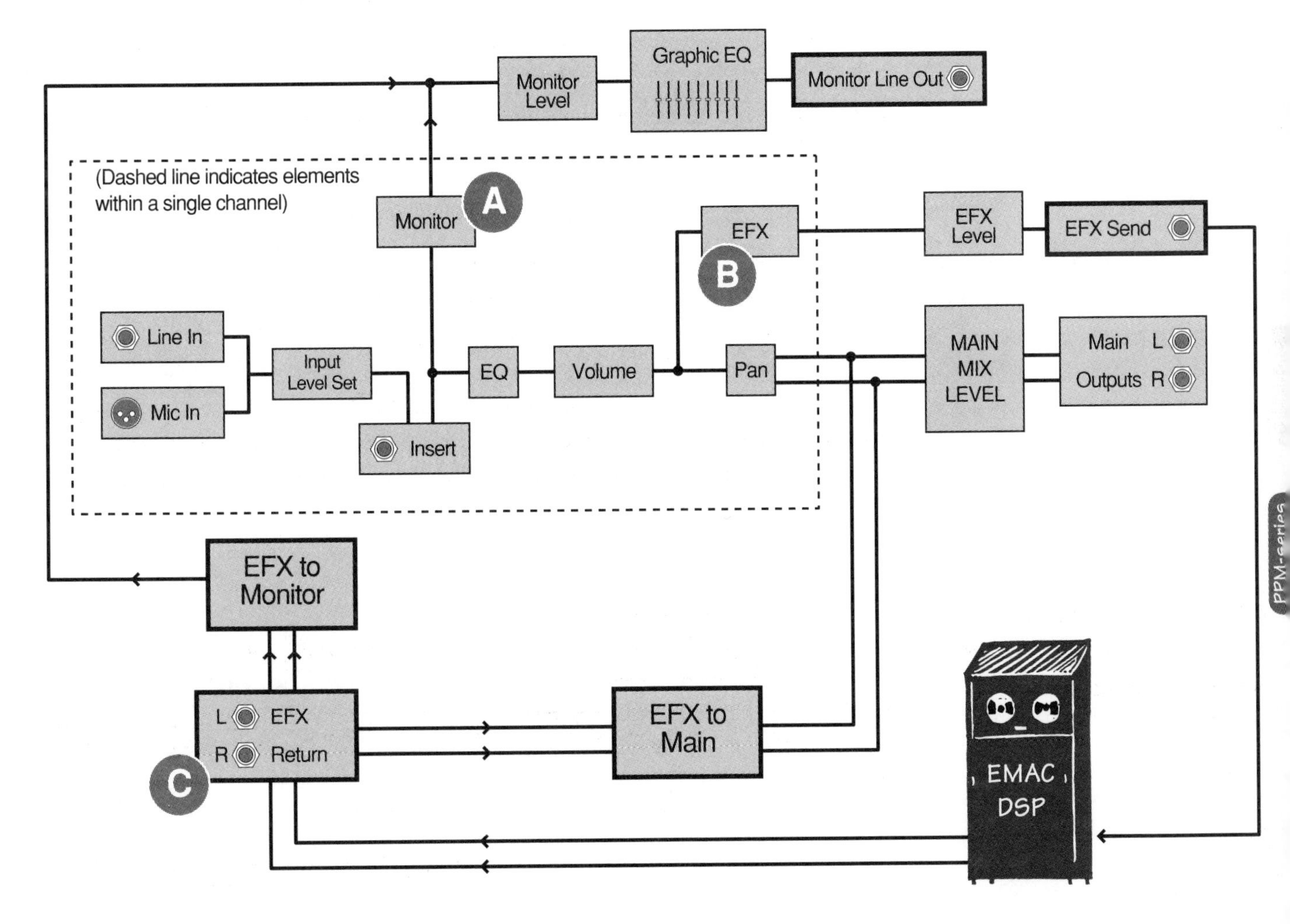

Sends, EFX

The EFX sends are post-fader, post-EQ (see letter B in the illustration on the previous page). The combined signal of all channels' EFX sends goes through the EFX Drive Level knob and is routed internally to the input of the emac effects processor (described in the previous chapter). In addition, the EFX aux send appears at the Effects Send jack.

But here's a wild little twist: the Effects Send jack is actually wired as an insert connector. This unusual feature also lets you tap a copy of the EFX send and route it to a second effects unit while still using the internal emac processor (although both effects units will be getting the identical same aux mix). This is done by plugging a regular guitar-style cable part-way into the Effects Send jack. This is called a "split out," and you can read more about it in the insert discussion the previous chapter.

If you try this, you may want to bring the outputs of the second effects unit back into the mixer, either through an unused input channel or alternately, the Tape In jacks (using appropriate adapters).

Connecting an insert cable to the Effects Send jacks lets you place an external device in the signal path before the emac. This makes the Super Cool EFX Volume pedal described on pg. 95 possible. For example, you could insert an equalizer into the effects send jack. The output of this would then feed the emac, so the resulting effect would be EQ-adjusted version of the emac effect.

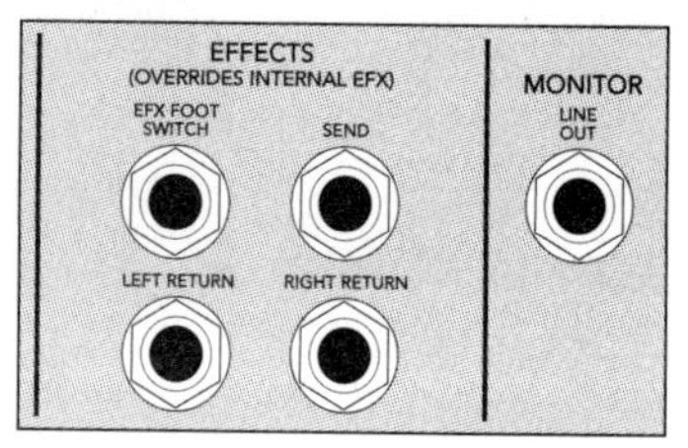

The knobs and jacks associated with the PPM's sends and returns are illustrated above (a stereo model PPM is shown, mono versions will differ slightly). The signal path relationships between these elements is shown in the illustration on the previous page.

Returns

The PPM has a single effects return, however, plugging anything to it disconnects the internal emac effects processor (see letter C on the prior page). Remember that an effects return is nothing more than a line input, so you could use the Tape In jacks or an unused input channel if you needed a second effects return, but didn't want to disable the emac's output.

Regardless of whether you are using the internal emac or an external effect connected to the effects return jacks, the EFX To Mon and EFX To Main knobs control the level of the effects mixed back into the main and monitor mix.

Note that stereo model PPMs have left and right effects returns; mono PPM-series mixers have a single mono effects return input.

Inserts

Each of your PPM's mono input channels has its own insert connector. They allow you to connect an external device, such as a compressor, into the signal path of an individual input. PPM Stereo inputs don't have inserts, as that front panel space is occupied by individual left and right input connectors.

From a signal flow perspective, PPM inserts come after the pre-amp. This type of insert is called *pre-fader* and *pre-EQ*, which means that twiddling a channel's level and tone controls won't affect the signal going out your mixer's insert, regardless of whether you are using it in the *insert*, *direct out*, or *split* hook-up configuration.

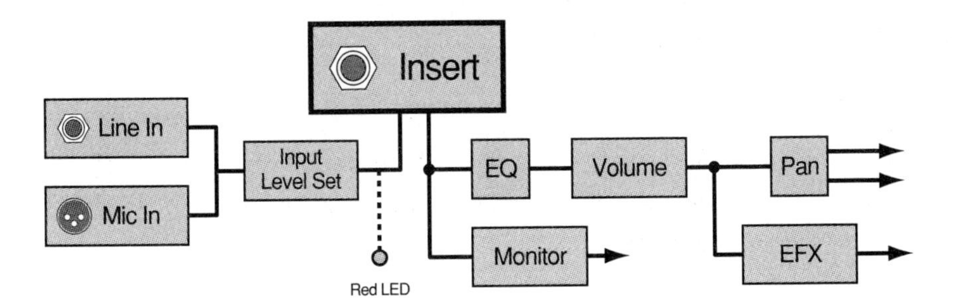

CFX Series

The CFX series mixers have four aux sends, a couple of returns and a built-in emac effects processor. While the CFX's sends and returns work just like other Mackies, their front-panel layout can be a little more confusing because of the extra controls for the emac's internal signal routing capabilities.

The knobs and jacks associated with the CFX's sends and returns are illustrated below. The signal path relationships between these elements is shown in the illustration on the opposite page.

Sends (Aux)

While CFX mixers have four aux sends, they aren't numbered aux 1 through aux 4. Instead, they are labeled Aux 1/2 and EFX 1/2.

The upper-two "Aux" sends are switchable between pre- and post-fader operation, via the Pre Fader switch on each channel. This gives you a choice. Set Aux Send 1 and 2 to post-fader (button up) if you don't need any monitor or headphone mixes (for instance, when doing a multitrack mixdown). For live performance with a stage monitor system, or when creating a headphone mix for musicians while recording, press the Pre Fader button down. It is possible to mix and match, setting some channels with Aux Send 3 and 4 to pre- and other to post-fader, although all the signals still end up at the same aux send output jacks (which are between the Utility Out and Sub Out 1-4 jacks on your mixer).

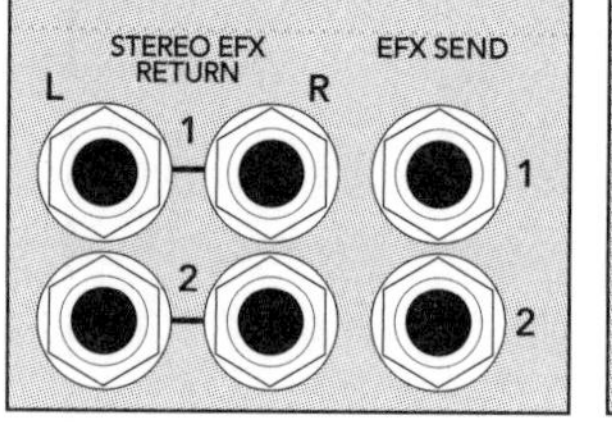

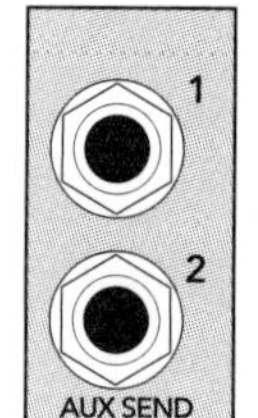

Note the location of the Aux 1-2 Pre in the signal path (see letter A in the illustration on the facing page). As you can see, the sends are post-EQ, even when the Pre button is depressed. This means your monitor mixes will reflect changes in the EQ settings of individual channels. In a live situation, you may be able to use individual channel EQ to reduce feedback problems on individual microphones, however, this may compromise the tone of that mic in the audience mix.

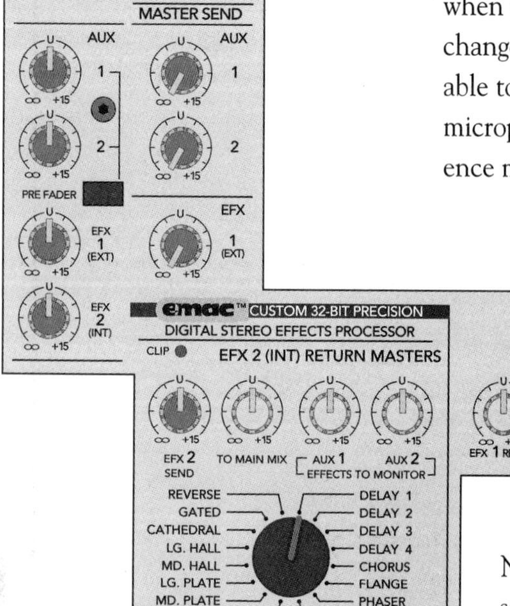

Sends (EFX)

EFX 1 (ext) is always a post-fader send, best suited for use with effects. The "ext" stands for external effects unit, which you'll need to hook up yourself. All signals sent to EFX 1 end up at the EFX Send jack which is right next to the Tape Output jacks on your mixer.

Now, how about EFX 2, which is marked (int)? If you guessed that it had something to do with the internal emac effects processor, you'd be right. Turning up the EFX 2 knob on any channel routes that input to the internal emac processor.

The CFX-series mixer's send and return system is illustrated below. The section within the dashed line indicates a single input channel, so imagine that part duplicated for each of your mixer's main inputs. The knobs, switches and jacks shown here are pictured on the opposite page as they appear on your mixer's front panel.

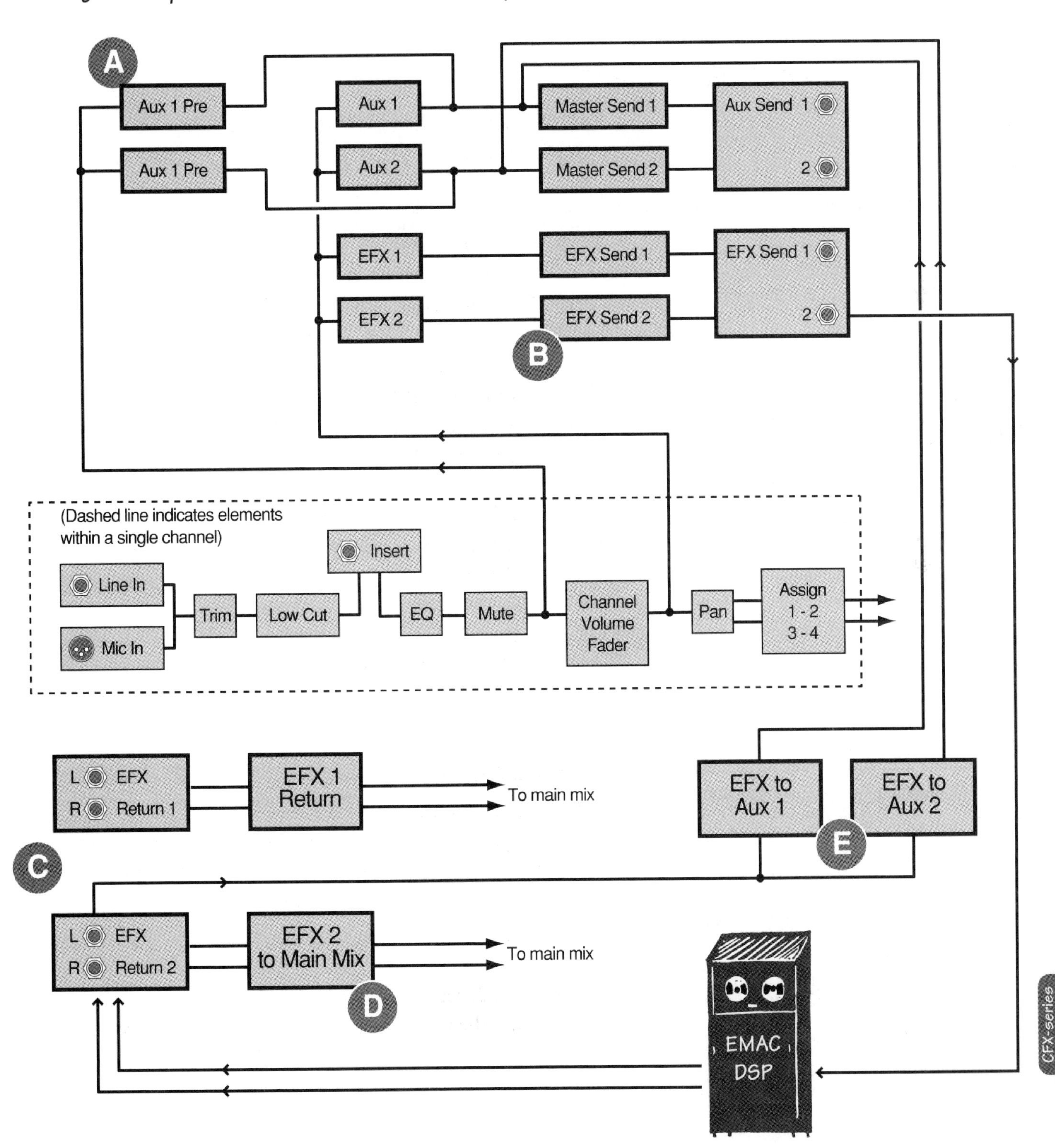

CFX-series

In addition to the internal routing of EFX 2 to the emac, your CFX mixer also brings EFX 2 out to a connector on the front panel (not surprisingly labeled EFX Send 2).

But here's a wild little twist: the EFX Send 2 jack is actually wired as an insert connector. Connecting an insert cable to the Effects Send jacks lets you place an external device in the signal path before the emac. It also makes the Super Cool EFX Volume pedal described on pg. 95 of the last chapter possible (you did read the last chapter, didn't you?)

You could also use this feature to add another signal processor in-series with the emac. For instance, you could insert an equalizer into the effects send jack. The output of this would then feed the emac, so the resulting effect would be an EQ-adjusted version of the emac effect.

Aux Masters

All four of the CFX's sends have master controls (see letter B on previous page). On your mixer, Master Send Aux 1, Aux 2 and EFX 1 knobs are located to the left of the graphic EQ. Immediately below EFX 1, in the emac control section, you'll find EFX 2 send. As you would expect, the EFX 2 Send controls the master level being sent to the internal emacs effects. If the Clip light blinks, turn EFX 2 Send down.

Returns

CFX mixers only have a pair of stereo Aux Returns (see letter C on previous page). Furthermore, Aux Return 2 is also used by the internal emac effects processor. If you plug anything into the Aux Return 2 jacks on your mixer, you'll be disabling the output of the emac. So, if you need to connect the output of more than one effects unit to your mixer, do so using other inputs. The stereo input channels are one option, the Tape In jacks another.

From a practical standpoint, if you're using Aux 1 and 2 as monitor sends you'll have enough dedicated sends and returns to run two effects units, which should be enough for most live situations.

EFX Return 1 has a level control next to the main level meters. EFX 2 (int) has three return level controls, all located in the emacs section. The first, To Main Mix (letter D on previous page), sets the overall level of the emac's effect output in the main stereo mix. The second pair of EFX 2 Return Masters (see letter E) lets you put varying amount of effects in the two Aux monitor mixes. This makes it very convenient to add some reverb or echo to the main PA as well as a bit of effect in either or both monitor mixes.

Inserts

Each of your CFX's mono input channels has its own insert connector. They allow you to connect an external device, such as a compressor, into the signal path of an individual input. CFX stereo inputs don't have inserts; instead, that front panel space is occupied by individual left and right line input connectors.

From a signal flow perspective, CFX inserts come after the pre-amp and low-cut switch. This type of insert is called *pre-fader* and *pre-EQ*, which means that twiddling a channel's level and EQ controls won't affect the signal going out your mixer's insert, regardless of whether you are using it in the *insert*, *direct out*, or *split* hook-up configuration. However, since the low-cut switch comes before the insert, its status does affect the signal present at the CFX insert jack.

CFX mixers also have a pair of inserts at their main stereo output. These come before the graphic EQ and main faders in the output section signal path. The most likely device to insert at this point would be a compressor or limiter. You could use these as a split out (as described in the previous chapter), note that the CFX's RCA Tape Output jacks are positioned at exactly the same point in the signal path.

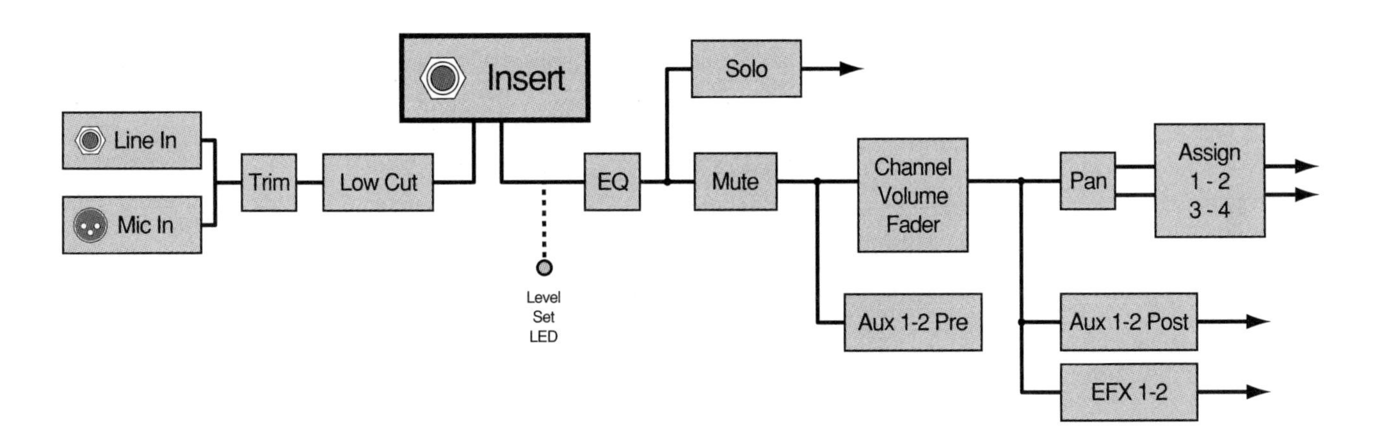

The CFX channel inserts come after the trim and low-cut switch.

CFX-series

CHAPTER 9

Output Section Concepts

Congratulations! You've just about made it to the end of your journey. This chapter and the next will complete your path from mixer inputs, through sends, inserts and routing, to the final destination—the output section of your mixer (sometimes called the "master section").

The output section includes controls which affect all the signals passing through your mixer, as opposed to individual channel controls which just affect a single sound. The output section is the final stage in the signal path before sounds exit your mixer and move on to the rest of your sound system.

It's easy to find the output section on any mixer—just look for some knobs and buttons that don't share the same vertical grouping as the individual input channel strips. On Mackies and most other mixing boards, the output section is on the right.

As in previous chapters, we'll begin our discussion with the basic concepts behind the mixer output section—that's in this chapter. The next chapter will take those generic concepts and apply them to each model Mackie mixer. So, you'll want to read all of this chapter before reading the next.

Output Section Road Map

The most basic function of a mixer's output section is to provide a master volume control and main outputs. Most or all of your mixer's input channels will end up being summed together on the main internal stereo "mix bus." In turn, this mixed signal passes through a master volume control and then on to the mixer's physical output connectors. Many Mackies include multiple sets of outputs, beyond just the main left and right output connectors of a basic stereo mixer. If your mixer has ALT 3/4 or assignable buses, there will be controls to route input channels assigned to these alternate buses through the main output, in some cases with level control. This should sound familiar to you, as we covered it as part of Chapter 5.

Other additional outputs within the master section include tape outputs and "control room" functions, which provide a convenient way to adjust your listening levels without affecting the signals feeding your audio recorders or audience sound system. There are other controls that end up physically located in the output section, even though they aren't really related to the main outputs from a signal flow perspective. For example, there are often controls related to solo (i.e. AFL vs. PFL operation and the RUDE solo LED). You'll also find auxiliary returns, which actually are inputs, but since they aren't complete input channels, their controls end up being placed in the mixer output section. Some mixers offer aux send master controls. Arguably, these *do* belong here, because they are also output controls. If present, they'll be grouped near the aux returns (which were covered in Chapter 7).

Overall, you can see a pattern emerging here. The output section of a mixer does have a lot of output-related controls in it, but there's another way of looking at it: The output section is the place where everything *except* mixer input channels end up. That's why some things in your mixer's output section don't really relate to outputs at all. So be it. In this chapter, we'll go through the output-related controls in detail. Following this conceptual overview, the next chapter will go through each model Mackie and explain its output section in detail.

Master Volume and Main Outputs

The single most important part of any mixer's output section is the master volume control. This is the final point at which you can adjust your signal before it goes on to the next device in your system, be that an audio recorder or amplifier and loudspeakers for a live performance. All signals passing through your mixer's main stereo bus flow through the master volume and turn up or down as a group as you operate the control.

Some mixers provide independent left and right level adjustments, while others offer just a single knob or slider that affects both left and right halves of the stereo output together. As you have no doubt gathered, the master volume is the perfect control for performing fade-ins, fade-outs and general "turn it up, dude" chores. I'm sure you also remember that if you want to turn just one part of your mix up—for instance, the lead vocal, you'd make that adjustment using the volume control on that individual channel and leave the master volume alone.

The overall signal level present at your mixer's main left and right outputs is controlled by the main mix fader (also called "Master"). This control will be a knob or one or two sliders. It adjusts the level of all signals assigned to the main output.

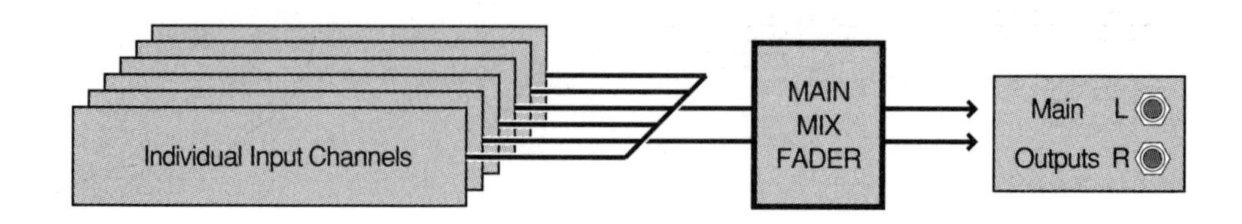

Tape Out

A Mackie tape out is best referred to as a convenience outlet. It carries exactly the same mix as the main stereo output. However, it makes for an easy connection to many cassette decks, Digital Audio Tape (DAT) or audio CD recorders, because of its phono-style RCA connectors. Note that the signal level that Mackie tape outputs can generate may be more than consumer or semi-pro recorders can handle without distorting, so watch your recorder's input level meters.

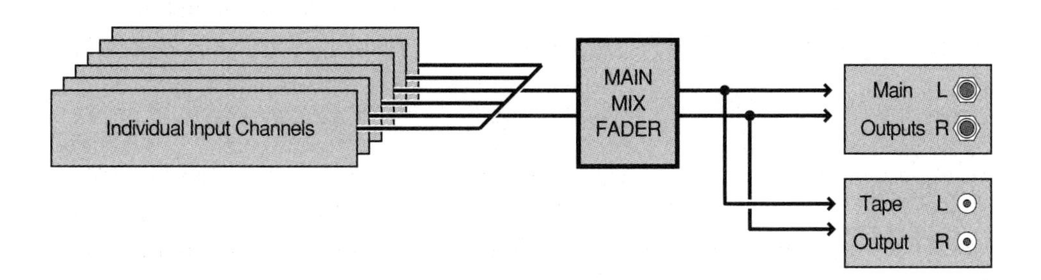

Tape outputs carry the same signal as the main output. The tape output level is typically controlled by the main mix fader, as shown here.

Headphone Outputs

So, the signal from the main left and right outputs (or tape outs) are connected to the next device in your system. When working on a recording project, this generally means your mixer's left and right outputs go to an audio recorder. But how are *you* going to hear what's going to the recorder? Just about all Mackies include a headphone output which lets you hear the same signal that is being sent to the main outputs of your mixer. As a further courtesy, most mixers will provide a level control to turn the headphones up or down.

But why should the headphones need their own volume? If they're too loud, couldn't one just turn down the master volume? On most mixers, this would indeed make the headphones softer. But there would be an unintended consequence: the signal level being sent to the audio recorder (or whatever else is connected to the main left-right outputs) would get turned down too! Clearly, this is not a practical solution.

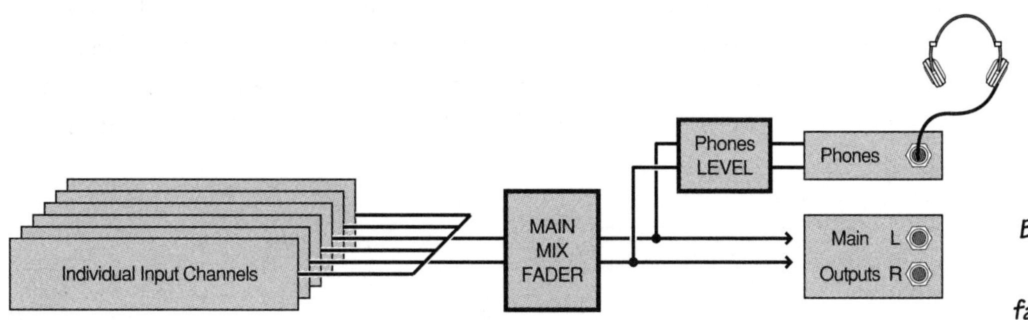

By adding a headphone level control after the main mix fader, you can monitor fade-ins and fade-outs over the phones.

A better approach is to provide a separate volume control for the headphones that won't affect the levels of the main left-right outputs (see illustration on previous page). Not surprisingly, this is often called a headphone level control. If the signal for the headphone level control comes from a point in the signal path after the master volume control, you'll hear fade-ins and fade-outs through the headphones. Alternatively, if the headphone level control takes its signal from a point before the master volume, the signal level in your phones won't be affected by the position of the master volume (as is the case with the original CR-1604).

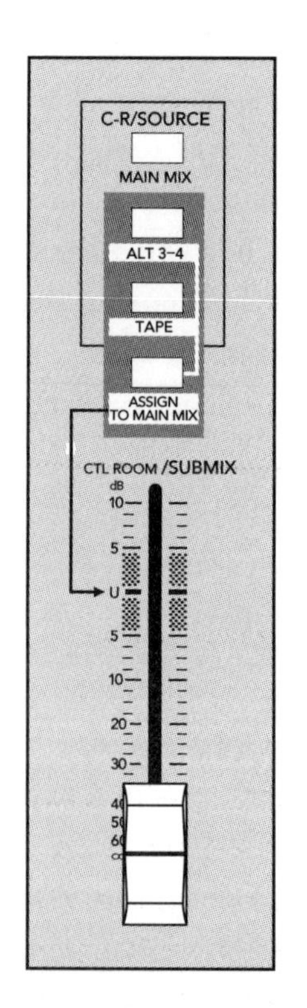

Control Room Outputs

OK, but what if you want to listen over loudspeakers, rather than headphones? You could plug the headphone outputs of your mixer into the line inputs of an amplifier and in turn, hook up the outputs of the amp to the speakers. These would be called the "control room" monitor speakers. (In a recording studio this would be the room where the engineer and producer sit.) The control room has its own set of loudspeakers that need to be turned up and down without affecting the levels of the recording in progress. Many Mackies have a dedicated set of control room knobs and buttons to help you hear what you want, without disturbing the level or content of the main stereo mix. The VLZ series mixers have extensive control room sections, while the SR and CFX have relatively limited ones. The original MS1202 and CR-1604 just have headphone outputs and the PPM series have no control room provisions at all.

All control room and headphone outputs switch to follow the mixer's Solo buttons. In other words, hitting a Solo button on a channel will cause that sound to appear in the control room/headphones output (the rude solo LED will also begin blinking to remind you that something is soloed). In addition to appearing in the head-

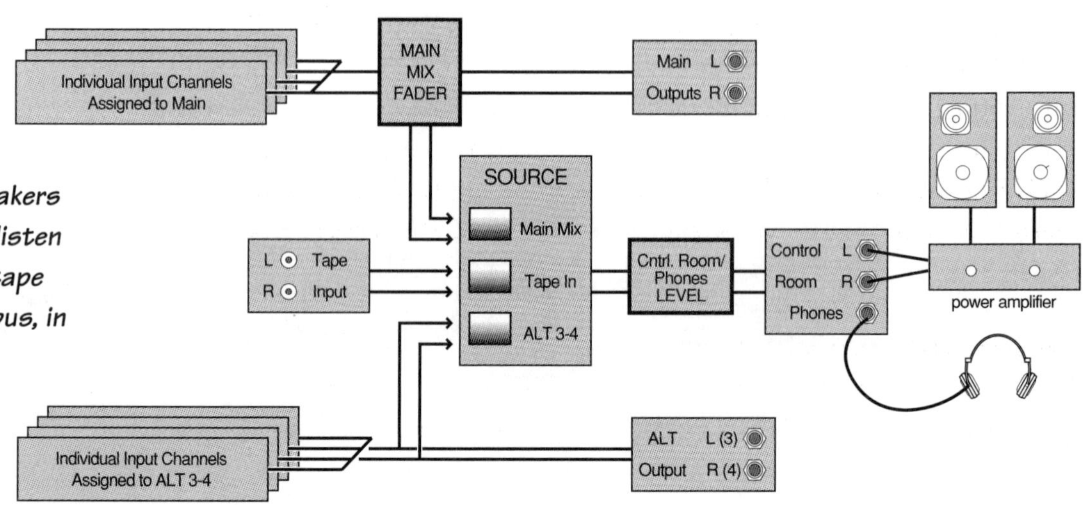

VLZ-series mixers have extensive control room routing options. In this simplified view of the 1202- and 1402-VLZ (pictured above), you can see that the monitor speakers and headphones can listen to the main mix, the tape input or the ALT 3-4 bus, in any combination, without disturbing the main left-right outputs.

phone/control room output, some Mackies let you route the soloed signals to the main output, but this is not often necessary.

OK, so the control room/headphone output can let you hear either the main stereo mix or the currently soloed channels. But the VLZ series control room sections go a step further by giving you "control room source" push-button access to multiple sources, including an external stereo tape recorder input, the main mix and the ALT 3/4 mix (or buses 1-2 and 3-4 on the 1604-VLZ). Pressing multiple buttons at once combines those sources for simultaneous monitoring. These source controls give you a quick and easy way to hear what's happening in different sub-sections of your mixer.

We'll talk more about control room features in the next chapter which describes features of specific model mixers. For now, just remember this: The control room output section of a mixer is designed to let you choose what you, the system operator, are hearing without changing the main stereo mix signal being provided to the audience or audio recorder.

Metering

Most Mackie boards have two rows of blinking LEDs in their output sections. These *level meters* provide a visual indication of operating levels of signals in your mixer.

Your mixer's meters let you keep an eye on current operating signal levels. The Level Set marking is used as a reference when soloing signals during the trim setting procedure. See your Mackie manual for details.

By default, your level meters will show the overall levels of the main stereo mix. In most cases, you can also use the meter to check the level of individual channels. This most often happens automatically when a solo button is pressed. By soloing a single channel, you can adjust its Trim control using the meter as a visual guide. See your Mackie manual for details on the optimum level-setting procedure for your mixer.

Generally speaking, meters in the green range mean your levels are fine (although only lighting up the bottom few greens means you may be running the board too soft). Hitting the yellow lights means caution, as you are getting close to the peak range your mixer can handle. The red light goes off when you are in danger of audibly distorting part, or all, of your mix. Note that hitting the red doesn't damage your mixer, although distorted mixes can potentially damage speakers at loud levels.

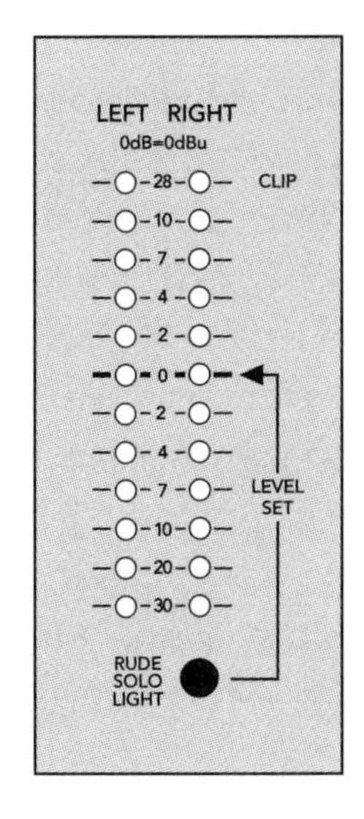

You should also remember that getting nice "in the green" levels on your board doesn't guarantee that the equipment connected to your mixer is operating at its optimum level. For example, when mixing to an audio recorder, be sure to keep an eye on your recorder's level meters too, because these won't necessarily match those of your board. Unintended distortion at any stage in your signal path is not your friend!

Here's a tip to help you get good levels to your audio recorder. Find a "test tone" CD or a synthesizer that can play a steady pure tone. Run that tone into an input channel, set your Trim properly and set the Channel Fader and Master output control so that the output meter reads "0." Now, put your mixdown deck into Record and use its record level input controls so that its meters read 0 also. If you want to be conservative, you could set it so that it reads a little lower, -6, for example. If you are using a digital recorder, be much more conservative, since with digital audio recorders, anything above 0 will clip badly. With 0 on your Mackie, you may want to be as low as -12 or -18 lower on your digital deck's input meters—ultimately, the level variations or "dynamic range" of the material you are recording will dictate a setting with an appropriate margin of safety or "headroom."

Now, when you are mixing, even if your record deck isn't directly in view, you can get a good idea of where its meters will be reading since you have calibrated them to those on your console. This way, if you see a really loud peak on your Mackie, you'll know that the tape machine went "over" too. But hopefully, your mixer levels will stay in a safe range and you'll know that the tape levels are OK.

Master Section Graphic EQ

In a few cases, the master section includes some EQ as well as level and routing controls. CFX, DFX and PPM series mixers include built-in graphic equalizers, which is another type of EQ. Not only does this look different from the EQ knobs on each of your mixer's channels, it affects the whole mix, not just individual inputs.

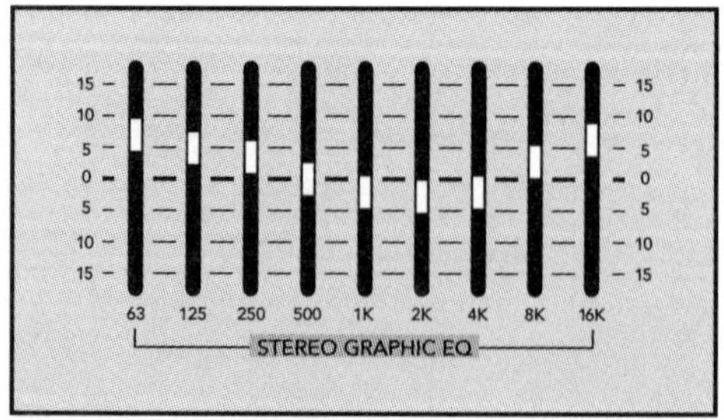

A graphic EQ lets you shape the tonal balance of an overall main or stage monitor mix.

Each slider of a graphic EQ can be used to boost or cut a particular range of frequencies. As you recall from our discussion of EQ in Chapter 3, this is similar to the mid EQ control, which also affects only a specific band of frequencies. (This type of EQ is called a *peaking* filter.) To operate a graphic EQ, simply push individual sliders up or down and listen to the results.

While graphic EQs can be used in a recording situation, they are most common in live sound applications. A graphic equalizer has two main purposes. First, it gives you much more creative control over the overall balance between the different frequencies your sound system produces. This means you can take a system that sounds too boomy and thin out the problem frequencies. Or, you may find yourself in a room with heavy carpet and drapes and need to brighten up your PA. Again, the graphic EQ can give a little high frequency lift to your whole mix.

A graphic EQ is also an important tool in the fight against feedback. If you are lucky, the feedback you are trying to eliminate will happen to land smack dab on the exact frequency controlled by of one of your graphic EQ's sliders. If this happens, you can just duck that slider down a bit and the feedback will go away. However, the feedback might just as easily land right in between two adjacent sliders, making it difficult to eliminate the feedback without taking a big bite out of your system's overall sound. For more about fighting feedback and other live sound topics, I humbly suggest another of my books, *Live Sound for Musicians*. Read all about it at www.trubitt.com.

Subgroup/Bus Faders

Mackies with true assignable buses, including the 1604-VLZ, CFX and SR-series mixers, have *subgroup* (also called *bus master*) *faders*. At the beginning of this chapter, we talked about the Master Volume Faders and the main Left/Right outputs. The bus master faders perform a similar function. They control the overall level of all signals that are assigned to a particular bus. After passing through a bus master fader, those signals continue on to one of the bus output jacks.

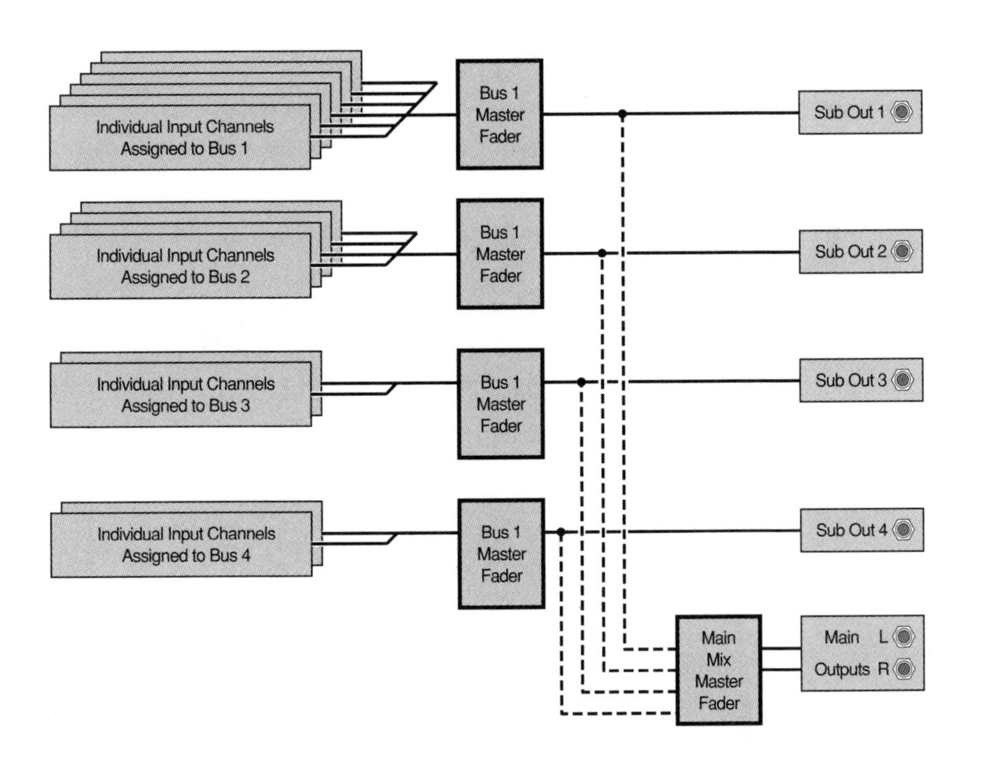

The illustration at left shows the signal flow of a four (or more) bus mixer. In order to mix with subgroups, the signal flow shown by the dashed lines must be enabled via the bus assign switches (see "Assign to Main Mix" on following page).

Using the submaster faders as illustrated in the signal flow diagram on the previous page requires the use of the Assign To Main switches. The 1604-VLZ's front panel is shown below; other Mackies will provide similar controls.

In addition to providing these extra physical outputs, a common use for these controls is to create *subgroups* while mixing. We'll go into this in more depth in Chapter 12, but for the moment, refer to the illustration on the previous page.

Conceptually, you can see that Bus Master Faders can be used to adjust the level of a number of individual channels, but not *all* channels, as the main mix fader does. Instead, the Bus Master Faders control the level of all channels that are assigned to that particular bus. Other groups of channels could be assigned to other buses.

The end result is that you could assign, for instance, all the vocals to a particular bus, then you'd be able to adjust the level of all the vocal mics from a single Bus Master Fader.

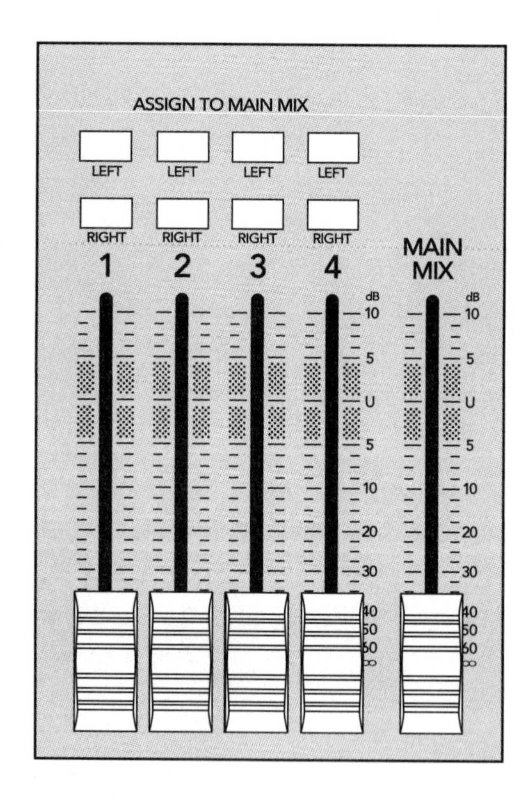

Note that making the internal connection from a subgroup fader to the main mix is done using each subgroup's Assign To Main switches, visible in the illustration at left. These switches operate much like the bus assign switches on your mixer's individual channels, allowing you to assign the output of each bus to the left, right or both channels of the main stereo mix.

Aux Masters

Several Mackies provide master level control over individual aux sends. While these controls are physically located in the master section of your mixer, they are really part of the aux send and return system. They are covered in Chapters 7 and 8, Auxes, Inserts and Effects.

CHAPTER 10

Mackie Output Sections

MS1202

The MS1202's output section is pretty gosh-darn simple (see the signal flow diagram on the following page). All input channels and aux returns end up in the main stereo mix. After passing through the main Master Volume control, the main mix splits and passes through the meters and on to the outputs, as well as feeding the Phones level control. While the mixer has two pairs of outputs (main and tape out), you can clearly see that exactly the same stereo mix will be present at both.

One item of note regarding the level meters: The 1202 meters normally show the output levels of the main stereo mix. However, when the Input Ch. Metering button is pressed, the combined signals of inputs 1 through 4, mic or line inputs is shown in the left meter. This metering option is provided to help you set your Trim controls properly. When this option is engaged, the right level meter will display the combined levels of all 12 inputs and the Aux Returns.

If you see metering peaks in the red, some level control is too high. If you get into the red in the left meter with the Input Ch. Metering button in, check your trim levels. If you get red peaks in the right with the Input Ch. Metering button in, or in either meter with the Input Ch. Metering button out, your individual channel gain or master knob is set too high. Try bringing down the master first, then individual channels if necessary.

Here are the output controls of an MS1202. See the following page for the signal flow diagram.

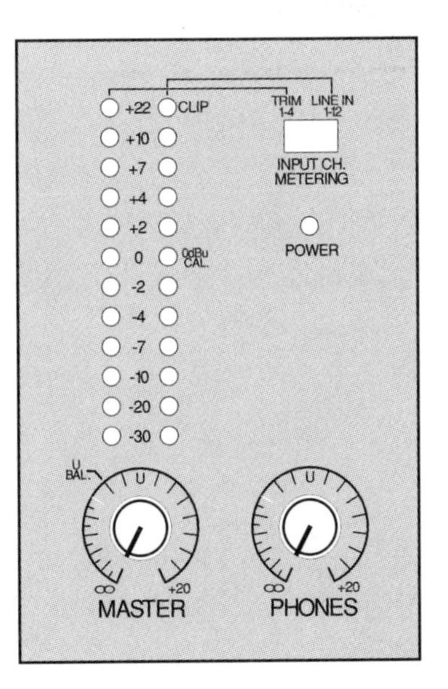

Remember that even if your Trim is set correctly, large EQ boosts can still cause an overload. In some cases, you can add a lot of EQ boost to a signal without causing an overload. The thing to watch out for is boosting EQ where a signal is already strong, like turning up the LO all the way on a kick drum sound. If you boost a part of a sound where there isn't much energy, you can usually get away with bigger boosts. A similar result can sometimes be achieved by turning other EQ controls *down*, instead of turning one up. Cutting frequencies with EQ will never cause an overload, and often is the better choice.

The signal flow of the MS1202 mixer's output section is about as straight-forward as they come.

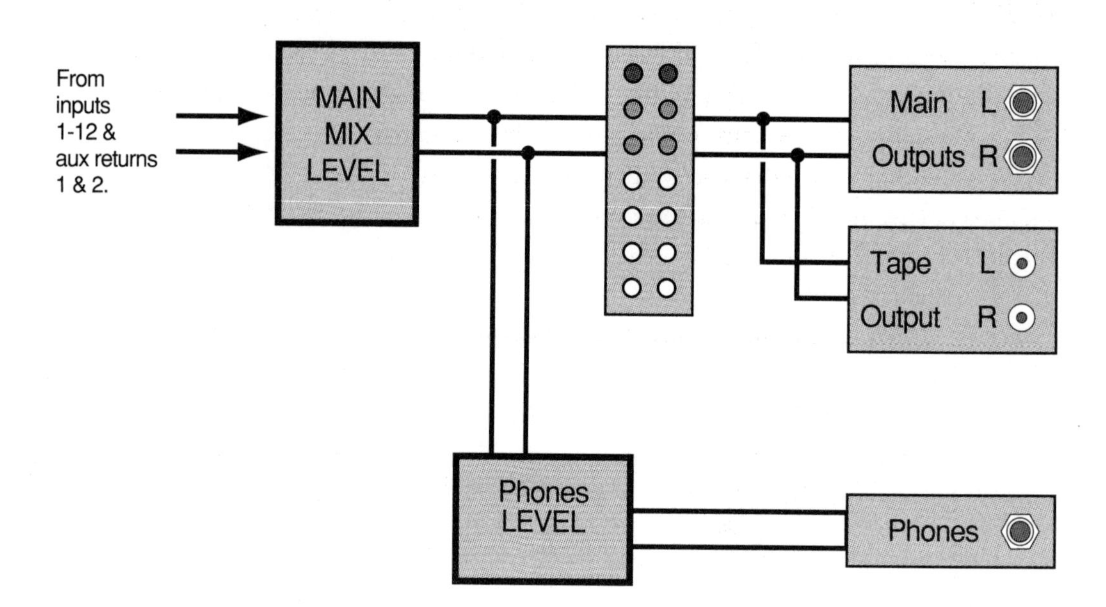

1202-VLZ & 1402-VLZ

The output sections of the 1202- and 1402-VLZ mixers (both regular and "PRO" models) are nearly identical.

Main Mix

Refer to letter A in the diagram on the following page. Here, input channels that don't have their Mute-ALT 3/4 buttons pressed and Aux Returns reach the output section and flow to the Main Mix fader (a knob on the 1202, a pair of sliders on the 1402). From that point, the main mix flows to three pairs of output connectors: Main L/R on 1/4" jacks, L/R on XLR connectors and the RCA-phono equipped Tape Output (see letter B). The +4/Mic switch is provided to cut the level of the XLR outputs down to the point where they can be connected to the mic-level inputs of another device that can't accommodate line-level inputs, such as some audio recorders, mixers, video camcorders, etc. Engaging this switch affects only the level from the XLR connectors, not the 1/4" or RCA outputs.

The main mix also feeds the Source/Control Room section, covered in a moment.

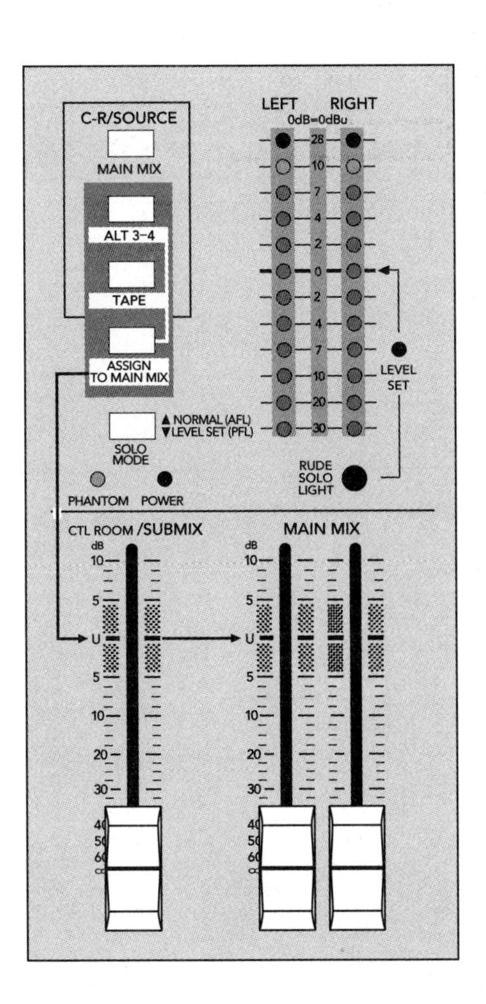

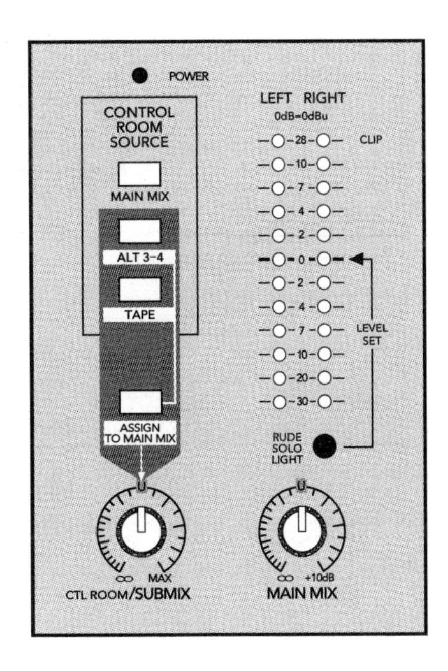

While their front-panel appearance differs, the 1202-VLZ and 1402-VLZ master control sections are nearly identical from a signal flow standpoint.

ALT 3-4

Any input channels with their ALT buttons pressed flow to the ALT 3-4 output connectors (letter C). There is no separate level control for the aux outputs as there is with the main left-right mix. The ALT 3-4 bus also connects to the Source/Control Room section.

Source/Control Room

The preceding chapter talked about the advantages of having a control room monitor section within a mixing board. The VLZ-series has an excellent one which lets you hear the main, alternate or tape mix, at any level, without disturbing the signal being sent to the main outputs.

The signal flow diagram of the 1202- and 1402-VLZ mixers.

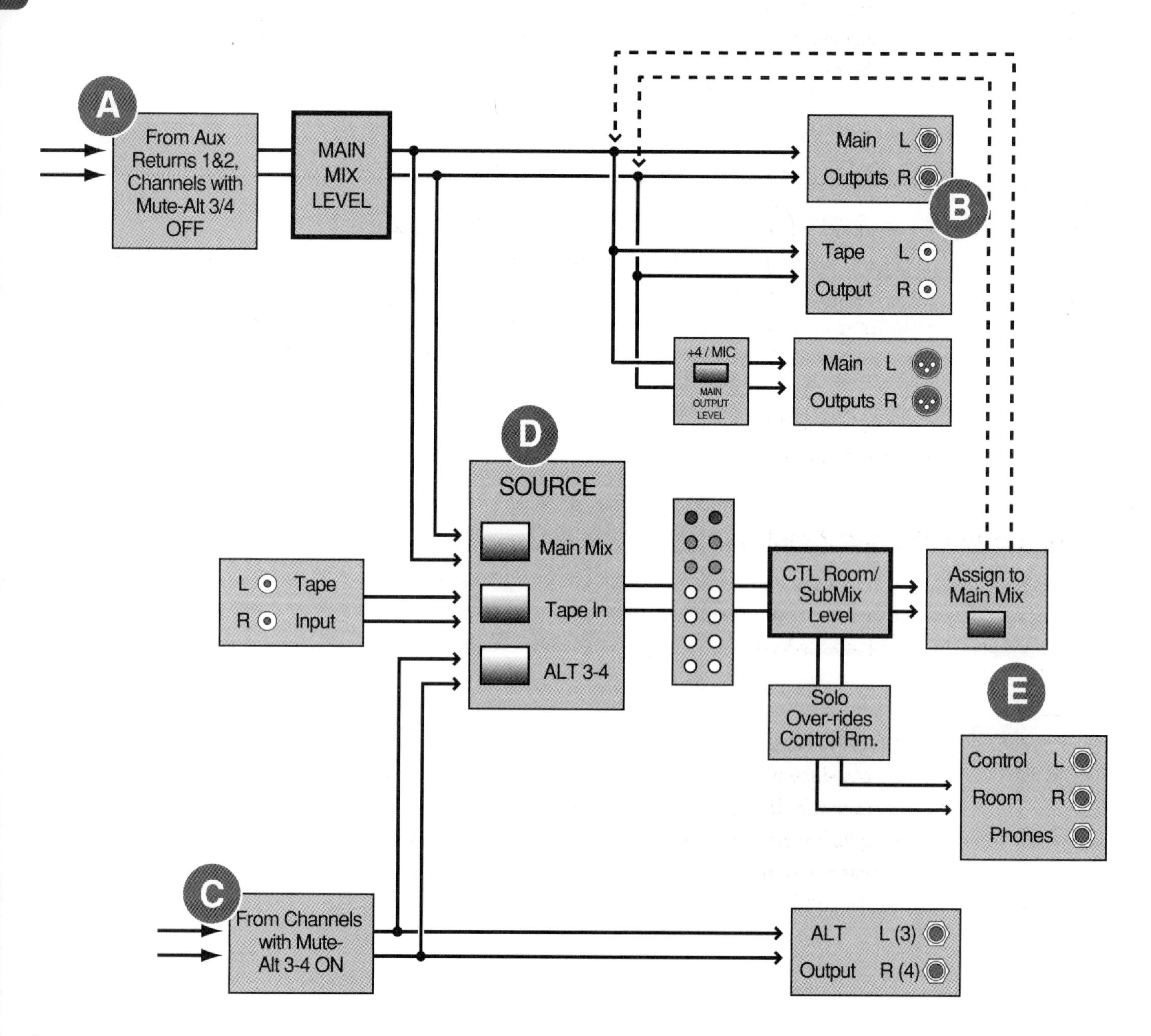

Note the three Source buttons: Main Mix, Tape In and ALT 3-4 (letter D). Any or all of these can be pressed, allowing the signals from one, two or three "sources" to be passed on to the level meters, and the CTL Room/Submix level control. This knob (1202) or slider (1402) adjusts the combined level of the selected Sources and passes them through to the Control Room and Phones output (just below letter E).

Note that engaging any of your mixer's solo controls will override the signal selected through the source buttons. Solo also commandeers the level meters, which helps when adjusting the trim levels of individual inputs.

Assign To Main Mix

Finally, we come to the Assign To Main Mix button (above letter E). As you can see from the signal flow diagram, this control lets you take the control room signal and connect it back into the main mix output.

The most practical application for this feature is when mixing a live performance, since you and the audience will both be listening to the Main Output. In this case, we can sacrifice the control room signal in favor of a "submix" or subgroup. Here's how: assign a number of related input channels to the ALT 3-4 mix. Then, select only ALT 3-4 on the Source button. Press the Assign To Main Mix and all the channels assigned to ALT will appear in the main mix output. And guess what? Using the Submix level control, you can turn them up and down as a group! We'll talk a bit more about submixing in Chapter 12.

Note: To play a tape during the intermission of a live performance, press the Tape In and Assign to Main Mix buttons.

Solo Mode (1402-VLZ Only)

The 1402-VLZ gives you two ways to solo—pre-fader or post-fader. Mackie has followed the common convention for referring to these options by calling them pre-fader-listen (PFL) and after-fader-listen (AFL). The 1202-VLZ provides PFL only.

The Solo Mode switch chooses between them. PFL is the correct choice when adjusting your Trim controls, since it takes the Channel Volume Fader out of the metering path. PFL is also good for live sound mixing, when you want to see if a mic or instrument is active before you turn it up in the main mix. AFL is nice when mixing, since it lets you hear one or more inputs at their post-fader mix levels, rather than turning them up to full when their Solo button is pushed.

CR-1604

The CR-1604 has a fairly flexible output section, although its lack of a Control Room section (described in the previous chapter) complicates certain tasks.

Main Mix

The CR-1604's Aux Returns and any channels *without* their Mute-ALT 3/4 buttons pressed will feed into the main mix faders, which are marked 1/Left and 2/Right. In turn, this stereo mix will flow through the meters and on to the main outputs.

There are two switches that will prevent the main stereo mix from reaching your 1604's output connectors. Note the position of the Master Mute. Pressing this button will prevent any and all signals from reaching your mixer's outputs. When the Solo To Main button is pressed, soloing any input channel or the Aux Returns will interrupt the main mix, replacing it with the soloed signals.

As you can see in the signal flow diagram, the main mix also flows to the headphone section, but only if the ALT Preview button is *not* engaged (more about that in a moment).

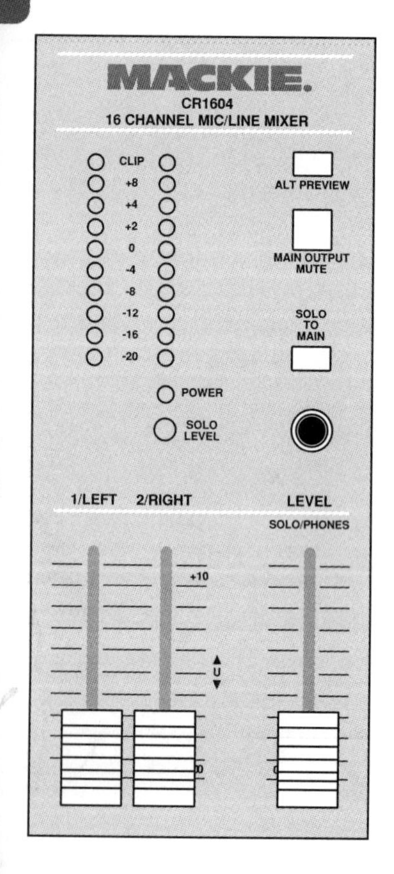

The CR-1604 master control section (above) and corresponding signal flow diagram (opposite page).

Level Meters

The 1604's meters normally show the operating levels of the main left/right outputs. However, if any Solo button is pressed, the meters show only the levels of the signals being soloed.

When inputs are soloed on the 1604, they maintain their stereo position (which is set by the Pan control). The meter follows along, showing levels in both channels. If the soloed input(s) are panned to the center, the left and right meter readings will be the same.

The meters are very useful for setting trim levels because they let you see exactly how close an input is to clipping. Your 1604 manual provides a detailed description of setting trims while watching your meter levels.

ALT 3/4

All input channels with their ALT buttons engaged will be routed through to the ALT 3/4 outputs. Of course, once you hit a channel's Mute-Alt button, you won't hear it in the main mix anymore. After all, the whole point is that Alt gives you a second, independent stereo mix, however, you'll need a convenient way to hear the Alt mix while you are working.

ALT Preview

For this reason, Mackie included the Alt Preview switch in the 1604's output section. Pushing this switch in will cause the current ALT mix to be played through the headphone out. Your main mix will continue playing through the stereo outputs uninterrupted.

Solo to Main

There's one more little option, the Solo To Main switch. Pushing this in makes the Main Left/Right output follow solo settings, just like the phones output. In other words, with this button pressed down you can solo over your main monitors if they are connected to the main stereo output of your 1604.

This is often quite convenient, but watch out! If you are in the middle of a mix to tape, or a performance for an audience, hitting a solo button will mute all the other inputs *if the* Solo To Main *switch is engaged!*

CR-1604 Mono Output

In addition to left and right, the 1604 also has a mono output. This balanced line out is a combination of the left and right signals. It's perfectly OK to use all three at once.

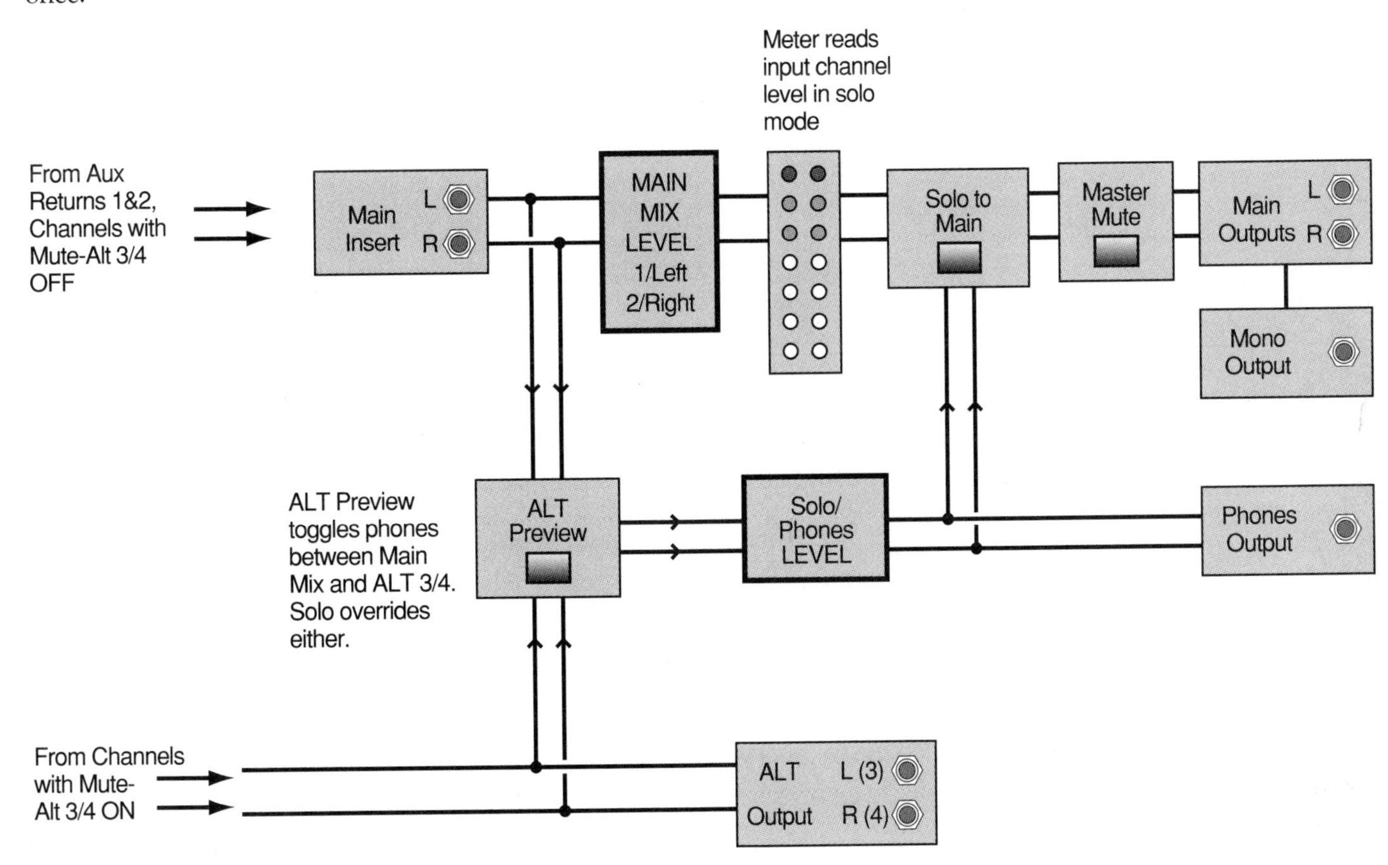

For instance, if your 1604 is mixing a live performance, you could use the mono out to drive the PA. Simultaneously, you could connect the left and right outputs to a tape deck to make a stereo recording of the performance. Using the Pan control on individual input channels would affect only the recording, not the live PA sound (in most cases, a mono PA mix is not only acceptable, it is often recommended).

CR-1604's Main Output Polarity

It's worth noting that the 1604's left and right balanced outputs are wired in opposite polarity to what one might normally expect. I'm going to explain what this means, but first, this is not something to get stressed out about! You can use your 1604 for years and never notice.

Here's the deal: Remember our discussion on the Nature of Sound? When a physical object vibrates, it sets up waves of changing air pressure. For example, when a bass drum is struck, the initial wave of air pressure is outward, in the same direction as the drum head moves when struck. Then the head pulls back, creating a low-pressure motion in the air, and so on. If this sound were recorded and then played back over a sound system, at the moment when the drum is first struck, you'd want the loudspeakers to push forward on the initial transient, just like the drum head did. However, it's possible to turn the sound wave "upside-down" electronically. If this happens, the initial strike of the batter against the drum head will cause the playback speakers to pull *in*, rather than push *out* on the attack.

While this is a subtle effect, evidence suggests that some trained listeners may hear a difference when this polarity reversal is made (sadly, I am not among this elite group of golden ears). You can wire up a simple cable to switch the main outputs' polarity back if you're concerned about it (just reverse pin 2 and pin 3 at either, but not both, connector ends). In my years of using 1604s, I've only come across one case where this polarity reversal caused a problem. In that case, I was using both the stereo main outs and a direct out, which were carrying the same signal to two different tape tracks. The sound on these two tracks ended up in opposite polarity and it took me a few minutes to figure out why.

Note that people commonly refer to reversed polarity as being "180° out-of-phase." Technically, this is misleading, since phase differences vary with frequency, but reversing the polarity of a signal affects all frequencies equally. Someday, I promise to write an article that explains phase so regular folks like us can understand it. Check my web site, www.trubitt.com, to see if I've gotten around to finishing it yet!

1604-VLZ, -VLZ PRO and 1642-VLZ PRO

The 1604- and 1642-VLZ models have one of the most flexible output sections of any compact Mackie mixer. However, all these options do take a moment to understand. Let us begin at letter A in the illustration on the following page.

Main L-R Mix

Any input channels with their L-R Assign buttons pressed reach your mixer's output section at letter A. Aux Return 1 and 2 also come back to this point. From here, they are joined by any sub masters assigned at letter D (more about that shortly). The main mix signals then pass through any device connected to the main inserts. If nothing is connected, the signal automatically flows on to the main mix fader at letter E. This single slider sets both left and right levels simultaneously.

From E, the main mix travels on to the main and tape stereo outputs, as well as the mono out with its own additional level knob tucked away next to the mono output jack.

The 1604-VLZ's master control section is shown below. The corresponding signal flow diagram appears on the following page.

Submasters

As you know, input channels and some Aux Returns can also be assigned to subs 1-2 and 3-4. Inputs assigned to these alternate stereo buses reach the master section at letters B and C, depending on the position of their Pan controls and which bus(es) they are assigned to. Each of the four buses has a submaster fader (see letter B). Once the bus signals pass through the fader, they split in two directions (letter F).

For the moment, ignore the split that goes to the Control Room Source (F). Instead, follow the path that leads to letter D, the Assign To Main Mix buttons. These buttons let you route any of the subgroups back into the main mix. Subs that are assigned back to main go through any signal processors connected to the main insert, and also through the main mix level fader before reaching the main outputs described previously. Regardless of the position of each sub's Assign To Main buttons, the sub signal continues on to the sub output at the upper-right edge of the illustration.

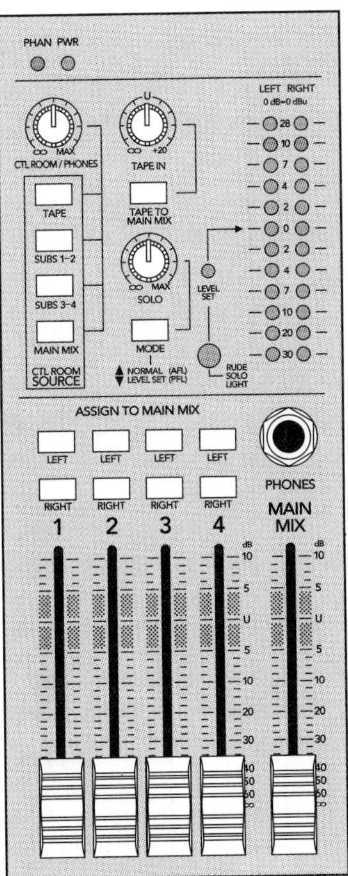

Assigning a sub to the main mix turns bus faders into sub masters, which are very handy when doing a mixdown session or live performance. You'll find out more about submixes in Chapter 12.

The 1604-VLZ master section signal flow.

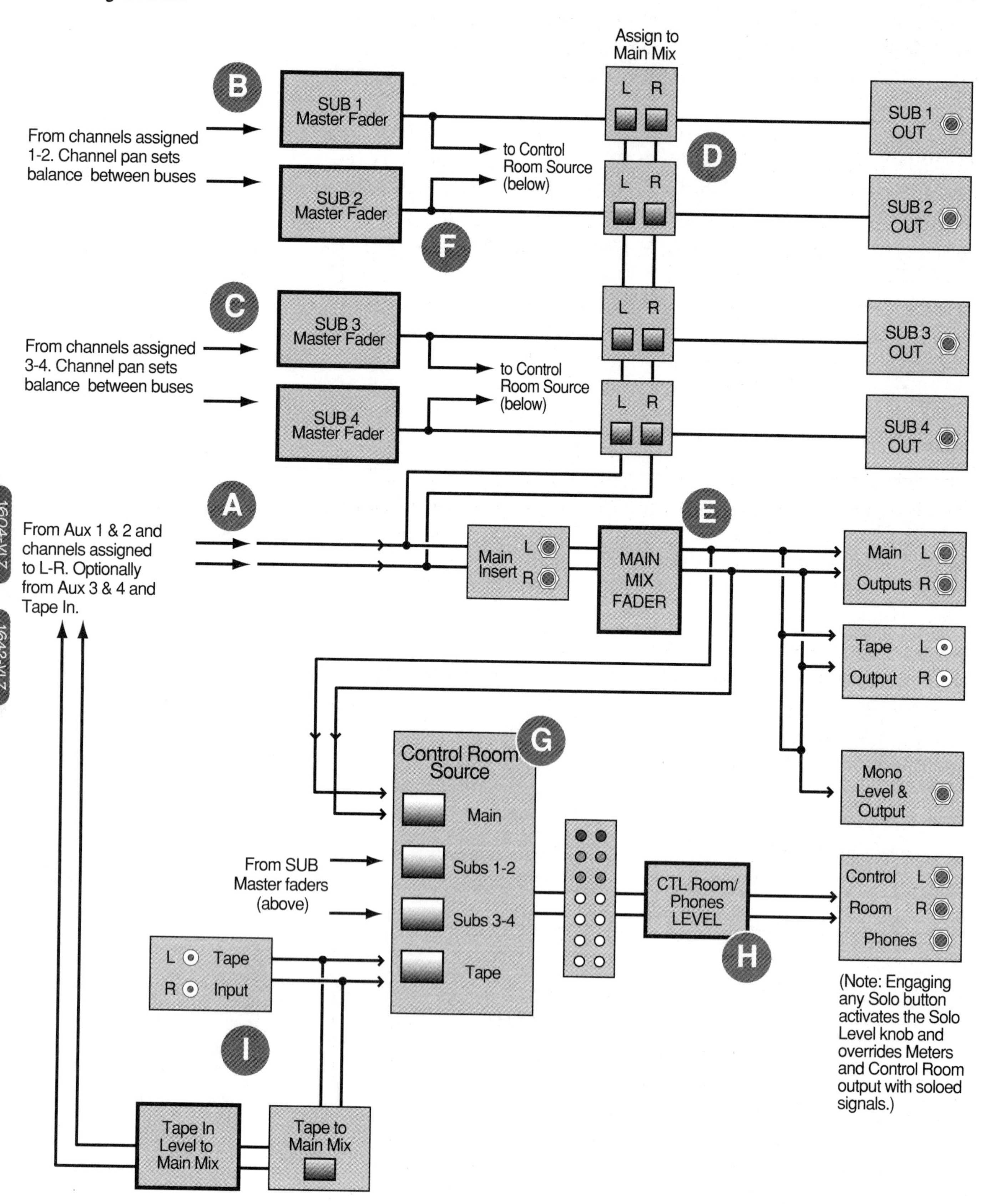

CONTROL ROOM SOURCE

The preceding chapter talked about the advantages of having a control room monitor section within a mixing board. The 1604- and 1642-VLZ PRO have an excellent one, which lets you hear the main, sub 1-2, 3-4 or tape input, at any level, without disturbing the signal being sent to the main outputs.

Check letter G and note the four Source buttons: Main Mix, Sub 1-2, Sub 3-4 and Tape In. Any or all of these can be pressed, allowing the signals from one, two or three "sources" to be passed on to the level meters, and the CTL Room/Submix level control. This knob (letter H) adjusts the combined level of the selected Sources and passes them through to the Control Room and Phones output.

Note that engaging any of your mixer's Solo controls will override the signal selected through the source buttons. Solo also commandeers the level meters, which helps when adjusting the trim levels of individual inputs.

Tape To Main Mix

When this button is pressed, the signal present at the Tape In jacks is routed through the Tape Level knob and back into the main mix (see letter I). The most obvious use for this feature is to play a tape between sets of a live performance, but I'm sure you can come up with other creative uses of your own.

1642-VLZ PRO XLR Outputs

The 1642-VLZ PRO does include one feature not present on the 1604-VLZ models: XLR-output connectors for the main stereo mix. This XLR output is switchable between +4 and -10 line levels.

SR24•4, SR32•4

SR-series master controls (below, right) and signal flow diagram (facing page).

The SR-series mixers have their share of unique output section features. While the control room functions are relatively simple, the SRs excel in live sound applications by including special "Air" EQ, subgroup inserts. And, the extra set of sub-outputs makes hooking an SR up to an 8-track recorder a snap.

Main Mix

Aux returns and individual input channels assigned to L-R reach the SR's output section at letter A in the illustration on the opposite page. From here, they are joined by any sub masters assigned at letter G (more about that shortly). The main mix signals then pass through any device connected to the main inserts. If nothing is connected, the signal automatically flows on to the main mix fader at letter B. This single slider sets both left and right levels simultaneously.

From B, the main mix travels on to letter C, the main stereo outputs (both XLR and 1/4"), the RCA-phono tape output, as well as an XLR mono out with its own additional level knob, tucked away on the rear panel of your mixer.

Phones/C-R Level

The stereo mix leaving the Main Mix fader also splits off to letter D, the Phones/C-R level control. This allows you to adjust your own headphone or monitor speaker level (connected to the Control Room or Phones out) without affecting the volume of the main output.

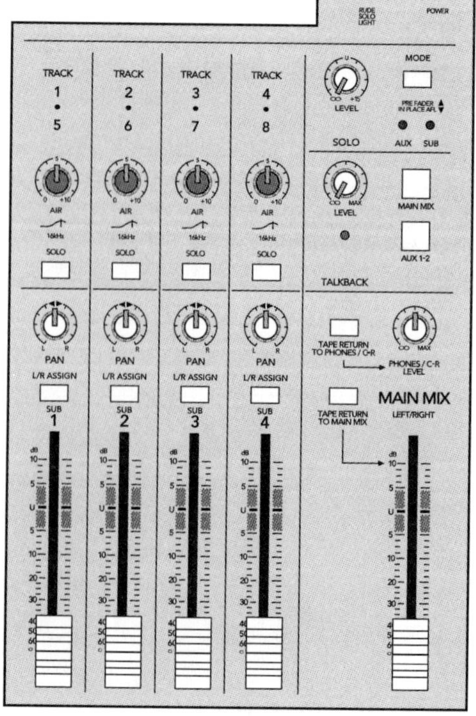

Sub Masters

As you know, input channels and some aux returns can also be assigned to subs 1-2 and 3-4. Inputs assigned to these alternate stereo buses reach the master section at letter E. The actual sub (numbered 1-4) to which a given channel is routed depends on the position of its Pan controls and which bus(es) it is assigned.

Once the signals assigned to a bus reach the master section, the first thing they pass through is the Air EQ circuit (see letter E). This is an EQ similar to the high-frequency control on each mixer channel. However, it is centered at a much higher frequency, 16 kHz. Dweebs such as myself may note that this EQ is a peaking filter, rather than the high frequency EQ which is of the shelving variety.

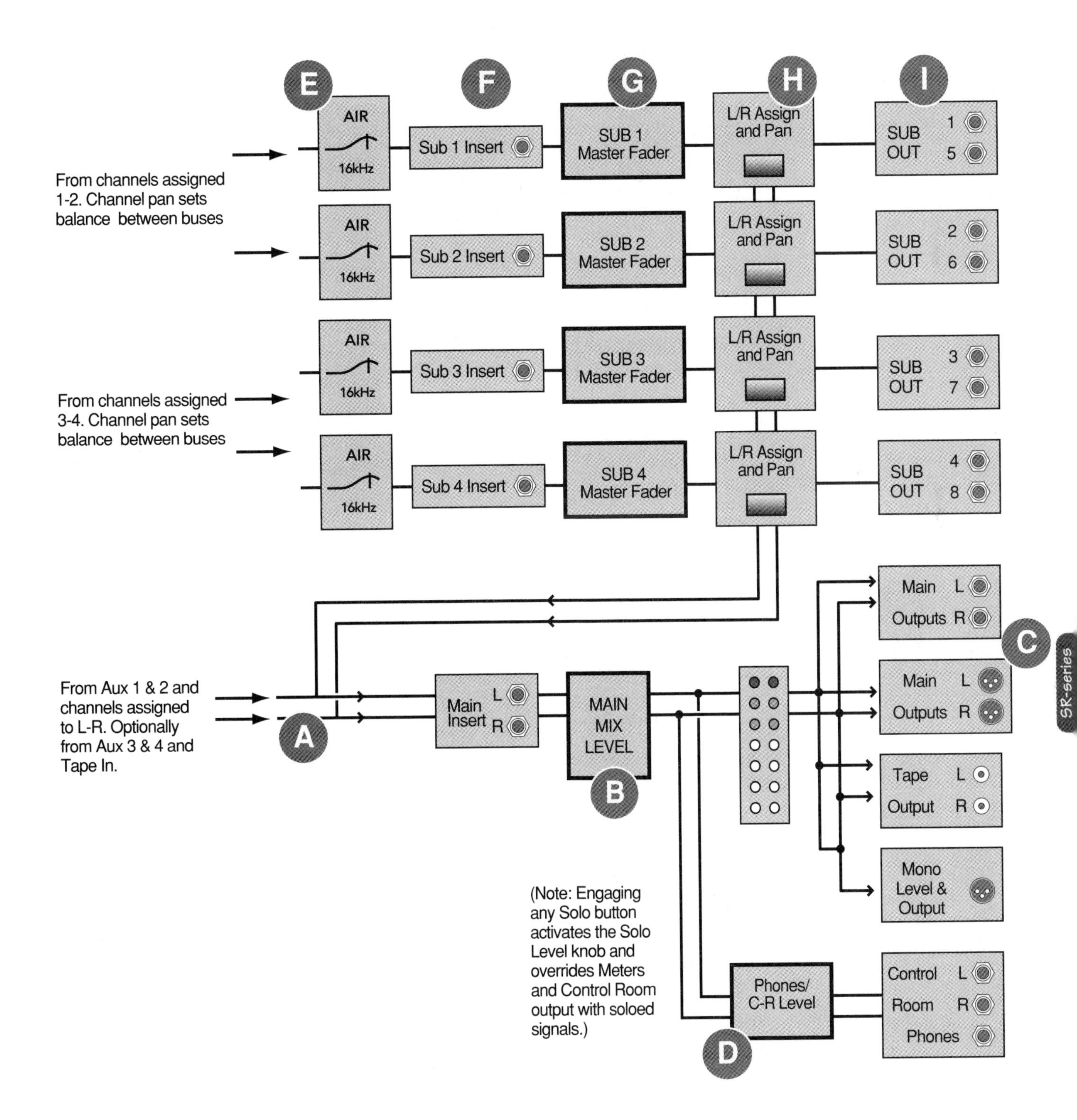

The purpose of the Air EQ is to add a bright "airy" shimmer to various sounds of your final mix. Mackie thought a special control for this would be welcome, as the powerful horn tweeters in typical live sound loudspeaker cabinets tend to lose these highest of high tones. By cranking up the Air, you can return some of the shimmer your mix would otherwise lose before reaching the ears of your audience.

The advantage of putting Air on each sub master is that you can selectively add this extra shine to just the instruments or vocals you want, rather than applying a blanket EQ to the entire mix.

Following Air, each sub master has an insert. This is ideal for adding compression to a group of vocals or instruments, since this allows a single channel of compression to be applied to multiple sources. While the result isn't quite as nice as dedicating an individual compressor to each mic input, it's a whole lot cheaper! Note that this works best when all elements assigned to the sub in question are similar. Even if the bass player happens to be the lead singer, I wouldn't suggest inserting a single compressor channel across both sounds, as the bass or vocal could cause the compressor to pump levels up and down in a way that would be detrimental to the sound of the other instrument.

Of course, it's possible to put a compressor across an entire stereo mix, so I don't mean to suggest you *can't* insert a compressor across different sounds at the same time. It's just that you'll probably need to set the compression controls a bit more cautiously. Like any insert, if nothing is connected to the jack, the signal automatically passes through to the next part of the signal path.

Letter G shows the sub master fader. As you have guessed, these are the four sliders marked Sub 1-4 at the bottom of the master control section. Following these, the signal moves on to the L/R Assign and Pan controls. These buttons let you route any of the subgroups back into the main mix. Subs that are assigned back to main go through any signal processors connected to the main insert, and also through the main mix level fader before reaching the main outputs described previously. Regardless of the position of each sub's Assign To Main buttons, the sub signal continues on to the sub output at the upper-right edge of the illustration.

Assigning a sub to the main mix turns bus faders into sub masters, which are very handy when doing a mixdown session or live performance. You'll find out more about sub-mixes in Chapter 12.

One final note: Each of the four assignable buses (Sub 1-4) shows up on *two* sub output connectors. As you can see in the illustration at letter I, Sub 1 and Sub 5 carry the same exact signal. The extra output (sub 5, in this case) is provided as a convenient way to hook your SR-series mixer up to a multitrack audio recorder. I'm sure you can come up with other creative uses of your own.

Tape In

While not shown on the signal flow diagram on the previous page, the SR-series does include a tape input which can be routed to either the main output, the control room out, or both. By pressing Tape Returns to Phones/C-R, you can monitor the tape inputs, which automatically disconnects the main mix from the control room monitor section. Similarly, pressing the Tape Return to Main Mix button puts the tape signal in the main, *but disconnects the rest of the mix from the main outputs!* Nothing wrong with that, just be sure it's what you intended, especially if you're in the middle of a great mixdown pass or a live gig.

When you do press one or both of the tape return buttons, you can adjust the level of the signal coming from tape using the Tape Return knob, which is located to the left of the level meters.

Talkback

The SR-series includes a talkback circuit. As the name suggests, a talkback system is provided for the sound system operator to communicate with the musicians on stage or over a headphone monitor system when recording.

To use the talkback, you'll need to plug a microphone into the talkback input on the rear panel. Next, simply press the talkback Main Mix or Aux 1-2 buttons. When pressed, your mic will pass through the talkback level control and come out of the main mix or Aux 1-2. In a live situation, Aux 1-2 are most likely to be stage monitor mixes, so you could tell the band (but not the audience), "I think the noise police are here!" If you want to address the audience before the show, press the talkback Main Mix button and smoothly intone the magical phrase, "Now sit back...relax...and enjoy an evening with (your band's name here)!" Works every time.

PPM Series

The output section of the mono and stereo models of the PPM series is identical, save for the obvious difference that one is mono and the other is...uh-huh...stereo. So, we're going to cover both in this section. You'll find signal flow diagrams for both; the little letters referencing parts of the drawing are the same for both, so look at whichever picture you prefer.

Main Mix Sources

There are three input paths signals can take to reach your PPM's main mix. Start with letter A in the illustration on the opposite page. The first six input channels pass through the Rumble Filter on their way into the main mix. Assuming you are running vocals through these inputs, you can safely engage the Rumble Filter without losing any important low-end beef from your mix.

Channels 7 and 8 (not available on the 406M model) and the EFX to main signals bypass the Rumble Filter (letter B), so use these inputs for drum machines, synths and other sounds with significant low end.

Finally, the Tape Input at letter C has its own level control to help adjust its playback level independently. Note that when the Break switch is engaged, the inputs connected to A are muted; those connected to B and C are not.

Tape Out

PPM master control panel (mono version) and corresponding signal flow diagram (opposite page).

The Tape Out jacks come before the main mix fader in the signal path. This means that you can set recording levels during your performance and not affect them if you turn up the overall PA (or if you are asked to turn it down—way down!). On mono PPM models, the left and right tape outputs carry the same mono signal. On stereo models, as you might expect, a stereo mix is available at the tape out jacks.

Master Level and EQ

Following the Tape Out, the main mix flows through the Main Master Level control and the Graphic EQ and level meters. From there, the signal reaches the Mixer Line Output(s). Regardless of whether you connected anything to this output, the signal flows on to the Power Amp Routing switch.

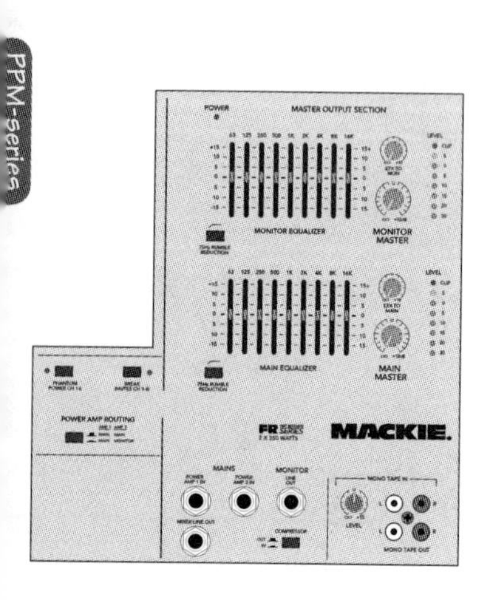

Power Amp Routing

With the Power Amp Routing switch (see letter H) in the up position, the main mix (upper half of illustration) will feed both channels of the power amplifier. With the routing switch pressed, the main mix will feed one amp channel and the monitors will feed the other.

On stereo PPM models, splitting the power amp between mains and monitors turns your main loudspeaker mix into mono. In most cases, a mono mix is better for live performances, so don't lament the loss of stereo. If you want to make a stereo recording of your gig (always a fun idea) you can use the Tape Out or Mixer Line Out connectors (letters D and E). These will still be in stereo, even if the main PA mix is mono.

Power Amp Section

These connectors are provided in case you are using a bigger external mixer and simply want to use your PPM as a power amplifier. Plugging anything in here disconnects the PPM's internal mixer from the amp inputs and speakers! The optional built-in Compressor operates on the signal just before it reaches the power amplifier circuit. Use this switch to protect your loudspeakers from excessive, possibly damaging levels. Finally, after the power amp, your signals exit the PPM at the speaker output connectors, letter J. Don't plug anything but loudspeakers into these outputs! If you want to connect your PPM mixer to another audio device with line-level inputs, use the Tape Out or Mixer Out jacks.

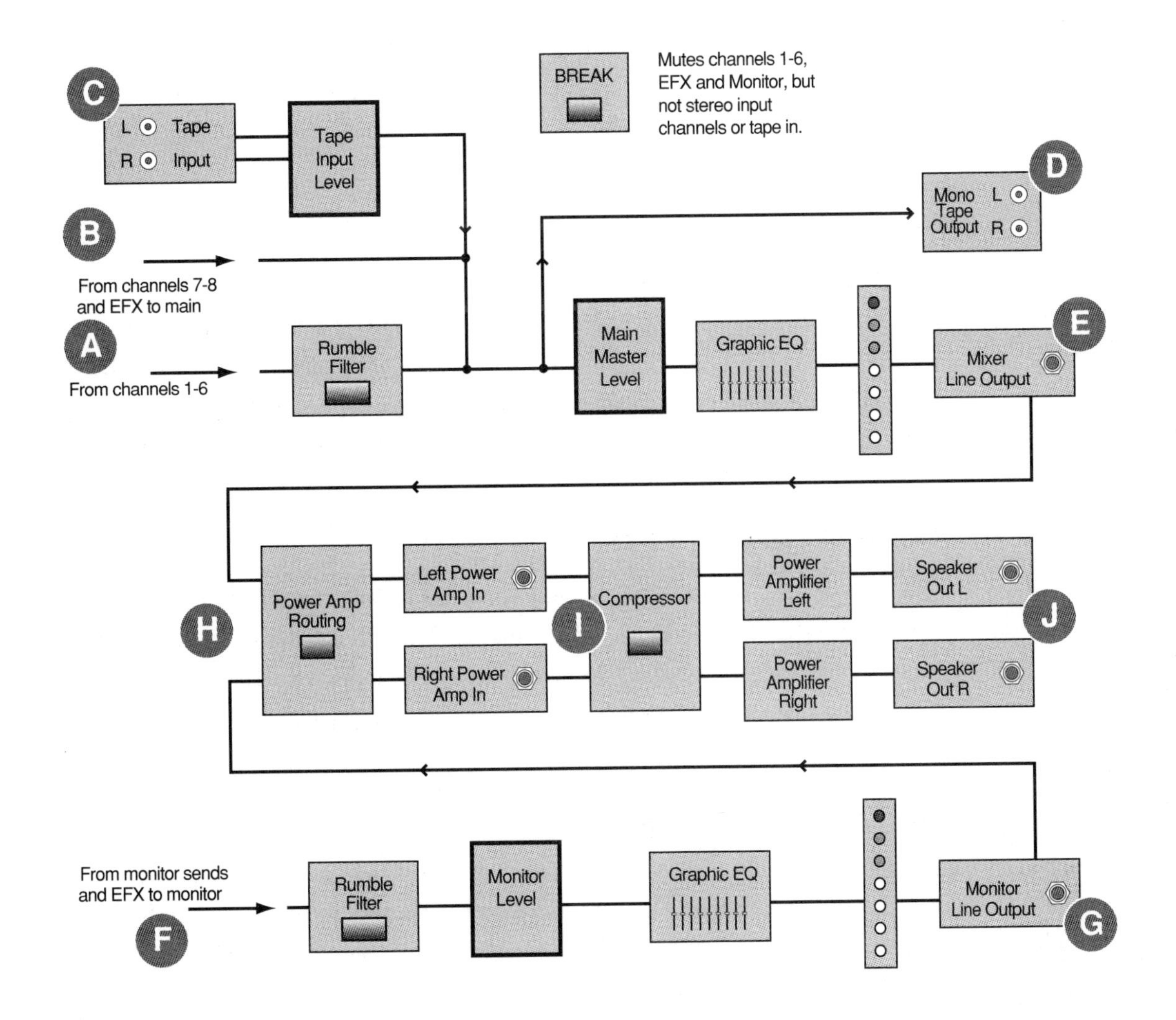

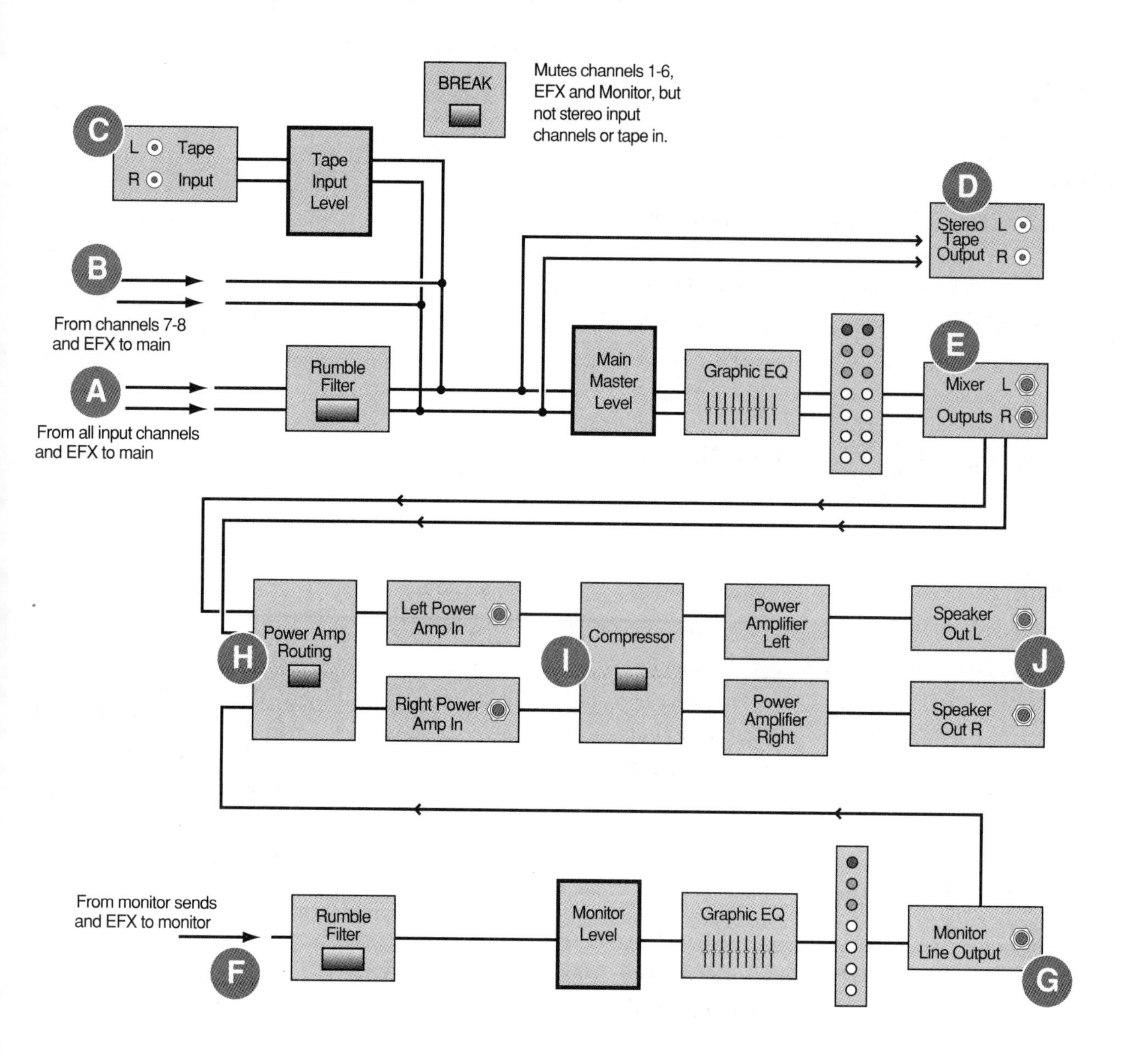

The PPM stereo master control section (right) and corresponding signal flow diagram (above).

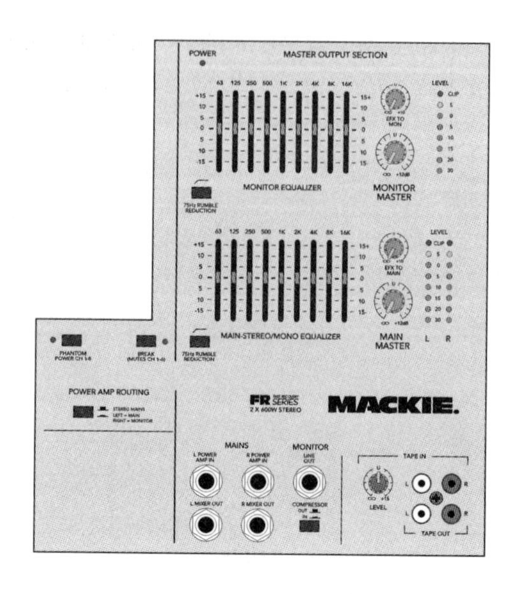

PPM-series

CFX Series

Sub Inputs

Input channels reach the CFX output section when they are assigned to bus 1-2 or bus 3-4. Inputs that are not assigned won't be present in the final stereo mix. Those channels assigned to 1-2 appear at point A in the illustration on the following page. Those assigned to 3-4 show up at point B. As you recall, each Pan control determines a signal's routing within the stereo bus, so, for example, channels assigned to 1-2 and panned hard-left will appear only in Sub 1.

Each sub mix passes through a sub master fader. This in turn flows to the corresponding Sub Output jack.

Sub Outs

The Sub Out jacks (letter D on following page) provide a way to get the individual sub-group mixes to an external device. For example, if you're making a multitrack recording, you might want to connect these outputs to individual multitrack inputs. You could also run these outputs to individual amplifiers and speakers for multi-channel surround-type effects, although your surround-panning options would be somewhat limited.

Assign to Main Mix

Each sub can be assigned to the main mix using the left and right buttons directly next to each sub master fader. Those subs assigned to the main mix flow along to point C in the illustration, where they are joined by the EFX to main returns (both internal and external).

Note: If you haven't connected anything to the Sub Out jacks, you probably want to press some of the Assign to Main Mix buttons!

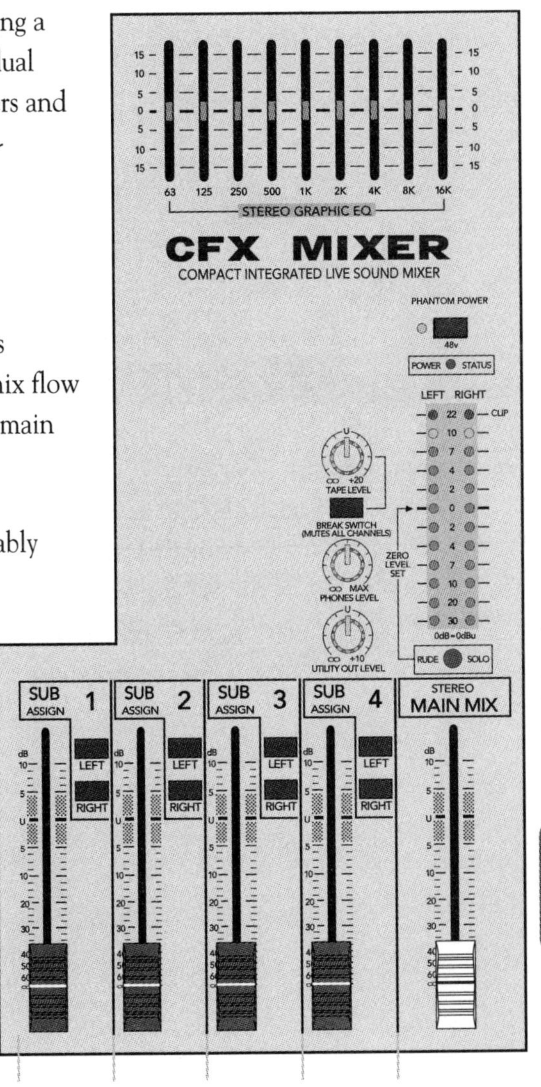

The CFX-series master control section. The corresponding signal flow diagram is on the following page.

CFX-series

The CFX-series master control section signal flow diagram.

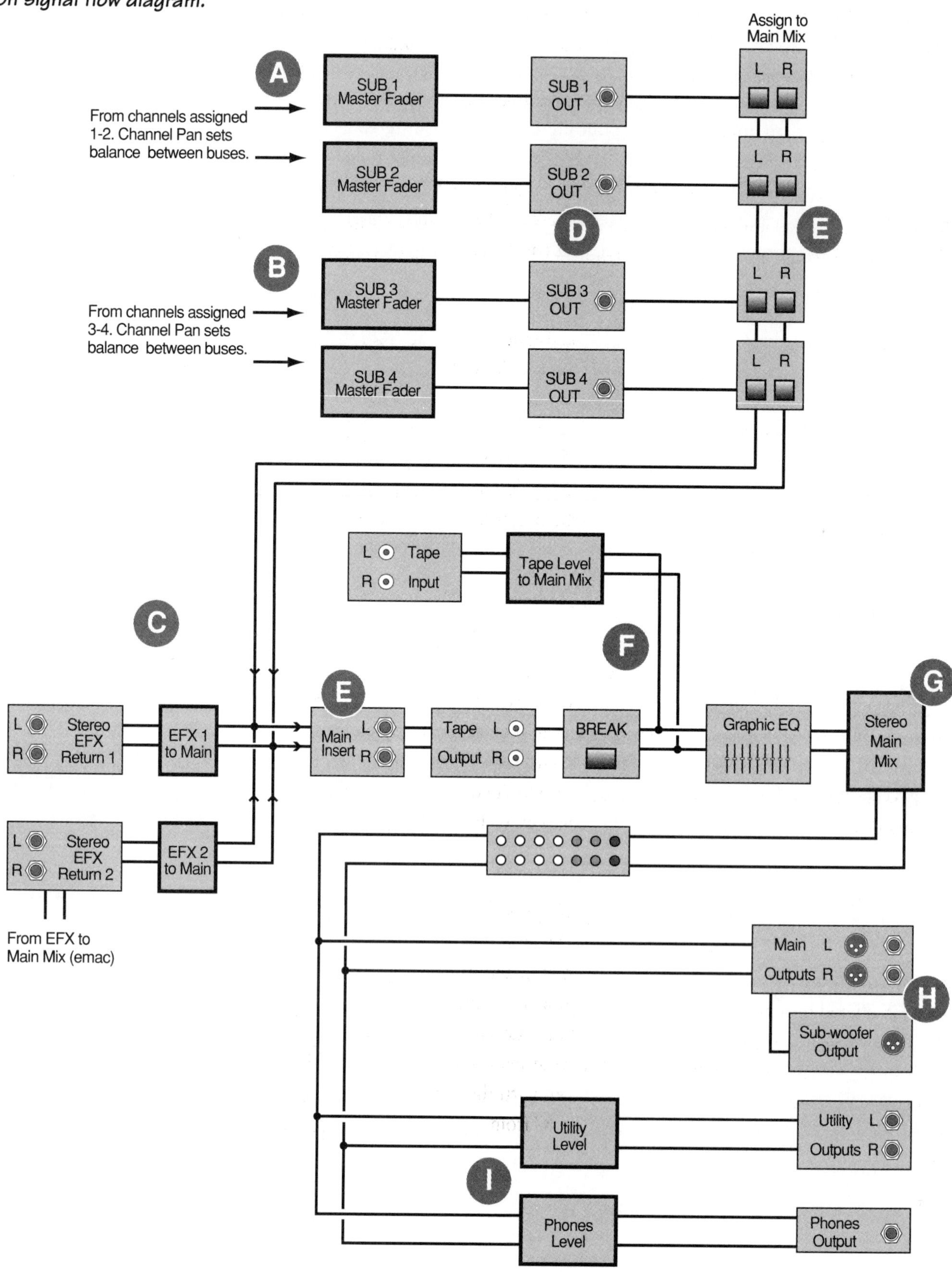

Main Mix Signal Path

The combined stereo mix including all subs and EFX returns passes through a main stereo insert. While you could add an equalizer to this point, it would be a bit redundant, since the main graphic EQ is coming right up. A compressor would be another candidate; just bear in mind that changing input levels upstream will change your compressor's effect.

Following the insert are the Tape Out jacks. Since these come before the graphic EQ and the main mix fader, you can set recording levels before your performance and not affect them if you turn up the overall PA (or if you are asked to turn it down—way down!).

Next comes the Break switch. Press this and you'll mute everything except the Tape Input and its Tape Level To Main Mix knob (see letter F).

Following the Break switch is the main graphic EQ. This is wired internally in stereo, although there is no way to set different EQ curves for each channel. Following the graphic is the Stereo Main Mix fader (letter G). Signals leaving this point are passed to the level meters and on to the outputs.

CFX Outputs

The main outs (letter H on pg. 158) come straight off the Stereo Main Mix fader. These outputs appear in both XLR and 1/4" connector types; both may be used at once if necessary. There is also a rather unusual feature—a 75 Hz Sub out. This output, if used, should be connected to a power amplifier and sub-woofer loud-speaker. All the lowest frequencies from the left and right stereo mix will be summed together and passed through this output. Don't confuse the "sub" in this output with the "sub 1-4" outputs—they are completely different, unrelated animals.

Note that this scheme, while convenient, doesn't provide quite all the benefits of an external crossover-based sub-woofer system. Specifically, the main outputs will still carry the below-75Hz signals, so your main amp and speakers won't be getting the break offered by a proper external crossover, which removes the lowest of the lows from the main speakers and feeds them *just* to the sub-woofer via its amplifier. In this way, you remove some of the load from your main system, increasing head-room and lowering distortion. That won't happen with the CFX's Sub out connector, but it's a quick and easy way to "get more bass" from an existing sound system with minimal hassle.

The other two outputs, Utility and Phones, each have their own additional level controls (see letter I). Aside from their variable level, these will carry the same signal as the main mix, *except* when a solo button is pressed. In this case, the main stereo mix will be switched off and the solo signal will appear in its place. There is no way to route the solo signal to the stereo main mix.

CHAPTER 11

Cables and Connectors

One thing you'll hopefully learn to love, or at least learn to live with, is wires, because you're going to be seeing a lot of them! Since cabling is such a ubiquitous part of the audio process, this chapter touches on a few important points about cables and the different kinds of audio connectors you'll find on your mixer.

Keeping Organized

Even with a small mixer like a 1202, it's easy to end up with 20 or 30 cables connected to your board. Besides the cost factor (good cables are not cheap and cheap cables are often less than good), a mammoth pile of spaghetti behind your mixer is trouble when trying to trace the inevitable missing connection.

So, rule number one of cable management is *staying organized*. Early on, I started labeling cables at both ends with tape, so I could tell which connection went where. Unfortunately, as anyone who has tried this learns, the result is a sticky mess in a surprisingly short amount of time.

A better solution to the labeling problem can be found in most industrial-strength electronics stores: Nylon wire ties with little write-on tabs (Panduit and Dennison are two manufacturers that come to mind). These handy ties (shown at right) hold related cables together in bundles. And, you can write on the tabs with Sharpie® pens. When you change things around, just snip the ties off with wire cutters—no sticky mess!

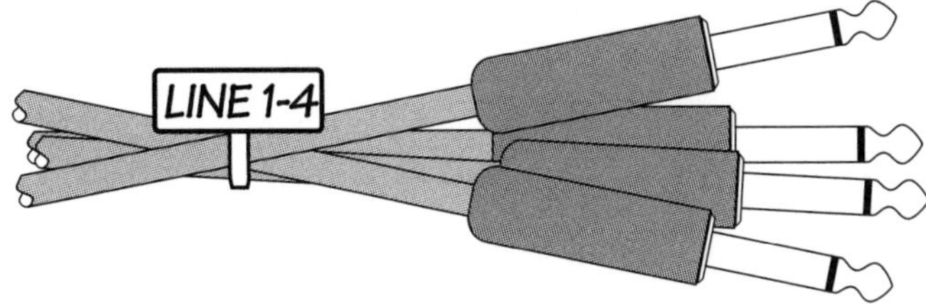

Use cable-ties to keep organized!

Kinds of Cable

Using the wrong cable for the job is a common mistake. As was mentioned way back in Chapter 3, audio signals come in a variety of strengths. Basically, the smaller the amplitude of the electronic signal, the thinner its cable may be.

Practically speaking, there are only two thickness categories of concern to us: speaker wire (cables for hooking up amplifiers to speakers) and everything else (meaning mic and line-level cables). High-power speaker connections require much fatter wire to avoid losing power in the cable before reaching the speakers.

The thickness of the rubber insulating coating on cables can be deceiving—it's the thickness of the wire inside that matters, not the rubber wrapper! Wire thickness is expressed in terms of "AWG," the American Wire Gauge, where higher numbers mean thinner wire. (For instance, 22 AWG can be appropriate for mic and line signals, but larger gauges such as 14 or even 12 are better for speaker connections.)

The level of an audio signal has an additional effect on cable requirements. Lower amplitude line and mic level signals are more likely to have problems with interference from radio or electrical equipment, such as motors or lighting dimmers. To keep this interference out of the signals connected to your mixer, the wires inside the cables are wrapped in a protective wire mesh, called a *shield*. This shield is connected to one or both of the outer shells of the cable's connectors. Then, if any interference is present, it gets picked up by the shield and sent safely (and silently) to the mixing console's *ground* connection.

One important point to remember is this: *Never use speaker cables for line-level signals*. Since speaker signals are at such a high level, background radio and other forms of interference aren't a problem—a good, beefy electrical connection is what's important. As a result, speaker cables *don't have a shield*. If you use speaker cables for line-level signals, you will be likely to pick up *a lot* of buzzing and radio-frequency interference. Don't do it!

The reverse is also true: *Don't use shielded instrument cables for speaker connections*. The thinner gauge wires in line or mic-level cables will rob power from your amplifier, dissipating it as heat before it gets to your speakers. This means part of the money you spent on that power amp is wasted, because you aren't using the right cabling!

Finally, as the length of speaker cables increases, it is appropriate to consider even thicker speaker wire (heavier wire means *lower* numbered gauges). This is because increasing a cable's length also increases its resistance to the flow of current. Increasing the thickness of the wire reduces resistance, thus compensating for longer cable runs.

A Balanced Solution

Keeping interference out of your mixer and the rest of the sound system is really important. Few things are as distracting as buzzes, static or, worse yet, some radio station playing in the background of your mix.

Shielded cables offer good protection against these problems. However, a shield may not be able to block 100% of the interference from reaching the inner signal-carrying wire. Protection from interference is further improved by the use of *balanced connections* of audio equipment.

All Mackies support *balanced* and *unbalanced* audio connections. Although you may be unfamiliar with these terms, you have certainly used unbalanced connections in the past. Electric guitars, home stereo equipment including CD players, tape decks and more all use unbalanced connections.

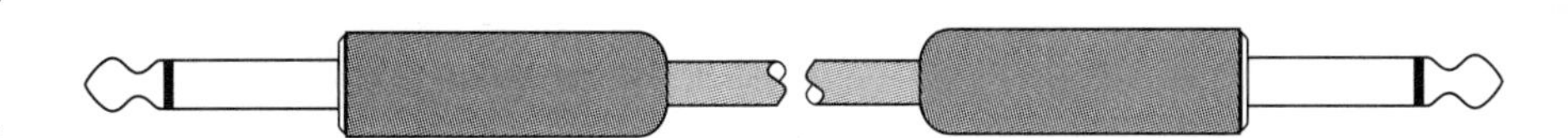

A standard "guitar cord," which uses two-conductor 1/4" tip-sleeve connectors, is an example of an unbalanced cable. You can use this to hook up balanced gear, but you'll lose the noise-cancelling qualities of a balanced connection.

An unbalanced cable uses two connections—shield (also called *ground*) and signal (also called *hot*). However, balanced cables need three connections—ground and two signal wires (called *hot* and *neutral*), each carrying the same audio signal at the same level (or voltage) but of *opposite polarity*. Opposite polarity means that as the signal on one wire is going positive, the second wire is going negative by exactly the same amount. (Note that we are talking about *monaural* balanced connections here. Three wires can also be used for unbalanced stereo signals, as is done with stereo headphones.)

When a balanced line reaches its destination, the signals of opposite polarity are combined in such a way that any interference that gets past the shield literally cancels itself out, while the original positive and negative signals add together for a signal with twice the amplitude! An unbalanced audio connection can't do this nifty trick and instead passes any accumulated noise along to the next stage of your signal chain.

So, if you have audio equipment with balanced connections, you should use them in a balanced configuration when wiring your studio. If you find yourself without the proper cables, it is OK to plug a *balanced* output into an *un-balanced* input. It's also OK to use a regular tip-sleeve "guitar cord" to connect a balanced device to one of your Mackie's balanced connectors. Just remember that doing either of these will result in the loss of a balanced connection's advantages in rejecting interference.

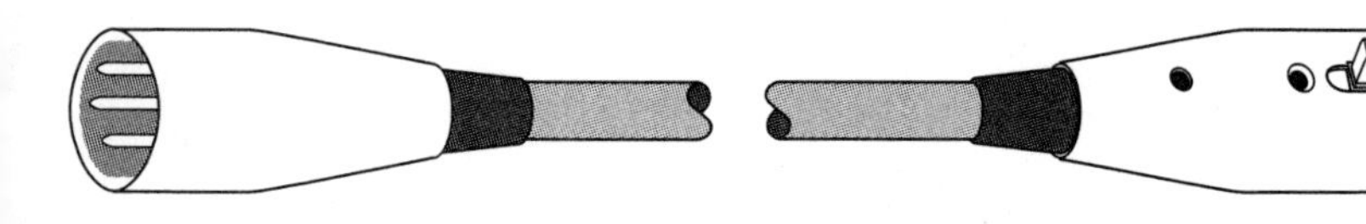

Mic cables, using three-conductor XLR connectors, can be used to create a balanced connection.

Also note that using an unbalanced cable reduces the level of the signal passed between the two units. Technically speaking, the unbalanced connection will be "6 dB" lower in level. This difference reflects a halving of the signal level, because only one of the two wires in a balanced cable is present. This somewhat lower level isn't a problem in itself—simply turning up either of the units can usually make up the difference, if necessary.

In case you were wondering, a dB is short for decibel. The decibel is a technical term that describes the *difference* in amplitude between two signals. When a scale is specified, a dB can also refer to an absolute level. For instance, dB SPL is used to measure absolute volume (sound pressure level) of outdoor concerts and industrial noise, to name a couple. The scale of your mixer's LED meters is marked in dBu, a standard scale for measuring the amplitude of line-level audio signals.

While understanding decibels is important, we're not going to get into them here, as doing so would violate this book's 100% Math-Free pledge. You can learn more about decibels in most technical books on sound systems. OK? OK!

Audio Connector Types

This section will introduce you to the different types of connectors you'll find on Mackie (and just about all other) mixers. As you'll notice, all these connectors are designed as pairs. By convention, the two halves of a connector pair are referred to as "male" and "female." A male connector is also called a "plug" and the female a "jack."

XLR Connectors

Also called "Cannon" connectors, this three-pin connector is the industry standard for balanced microphones and professional line-level signals.

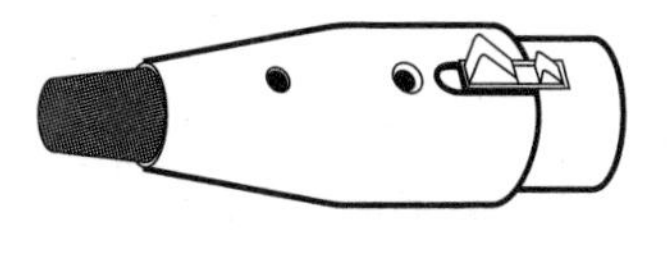

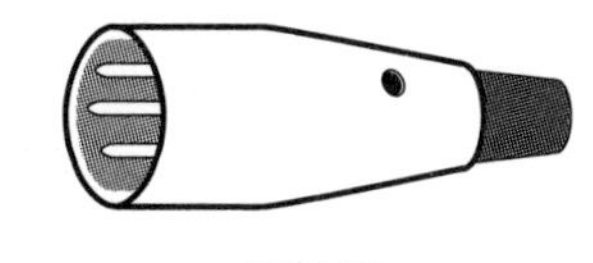

Pros: Durable, reliable and widely used. To ensure a nice solid connection, these connectors often have a latching mechanisim to hold them firmly in place when connected. Note that Mackie's XLR connectors don't have the latch, but the connection is still snug.

Cons: Expensive and somewhat large—XLR connectors use up a lot of panel space. There often isn't enough room to use lots of these on smaller mixing boards.

1/4" Phone or "Guitar Cord" Connectors

The most common audio connector for musical instruments is the 1/4" phone plug, so-called because they were originally used in old-fashioned telephone switchboards. This connector will be immediately familiar to all electric guitar, bass and keyboard players in the world. In fact, this sort of cable is also known generically as a "guitar cable" or "guitar cord." Mackies use 1/4" phone jacks for most line inputs and connections to and from effects.

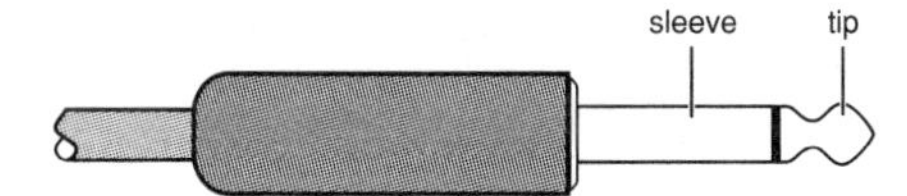

Pros: This is a relatively inexpensive and space-efficient connector, and it's very widely used. If you're short one, someone in the band probably has a spare you can borrow.

Cons: These aren't the most secure contacts; they lack the locking action of XLR connectors. In one of my first public performances, I was sitting in with a friend's band on electric guitar. When the moment of my first big solo arrived, I stood up without realizing my foot was on my guitar cord.

The effect was dramatic: Skinny kid stands up with a flourish, followed by a hideous buzzing squawk as the cable pops out of his guitar and falls onto the floor.
Moral: Don't stand on your cord, or skip your first few public performances—take your pick.

1/4" TRS, Stereo or Balanced Connectors

These are a variation on the 1/4" connector. Adding a "ring" contact provides an additional signal connection. This can be used for single-plug stereo (look at the end of any pair of headphones for proof). It can also be used to create a mono, *balanced* audio signal, just like an XLR connector. Both have three contacts.

1/4" tip-ring-sleeve or TRS connections are also used for channel inserts. In this case, a single TRS connector provides two independent connections: an *unbalanced input* and an *unbalanced output*.

Pros: Space-efficient and cheaper than XLR connectors.

Cons: TRS cables are much less common than 1/4" tip-sleeve guitar cables. Also shares the cons described in 1/4" Phone above.

Note: In a pinch, you can substitute a two conductor 1/4" cable for a balanced TRS connection. Note that doing so will result in a softer signal (actually half the amplitude) compared to using a proper three-wire connection, because you are only connecting *one* of the *two* audio signal voltages.

Here's how this happens: If you look at a TS and TRS plug side by side, you'll see that the Sleeve of the TS plug extends all the way up to where the Ring on the TRS plug ends. As a result, when a TS plug is inserted into a TRS jack, the TS's sleeve will make contact with *both the sleeve and ring connector* of the TRS jack! This is OK, but it does have an important effect. Since the sleeve is touching the ring connection, the ring contact is "shorted" to the sleeve, or "ground."

Shorting the ring to ground means only the tip carries audio. As mentioned, this results in a signal that is only half the amplitude of a balanced (two connection) signal. In technical terms, the unbalanced TS connection is 6 dB lower in level than the properly connected balanced signal would be.

Again, you can use TS plugs in TRS input or output jacks—just be prepared to deal with the resulting level reduction and the increased susceptibility to interference.

Phono Connectors

Not a misprint, these are *phono* (as in phonograph) with an "O" as opposed to phone with an "E." You'll certainly remember these from your last visit to the backside of your stereo or VCR. These connectors are *the* standard for consumer hi-fi. They are sometimes referred to generically as "RCA" cables and plugs.

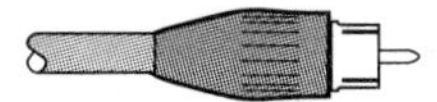

Pros: They're cheap. If you're a dweeb like me, you have a big box of these cables in your closet, spotted with the gummy residue of the masking tape from the time you were 15 and tried to label your stereo's wiring.

Cons: Phono cords are the connectors most often associated with molded plastic connectors, which are impossible to repair if they fail. Connector fit can also be problematic: They can be too loose, which can lead to pops and crackles, or too tight, in which case the typical close spacing from one to the next makes them nearly impossible to remove. I hate when that happens.

CHAPTER 12

Hooking Things Up

All right, it's time to take everything we've covered up to now and hook up your equipment. Rather than show you complete system hook-ups, this chapter breaks each group of connections down to single examples. This is because there's no way to tell what equipment you have in your system, so rather than try and invent a "typical" sound system (there isn't one), we'll just look at one piece at a time.

To begin the process, you may want to make a list of the equipment you have and then turn to each relevant section of this Chapter to see different ways to connect it to your mixer. Common sense tells you that you can't hook two things up to the same mixer connection. Therefore, as you are going through the different hook-up examples, you'll need to keep track of your needs and make sure that your plans don't depend on two pieces of equipment being plugged into the same jack on your mixer!

If you've already used up the connections required by one style of hook-up, you may be able to use another plan for the next piece of gear. Fortunately, there are usually several types of hook-ups that achieve the same result. In many cases, you'll find more than one diagram showing different wiring schemes.

Of course, this process will require some page-turning, head-scratching and gear-juggling on your part. Still, I hope you'll find this piecemeal approach more effective than big generic setups that can't possibly anticipate the specific equipment in your system.

Remember, the whole purpose of this book is to help you understand sound system signal flow *so you can figure these puzzles out for yourself*. Learning to make these plans on your own will help you deal with the unexpected situations that are an inevitable part of playing with audio equipment.

Mic Inputs

Remember that these illustrations show just one aspect of a hookup. You must combine several examples to connect your entire system.

Mackie mixers are designed to work with professional-quality microphones, including those that require *phantom power* (described in Chapter 3). Higher-quality mics typically use XLR, rather than 1/4" phone connectors.

Microphones with 1/4" unbalanced phone connector outputs may be connected to line inputs. However, since these inputs were designed for line-, rather than mic-level signals, you will probably need to turn the Trim control up most or all of the way to match the level of mics connected through XLR cables.

If you need more microphones than you have XLR inputs, you can connect additional XLR-mics to Mackie line inputs by using *line matching transformers*. These inexpensive devices convert the XLR connector to a 1/4" plug while fixing some of the mismatch problems that result from connecting a mic output to a line input.

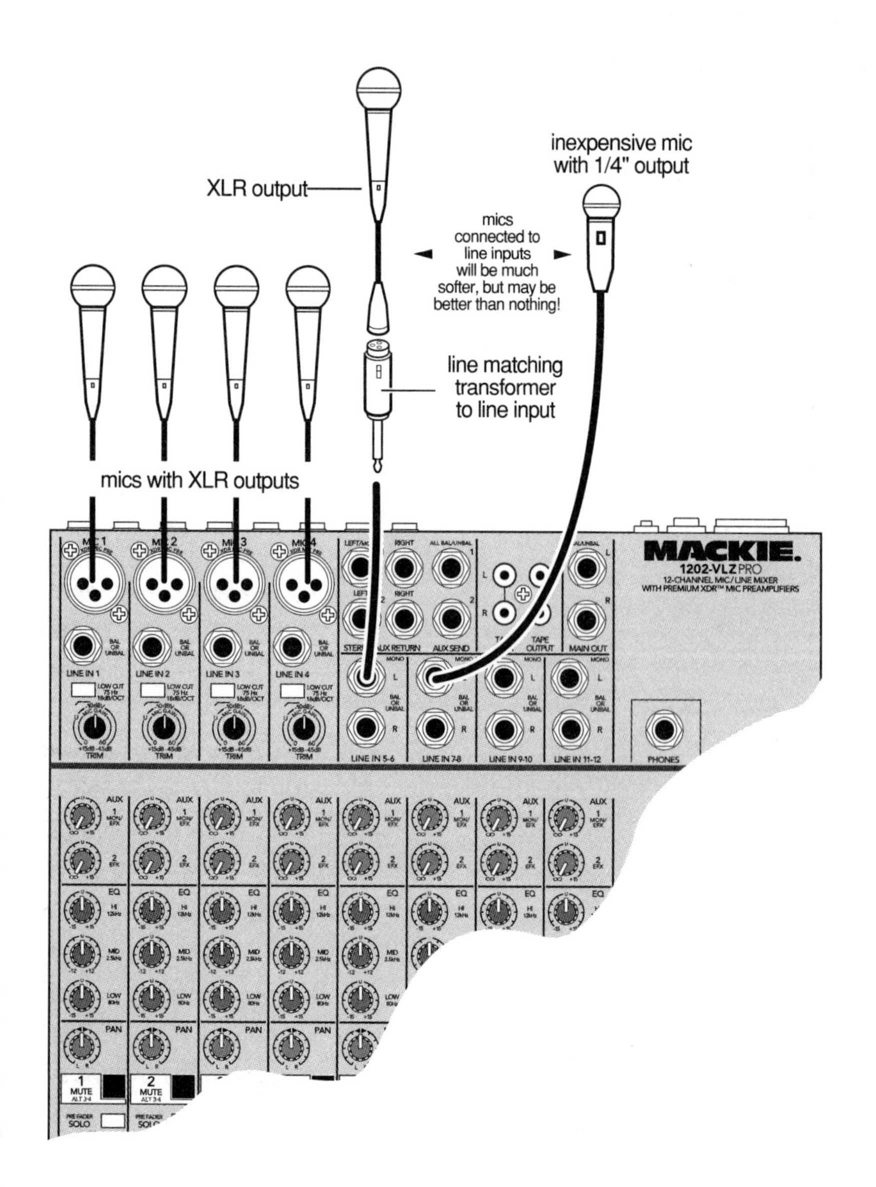

Connecting XLR mic outputs to mixer XLR mic inputs is always recommended. Mics with 1/4" outputs may be connected to a line input with a Trim control, if one is available. In this illustration, a line matching transformer and 1/4" output mic are shown connected to Trim-less 1201-VLZ inputs. This won't hurt anything, but you may not be able to get these mics loud enough.

Line Inputs

Line-level sources can be plugged into any Mackie's line inputs (remember that aux returns are line inputs too!). If the equipment connected to your Mackie has *balanced* outputs, try and use three-wire TRS cables (and any adapters required) to preserve the advantages of balanced connections.

Many models include a few stereo line input channels. Ideally, you'll want to connect these to devices with stereo outputs, like drum machines, stereo effects processors, CD players, etc. This makes it easy to control these stereo signals with a single set of controls.

Never connect amplifier speaker outputs to any line inputs. This will certainly result in severe distortion and possible equipment damage!

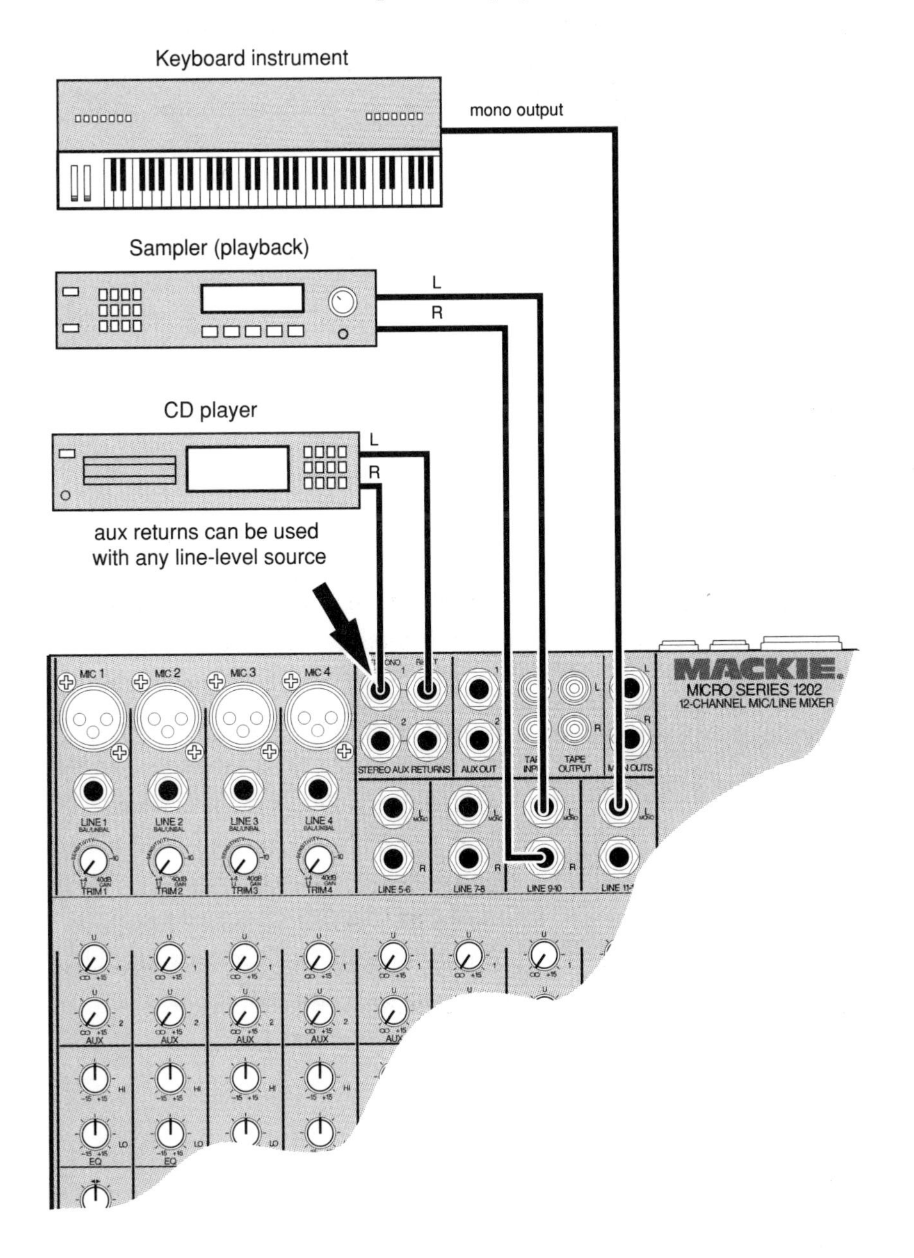

Any device with line-level outputs can be connected to your mixer's line inputs. Here, a keyboard, sampler and CD player are shown attached to an MS1202 mixer. Note that the CD player is connected to an Aux Return, rather than a main mixer channel. Remember that connecting an XLR line output to a Mackie XLR mic-level input may cause distortion.

Insert Examples

Your mixer's insert/channel access jacks allow you to connect an external signal processor directly into the signal path of an individual input channel. This means you can make adjustments to an individual signal's sound that would be impossible to achieve with your mixer's built-in controls.

The two most common external devices to connect are equalizers and compressor/limiters. Equalizers are used to alter the frequency balance or tonal character of a sound. Compressors and limiters are *dynamics processors*, meaning they affect the amplitude of audio signals (a brief introduction to compressors appears later in this chapter).

Equalizers and dynamics processors are available in mono (single input/output) or stereo (dual-input/output) configurations. In the following examples, typical stereo devices are shown. While most of the upcoming examples show devices inserted on a microphone input, equalizers and compressors can also be used with line-level sources, such as keyboards or audio recorder outputs.

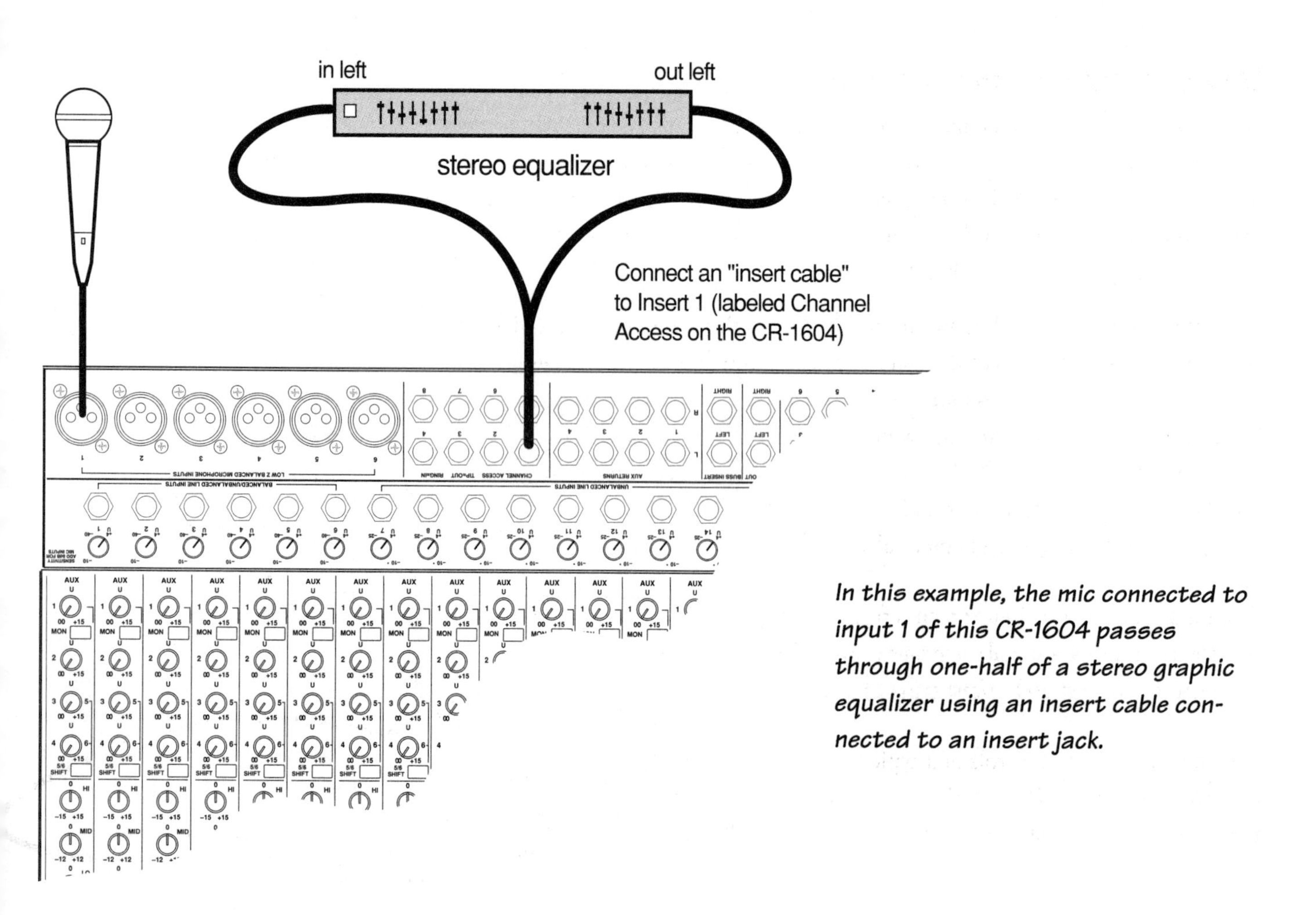

In this example, the mic connected to input 1 of this CR-1604 passes through one-half of a stereo graphic equalizer using an insert cable connected to an insert jack.

Both equalizers and dynamics processors can be used with any model Mackie. However, using dynamics processors on the classic CR-1604 is complicated by the location of the 1604's insert jack in its channel signal path (see Chapter 8, pg. 112 for a detailed explanation and suggested work-around).

Inserting Equalizers

With the hook-up illustrated on the facing page, the extensive tonal shaping possibilities offered by this stereo graphic equalizer can be applied to the sound of Mic 1. In this case, only the left channel of the equalizer is active, as the right channel is unconnected.

If desired, a second mic could be passed through the unused EQ channel in the same way. Then the EQ added to both mics could be adjusted independently, since the graphic equalizer is a stereo unit, providing two separate signal paths.

Note that when two identical mics are used as a stereo pair, you should start by setting both channels of the equalizer identically. This will help preserve the stereo balance between mics.

Understanding Compressors

When you listen to commercial pop records, you are hearing compressors in action. It is safe to say that *every* song on this week's charts uses compression on the vocals and more. Properly applied, compression imparts a smooth "professional" quality to vocal and instrumental sounds. When over-used, compression robs a sound of its vitality, turning it into a squashed little noise.

A compressor is a device that automatically adjusts the level of an audio signal by reducing the loudest peaks of the signal passing through it. Using a compressor or limiter (a device which does much the same thing) is the same as if you had a hand on your volume control and were fast enough to turn it down a bit every time the signal got too loud. Compressors can help you "hold" different parts of a mix in balance with each other and can also protect speakers (and ears) from damage if a really loud sound comes along.

The name compressor/limiter comes from the fact that these devices reduce (compress or limit) a signal's *dynamic range*, which is the difference between the loudest and softest parts of a sound.

The most important controls of a typical compressor or limiter are the *ratio* and a *threshold* control. The threshold control determines the level that you want the compressor to consider "loud enough"

Below, the mic's signal is passed through the left channel of a stereo compressor, via a PPM mixer's channel insert jack.

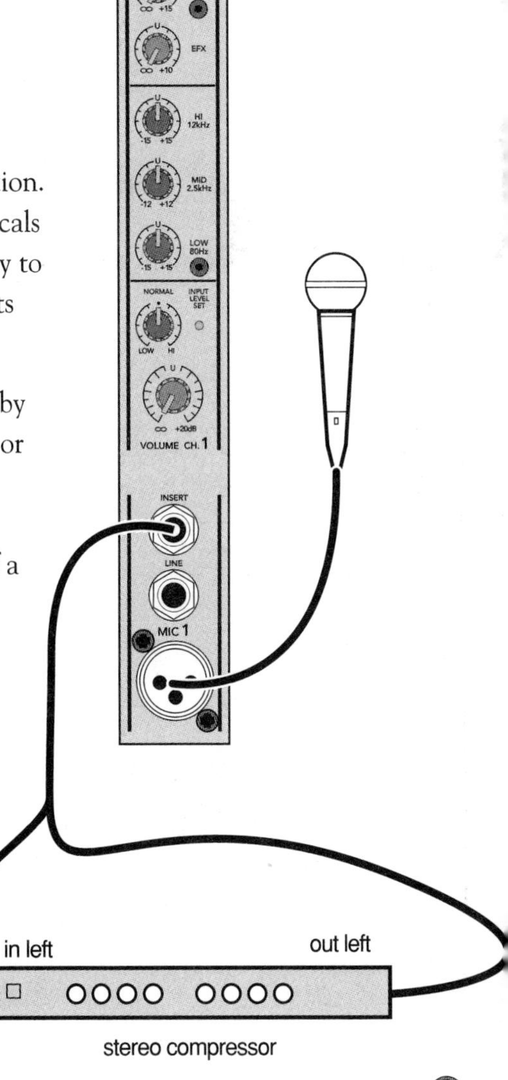

to begin doing its work. When the signal is below the level set by the threshold, the compressor/limiter does nothing. When the signal gets louder than the threshold, the compressor/limiter momentarily reduces the level of the audio signal.

The ratio control sets the amount of gain reduction that takes place when the signal exceeds the threshold setting. The higher the ratio setting, the more extreme the gain reduction becomes.

If you want to insert two devices on the same channel, use a normal "guitar cord" and an insert cable as shown below. Note that the order of the inserted devices will have an effect on the final sound (see text).

Compressors usually use moderate ratio settings, which let signals gently exceed the threshold setting. Limiters typically have much higher gain reduction ratios than compressors. Extreme limiter settings will hold signal peaks at or below the threshold level and prevent them from getting any louder, period.

Compressors are best used in conjunction with your mixer's inserts, because this routes a signal solely through the compressor. In contrast, using an aux send/return hookup would pass the un-compressed signal to the main mix *in addition to the compressed sound*, partially negating the effect of the compressor!

Finally, beware of over-compressing individual signals, or worse yet, an entire mix. Too much compression robs your music of its dynamics, or creates obvious artifacts—in other words, you'll be able to clearly hear the compressor shoving your signal levels around. Also be aware that in a live performance situation, compression increases the chances of feedback.

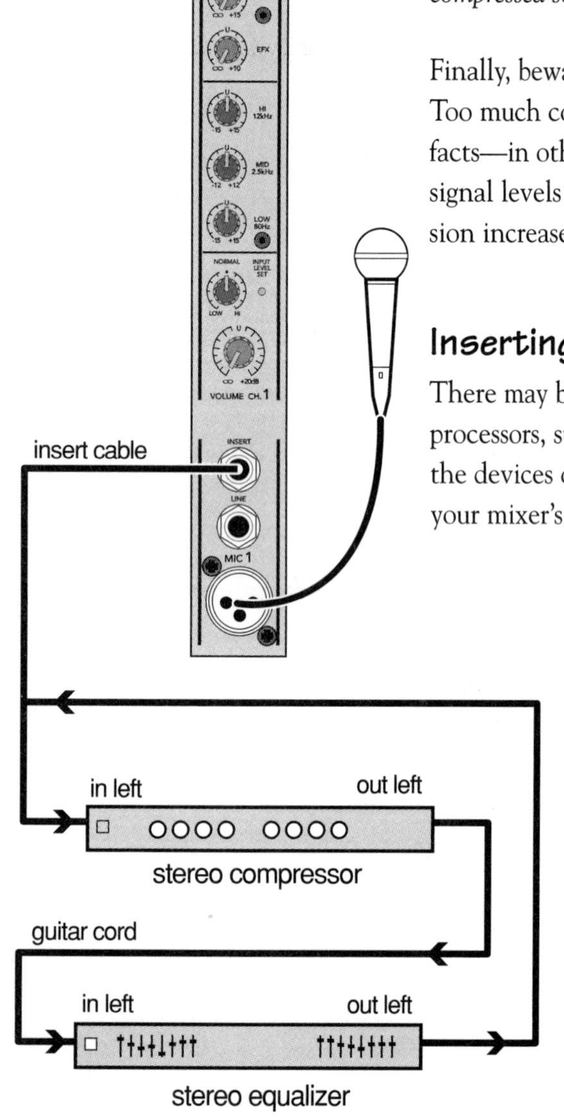

Inserting Two Devices on One Channel

There may be cases in which you want to pass audio through two external signal processors, such as a compressor and an equalizer. In this case, simply "daisy-chain" the devices one after another and then connect the last device's output back to your mixer's insert return, as shown at left.

When compressing and equalizing a signal, one might wonder which signal processor should come first. An excellent question, as the two choices yield slightly different results.

As a rule of thumb, if you are going to be cutting frequencies, especially lows, put the equalizer first. Cutting unwanted lows *before* the compressor will prevent the compressor from unnecessarily clamping down on low-frequency thumps and bumps that you're going to be cutting out anyway. On the other hand, if you plan on boosting lows with the EQ, try compressing first (as pictured at left). That way, the compressor won't be fighting to turn things down as you pump up the lows. In both cases, the goal is to minimize any unnatural "pumping" or "breathing" as the compressor responds to strong low frequency signals.

Aux Send and Return Hook-Ups

The previous section covered signal processing using your mixer's inserts. Insert connections are ideal for compressors, equalizers and similar devices that are applied to a single channel's signal.

However, to connect effects like reverb or echo units, you'll want to use your mixer's aux send and return connections. Using an effect with aux sends and returns allows each channel to have a different amount of effect on it, and many channels can *share the same effect unit*. We covered sends and returns in detail in Chapters 7 and 8.

When hooking up an effects unit as shown in the following examples, there is one very important setting to make in the effects unit itself. The effects unit will have level controls for both the "effected," as well as the unadulterated "dry" sound. This setting may be called *wet/dry mix*, *direct/effect level*, *balance*, etc.

Be sure to set the level of the dry, or unaffected signal *off* or to *zero*. The output level of the "effected" signal can usually be set at the default level. In units where a "balance" style control is used, both wet and dry levels are controlled together. In this case, make the signal 100% wet.

Failure to do so will make mixing difficult—when you change send, or return levels, you'll unintentionally change the level of the direct sound, throwing off your mix.

Mono Send, Stereo Return

In almost all cases, you should hook up a reverb or similar effects unit's input to a single aux send from your mixer (see illustration on following page). Because only a single aux send is used, you'll be feeding that device a mono signal. This is usually best, even if the device has "stereo" inputs. On less expensive effects units, these "stereo" inputs get combined to mono before processing anyway, so trying to send it a stereo feed just wastes aux send resources. On the other hand, the output of most effects units sound much more interesting in stereo than mono. Since your Mackie has stereo aux returns, you should typically use them to get the stereo output of your effects back into your main mix.

Remember that these illustrations show just one aspect of a hookup. You must combine several examples to connect your entire system.

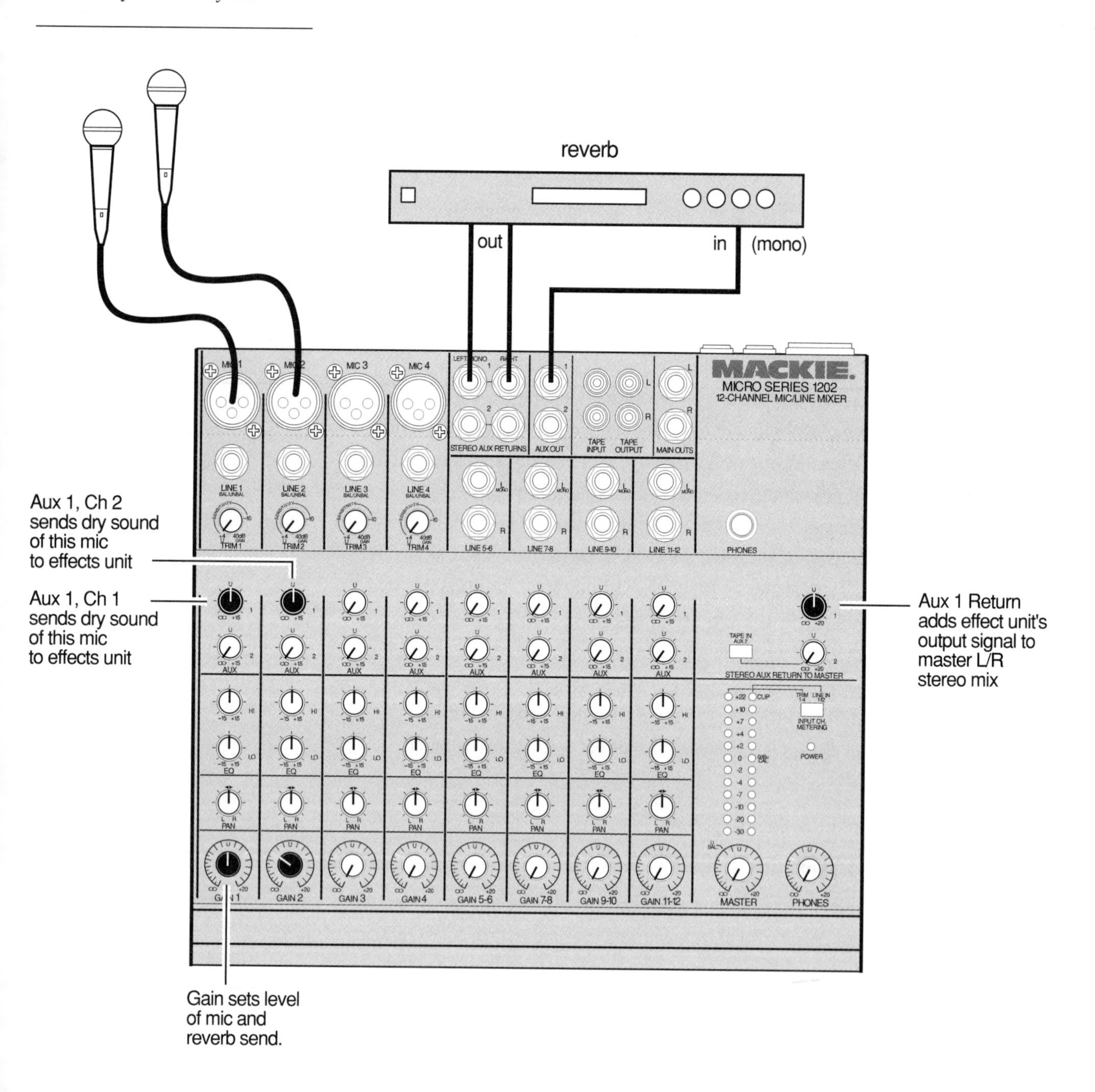

A single Aux Send is typically all that's necessary to hook up a reverb or other effects unit. Even if your external processor has individual left and right inputs, few units actually provide separate processing of left and right signals. Therefore, using a single, mono send is appropriate in most cases.

Creating Stereo AUX Sends

If you happen to have an effects unit that really *can* process stereo sounds internally, you can send it a stereo mix by using signals from *two* aux sends.

All the mixers described in this book provide mono aux sends and stereo returns. However, with a little extra knob-twiddling on your part, you can create a stereo aux send by using two mono aux sends. You'll be able to control each channel's aux send pan position, even though there is no pan control on the aux sends. This can be useful if you have an effects unit that has true stereo processing, or if you want to create a stereo stage monitor or headphone cue mix.

Here's how: You will be mimicking the effect of a Pan control by using two individual Aux controls. In this example, we'll assume that Aux 1 is connected to the Left, and Aux 2 to the Right input of the stereo effects unit. If on a given channel, Aux 1 and Aux 2 are turned up the same amount, you will have equal level in both channels of the effects unit. This is analogous to a Pan control set to 12 o'clock.

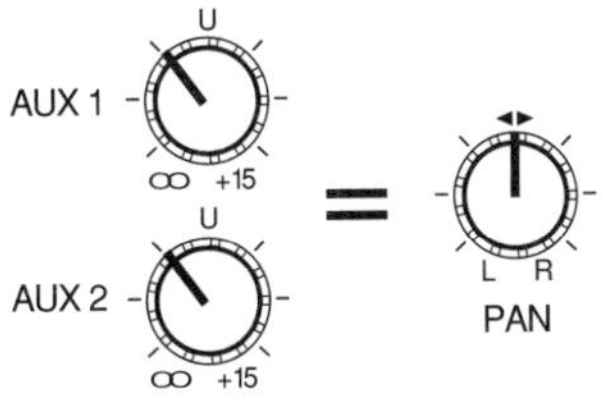

AUX 1 = PAN
AUX 2

To send an input channel's signal to the extreme left side of the effects unit, turn up Aux 1, and turn down Aux 2. This is like a Pan control set to 7 o'clock.

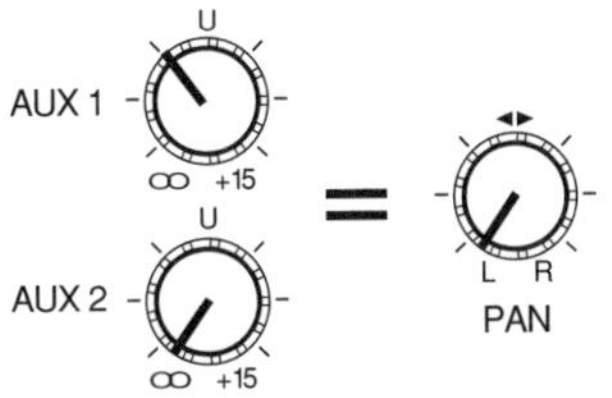

AUX 1 = PAN
AUX 2

To send an input channel's signal to the right of center to the effects unit, turn up Aux 1 a bit, and turn up Aux 2 a bit more. This is similar to a Pan control set to around 2 o'clock.

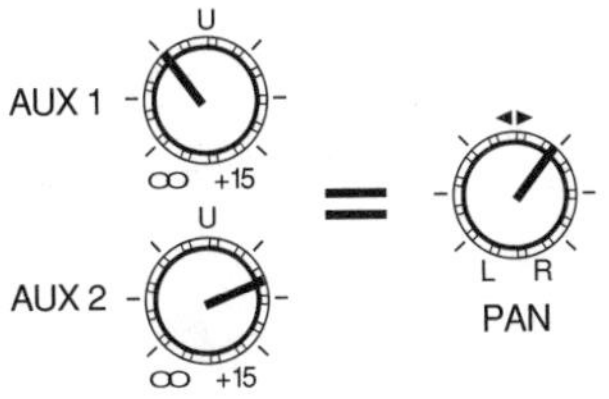

AUX 1 = PAN
AUX 2

By varying the balance between Aux 1 and Aux 2, any equivalent pan position can be achieved. Just make sure you don't make both auxes so loud that the input of the device you are sending to begins to distort.

Stereo aux sends offer new creative options. For example, you can give a channel's "effected" sound its own position in the stereo field. You may want to keep the dry and effected version of a given sound in the same stereo position. For example, if your dry signal is panned to the center, sending an equal amount of signal from both halves of the stereo send would keep the effected signal in the center of the mix. Alternately, you could put the dry and effected sounds in different places of the stereo spread, for instance, at opposite sides of the stereo image. This would create a more dramatic effect.

Adding a Monitor/Cue Mix

In addition to hooking up an effect unit's input to an aux send, you can hook up a headphone amplifier or a power amp and monitor speakers to create a separate mix for musicians for recording or live work. (On stage these are called "monitor mixes" and in the studio, it's a "headphone cue mix." From a signal flow standpoint, they are identical.) Remember that using a pre-fader aux send means the monitor mix will be unaffected by changes to individual channel levels.

This signal flow diagram illustrates how a single channel reaches both the main stereo output and monitor mix. See text at right for more details.

The illustration below shows the signal flow of a monitor mix. The channel volume fader (letter A) sets the mic level in the main mix, while the Aux 1 control (C) sets the monitor level. Both mains and monitors can have master volume controls (B and D). Finally, the main stereo mix and monitor mix exit the mixer from their own jacks. The 1202-VLZ mixer pictured on the facing page shows a real-world example of a monitor system hook up. Note that the same letters appear on both illustrations to help you identify which physical mixer controls relate to various parts of the signal flow diagram pictured below.

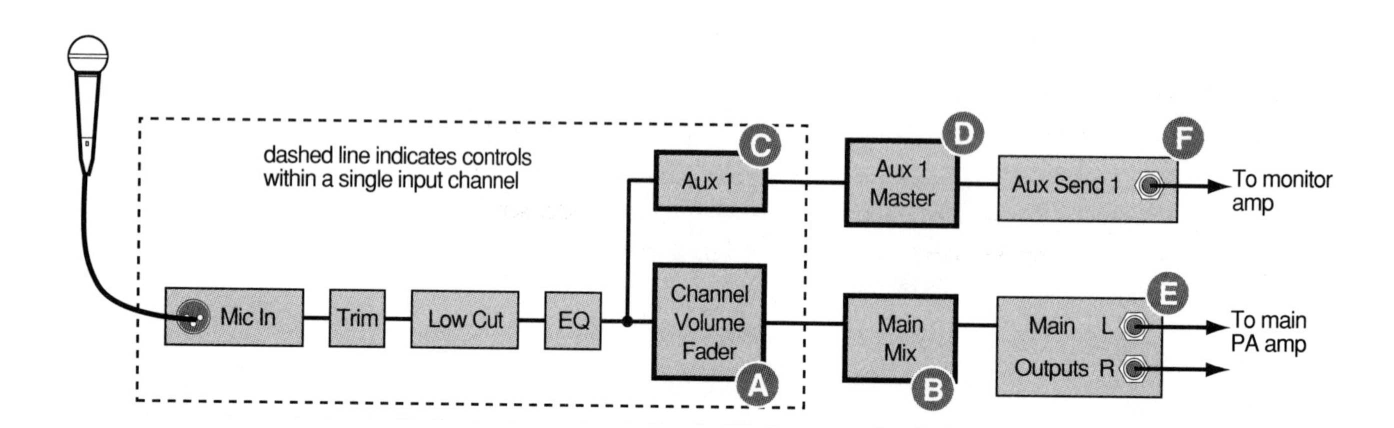

The 1202-VLZ shown below is running a wedge monitor mix as well as a main PA mix. Monitor mixes are easier to use with a pre-fader aux send. On this mixer, that is set by the Aux 1 Select button below the Aux 1 Master knob. The circled letters identify corresponding points in the companion signal flow diagram (facing page).

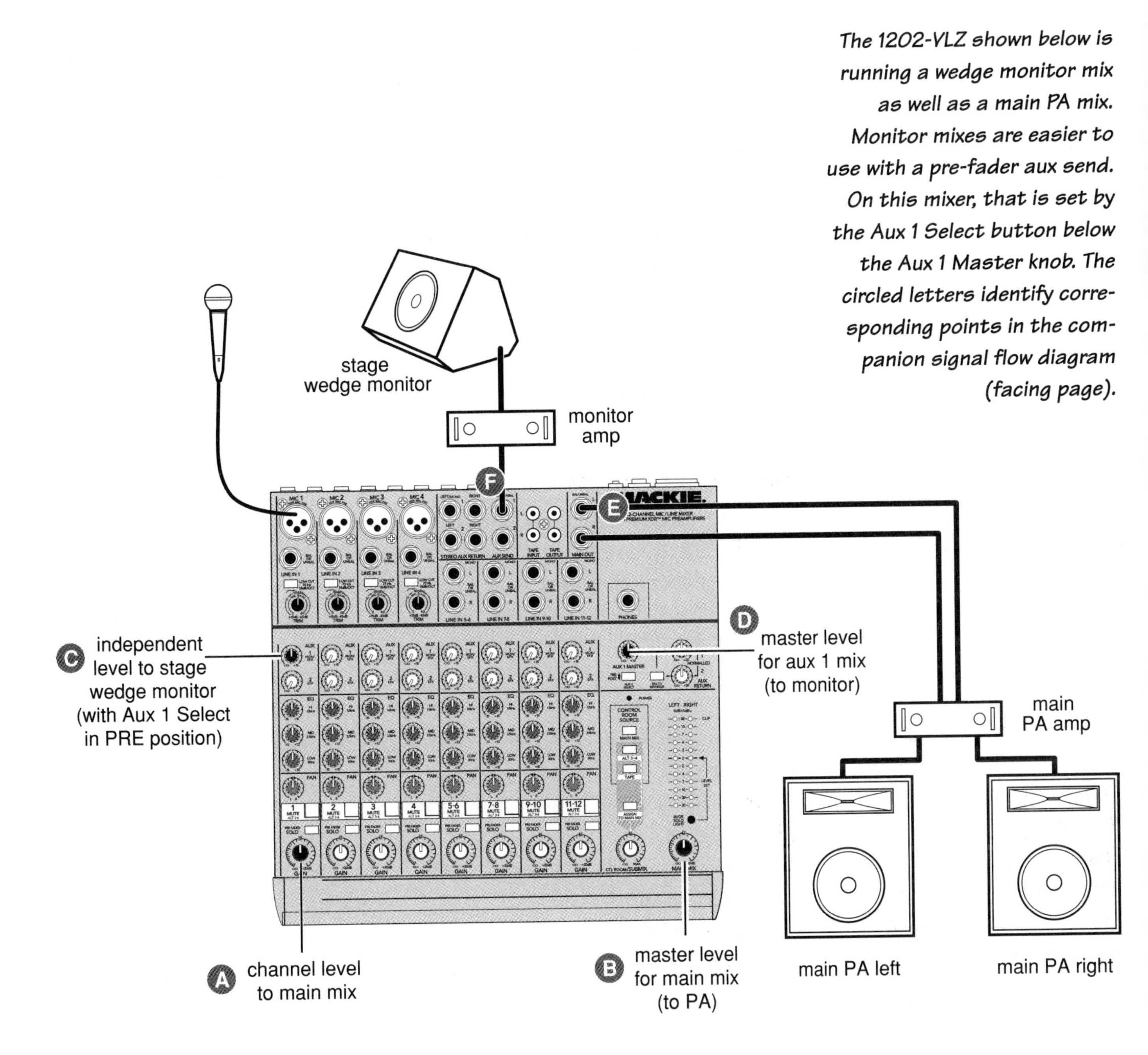

If you wish to create stereo stage monitor or headphone cue mixes, a pair of pre-fader sends would be ideal. However, this is an option only for 1604-VLZ, 1642-VLZ and SR-series users. If you create a stereo aux send using post-fader sends, any change you make to a channel volume fader will affect that channel's level in your stereo monitor mix.

Effects in the Monitor Mix

Most Mackies, including VLZ, CFX, SR and PPM-series mixers, have provisions to easily return effects to the monitors, as well as the main stereo mix. If you have one of these mixers, refer to Chapter 8 for a model-specific explanation.

If you have a classic MS1202 or CR-1604, read on: Adding effects to a monitor/cue mix is a two-part process. First, you must connect your effect outputs to normal Input Channels, not to the Aux Returns. Then, you can add effects to the monitor mix by bringing up the effect return channel's Aux Send that feeds the monitor mix. Don't turn up the return channel's effect Send or feedback will result.

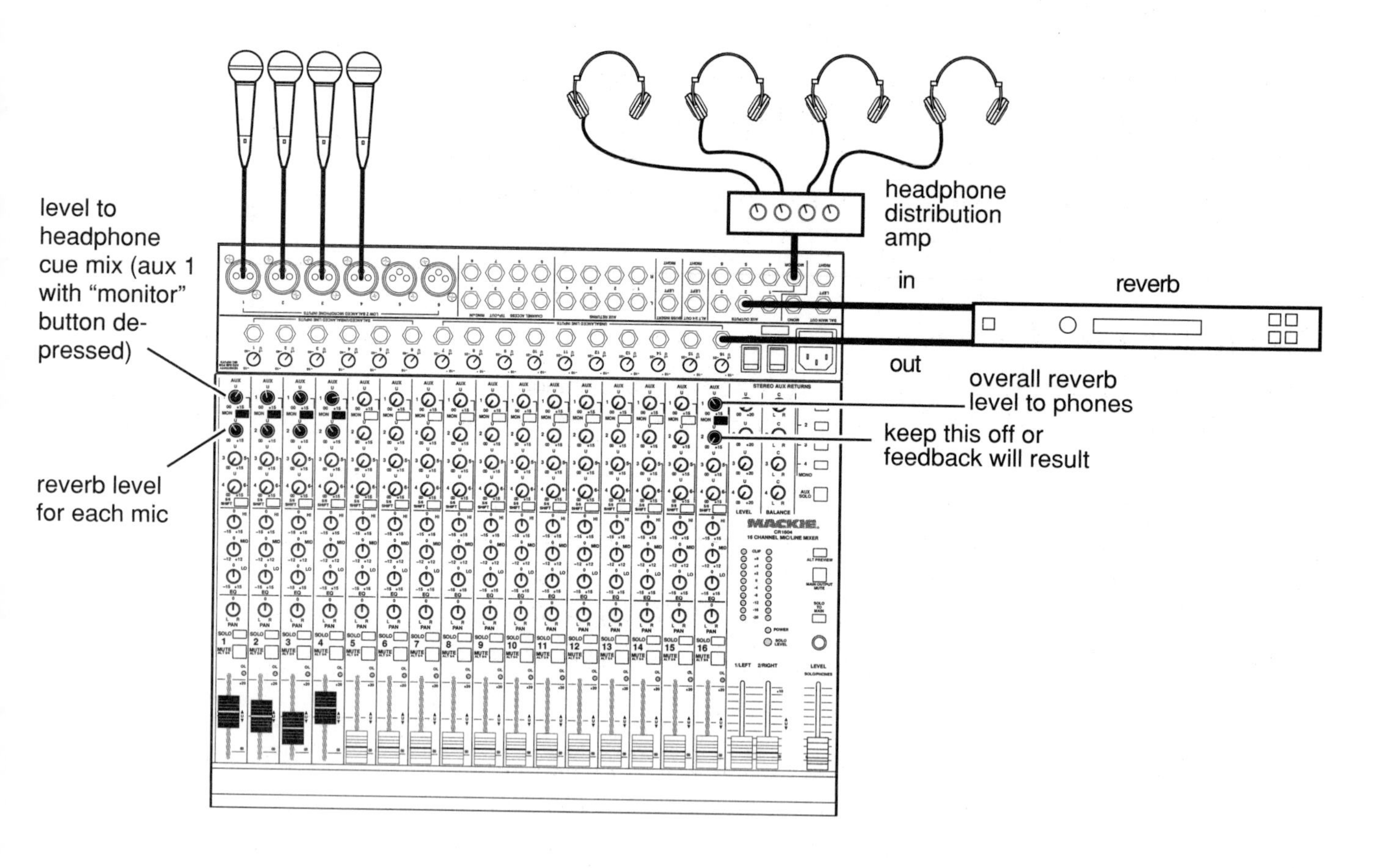

It's difficult, but not impossible to add effects to your headphone or stage monitor mix using a classic CR-1604—and the MS1202 hook up is similar. This trick is much easier with more recent Mackies. Check out Chapter 8 for details on your particular mixer.

Stereo Mixes

The most common use for mixers is to create a combined stereo output, or *mix* of all the connected input signals. A stereo mix typically goes to one of two places—either an amplifier and speaker system to present the mix to an audience (as is the case with a PA system), or the mix will be recorded on a two-track stereo recorder.

In some cases, you may want to try and create an audience PA mix and a mix to tape at the same time, but this is a pretty challenging proposition. While using a Y cord will allow you to connect the PA amplifier and your recorder at the same time, the aesthetic demands of the audience mix vs. the tape mix will always be different, so one mix or the other is bound to suffer.

Whether going to PA or to tape, the Main Out Left and Right connectors are the point where your stereo mix leaves the mixer and goes to the next device in the system. If the equipment connected to your mixer's output has balanced inputs, try to use three-wire TRS cables, TRS-to-XLR adapters, or both to connect them, thereby gaining the advantages of balanced connections.

Connecting a Mixdown Deck

Mixing to a stereo recorder (a.k.a. "tape deck") is among the most common of hookups. Whether mixing a live performance to a stereo recorder or blending the outputs of a multitrack, you're going to need to listen to the mix in progress as well as be able to play the tape back after each mix. Mackies with a separate control room section make this easy. Unfortunately, several models lack this useful feature, including the classic CR-1604 and MS1202 as well as the PPM-series powered mixers. This section describes ways to work around the lack of a control room section on these models and some general tape-monitoring issues applicable to all mixers.

But before we talk about specific model mixers, let's look at some simple generic examples. In the first example, your stereo mix is connected to the line inputs of the mixdown deck (DAT, cassette, MiniDisc, reel to reel tape, etc.). Then, the line outputs of the recorder are connected to your amplifier inputs and then on to monitor speakers. This allows you to mix while listening to the signal passing through the mixdown deck, which ensures that you are listening to exactly the same mix that is going to tape.

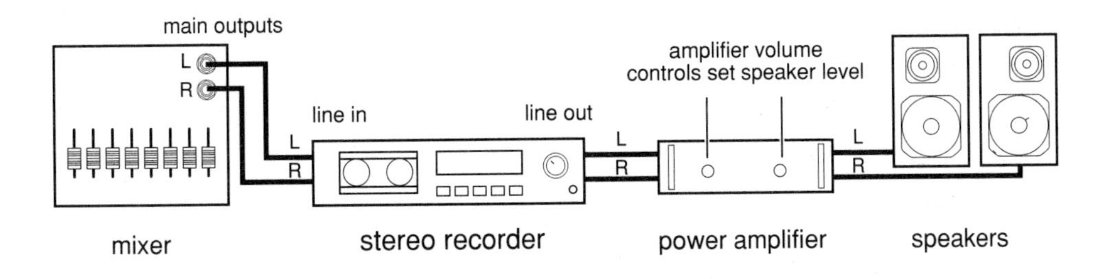

Although there are much better alternatives (keep reading), this hookup lets you hear your mixes as they are recorded and when played back from tape.

However, there are two inherent problems with this hookup: In most cases, the mixdown deck will have to be in record or record/pause in order to hear anything from your mixer. With DAT recorders, this can eventually lead to increased head wear, because the head spins even while paused. (Note that some decks have a "source" or "monitor" switch that allows monitoring the inputs without having to be in record.)

You'll also have to adjust the playback level over the speakers by using the input level controls of the amplifier. Depending on your amplifier and its location in your studio, this might be inconvenient (some two-track mixdown decks have line output level controls, which can solve this problem).

In this second example (below), signal-splitting "Y" cords connect your mixer output to an amplifier input as well as a two-track recorder for mixdown. Then, the output of the recorder is patched back into unused input channels or aux returns for playback (a cautionary note about this in a moment).

Here's a marginally better scheme that uses "Y" cables to run the mixer output to two destinations—the record deck and your amplifier/speakers.

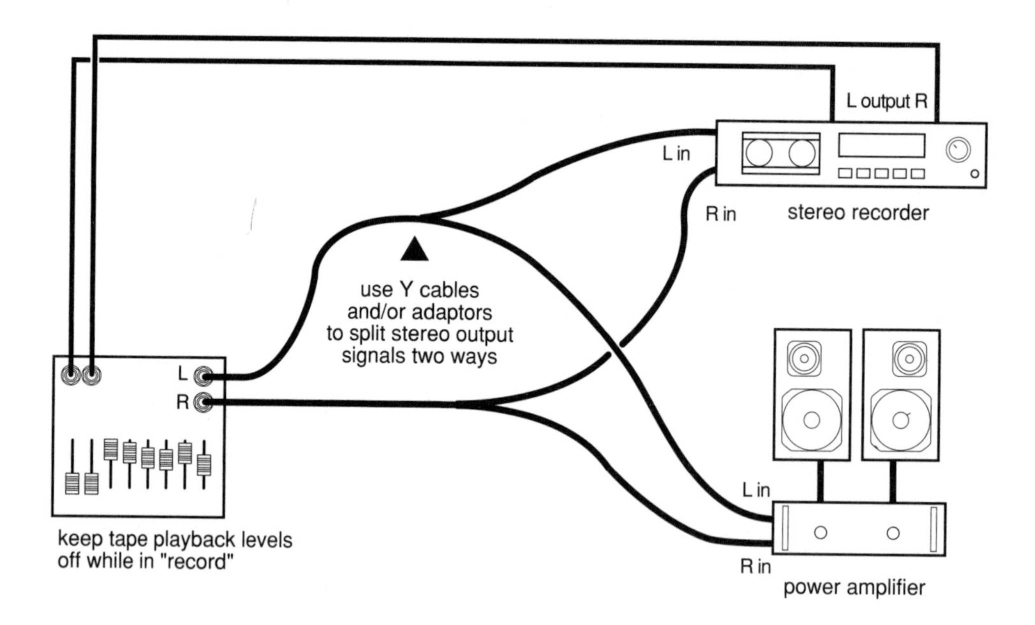

The advantage to this approach is that your deck doesn't have to be in record for you to hear what's going on. Unfortunately, you still have to use the input level controls of the amplifier to change the playback level in the speakers, since your mixer master controls will affect *both* the level to tape and the level to your monitor speakers.

After you've done a mix, you can hear the results by playing the tape and turning up the mixer inputs that the tape deck outputs are connected to. But now there's another problem to consider—feedback.

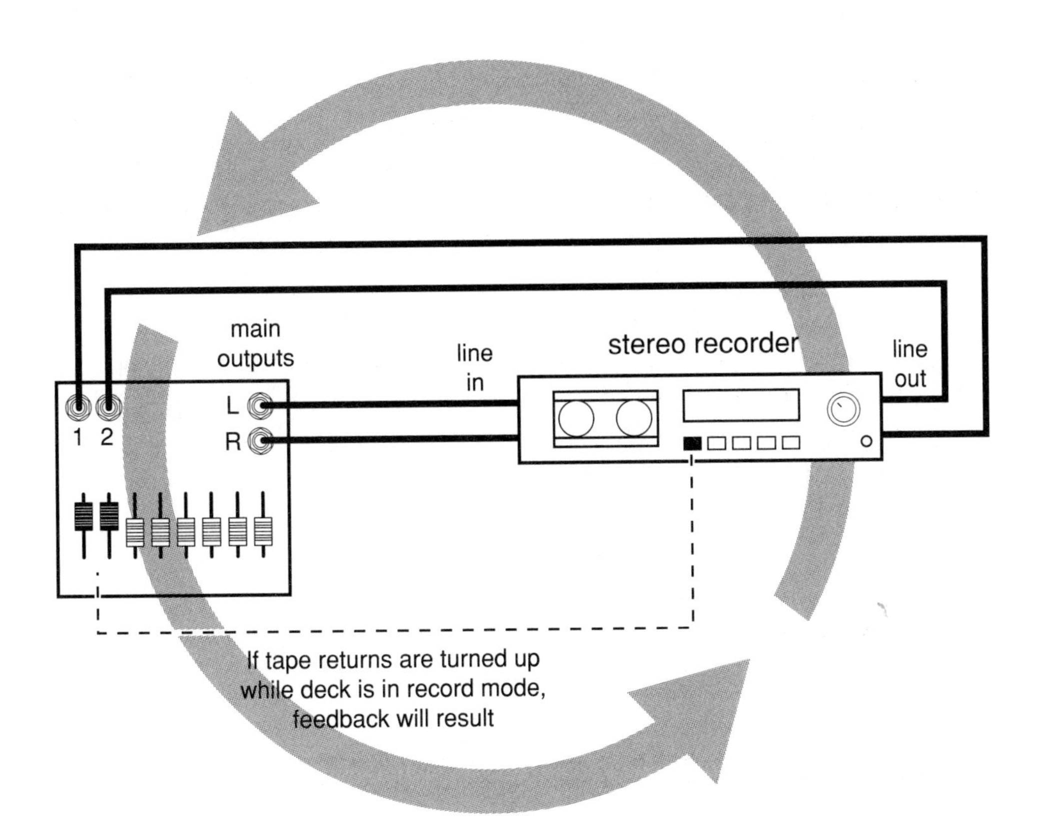

When your recorder has both its outputs and inputs connected to your mixer, be sure not to turn up the mixer channels connected to the tape deck's output when the deck is in record. If you do, it will feed back with an alarmingly loud howl!

When you connect a mixdown deck to your mixer's input *and* outputs, you can create a circular "feedback" path in which signals go around and around, almost instantly building to a furious howl!

This happens because, when the deck is in record, it will be sending your console's mix output *back into mixer inputs*. If these inputs are turned up, a feedback path is created—the mixer's output is connected to its input. So, if you hook up a tape deck as shown above, be sure to turn down the deck's playback level at your mixer *before putting the deck into record*.

Another thing to watch out for is "double-monitoring." Double-monitoring means you are hearing the same mix from two different sources. For instance, you could be splitting the output signal from one recorder to another's input externally, while inadvertently listening to the outputs of both recorders.

If the record deck is digital, the short delay time from its input to output can result in funny-sounding phase-cancellation between decks, creating a very metallic, "flangey" sound. You'll hear a similar effect when the feedback situation described above happens, but the mix levels aren't high enough to howl.

Control Room Monitors

As you have probably figured out from the previous examples, there are two sets of levels to be concerned with when mixing. The first is to ensure that you are getting good levels to tape. (Note that while I'm saying "tape," these comments apply if you are mixing to recordable CD, MiniDisc or a hard-disk recording system.)

The second signal level to consider is the volume of your monitor speakers while mixing. Many Mackies, such as the VLZ-series, include separate outputs and level control for the "control room" playback system. The classic MS1202 and CR-1604 don't include this feature. However, you may be able to get by if you use your mixer's headphone output instead, as shown below:

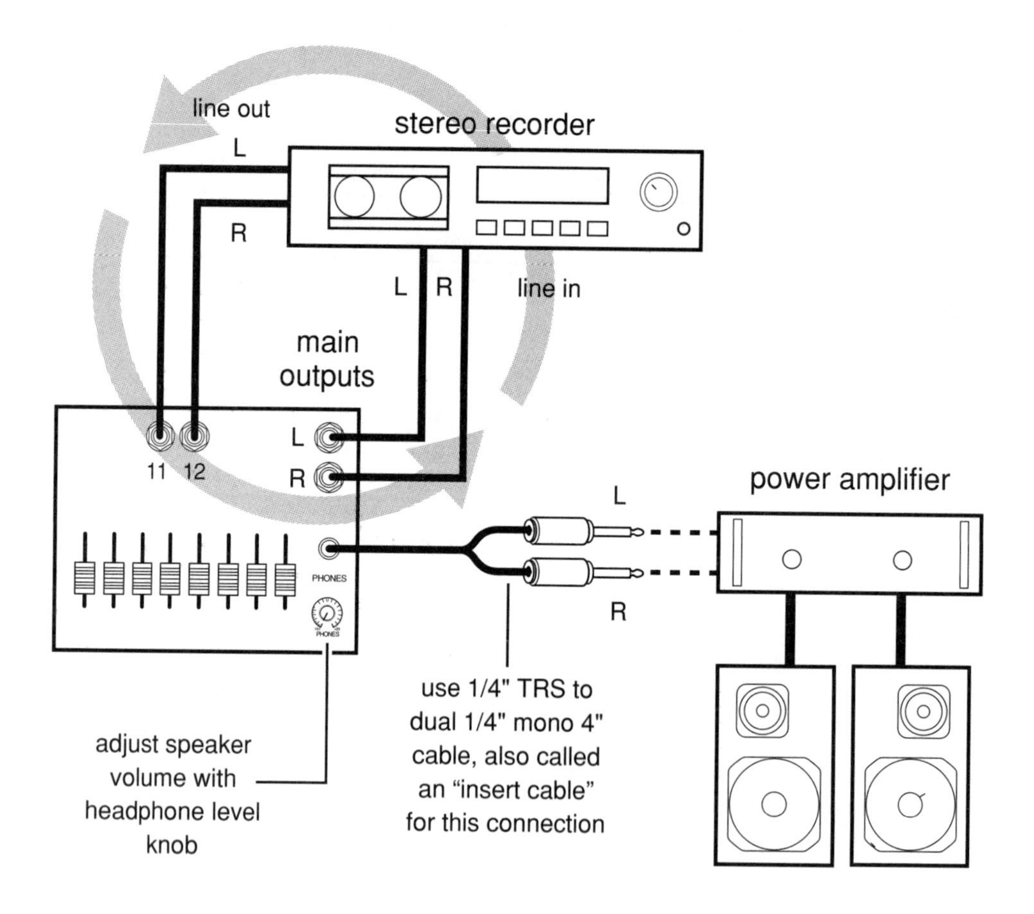

This hook up will give you an independent level control for playback over your monitor speakers without having to adjust your amplifier. Also, you'll be able to use the solo buttons (if your mixer has 'em) while mixing and the solo signal will only appear in the monitor speakers—you won't disturb the stereo mix being sent to the recorder.

And now, the bad news for CR-1604 users: A look at your mixer's block diagram will show that the Headphone output comes before the 1/Left and 2/Right faders. This means that if you are trying to end a mix with a fadeout, you won't actually hear the fade that's going to tape. In other words, fading the master will create the fade on your mixdown deck, but the mix coming through your control room monitors will be as loud as ever! Yuck!

If worse comes to worst, you could always switch to one of the previous examples, which will allow you to hear fadeouts as they are being performed.

CONTROL ROOM MONITORS (THE EASY WAY)

For those of you using a mixer with independent control room outputs (VLZ models, for instance), simply connect your main outputs to the tape deck's inputs and run your amplifier and loudspeakers from the control room outputs. On these models, the control room output appears *after* the main stereo fader, so your control room speakers will follow any fade-ins or outs you include in your mix to tape.

Mixing with Subgroups

During mixing of a live performance or a mixdown of a multitrack recording, you can use bus assigns to create submasters for improved mixdown control. It's slightly complicated, so I'm going to explain the benefits of this scheme before I describe how to do it.

When mixing, you'll often find yourself with a number of individual channels of related sounds, for instance, four mics for backing vocals, or six mics covering different parts of the same drum set, etc. During mixdown, you might find yourself thinking "...hmmm, drums could come up a bit here," but making this change means adjusting each of the drum channels individually. In doing so, it's quite easy to disturb the *relative* balance between drums, so that after you've turned all the drums up, you realize that the snare drum seems a little too loud, and the hi-hats might be a little softer, relative to the rest of the kit. Wouldn't it be cool to have a single fader that would turn up or down *all* the drum channels, or the *entire* backing vocal section *without* having to adjust their individual faders? This would be sort of like a special "master" volume, but one that *only* effects the drums, or the backing vocals.

In order to do this, you need a mixer with some sort of assignable busing. It's possible to pull this off with ALT 3/4-equipped mixers (CR-1604, 1202- and 1402-VLZ), but you get much greater flexibility when using mixers with four assignable buses (1064-VLZ, the SR- and CFX-series).

Let's mix a live show for a hypothetical blues band, that includes a three-piece horn section, three-piece rhythm section (piano, bass and guitar), a drum set with four mics and the band leader who sings and plays lead guitar. That's twelve inputs to deal with, and a lot of musicians. Rather than have to deal with each instrument and mic individually, let's set up several "submaster" faders to control related groups of instruments, as shown in the illustration below. In these examples, a four-bus mixer is shown, but the same concepts apply to those using ALT 3/4-equipped models.

Now you can begin to see the advantages of this approach. Instead of twelve individual faders to deal with, the whole band can be mixed on just five! Want more horns? Turn up the horn master. Drums too loud? Down goes the drum master. Of course, if the trumpet player is too loud, you adjust their individual input channel, but once your relative levels for each section of the band are set, you can really just concentrate on the music, rather than chase a lot of individual faders all over the place. The lead vocal and guitar don't have their own master, but these are easy enough to adjust individually. And, when the club owner makes his inevitable request to turn the whole thing "*way* down," you've still got a single, overall master volume slider.

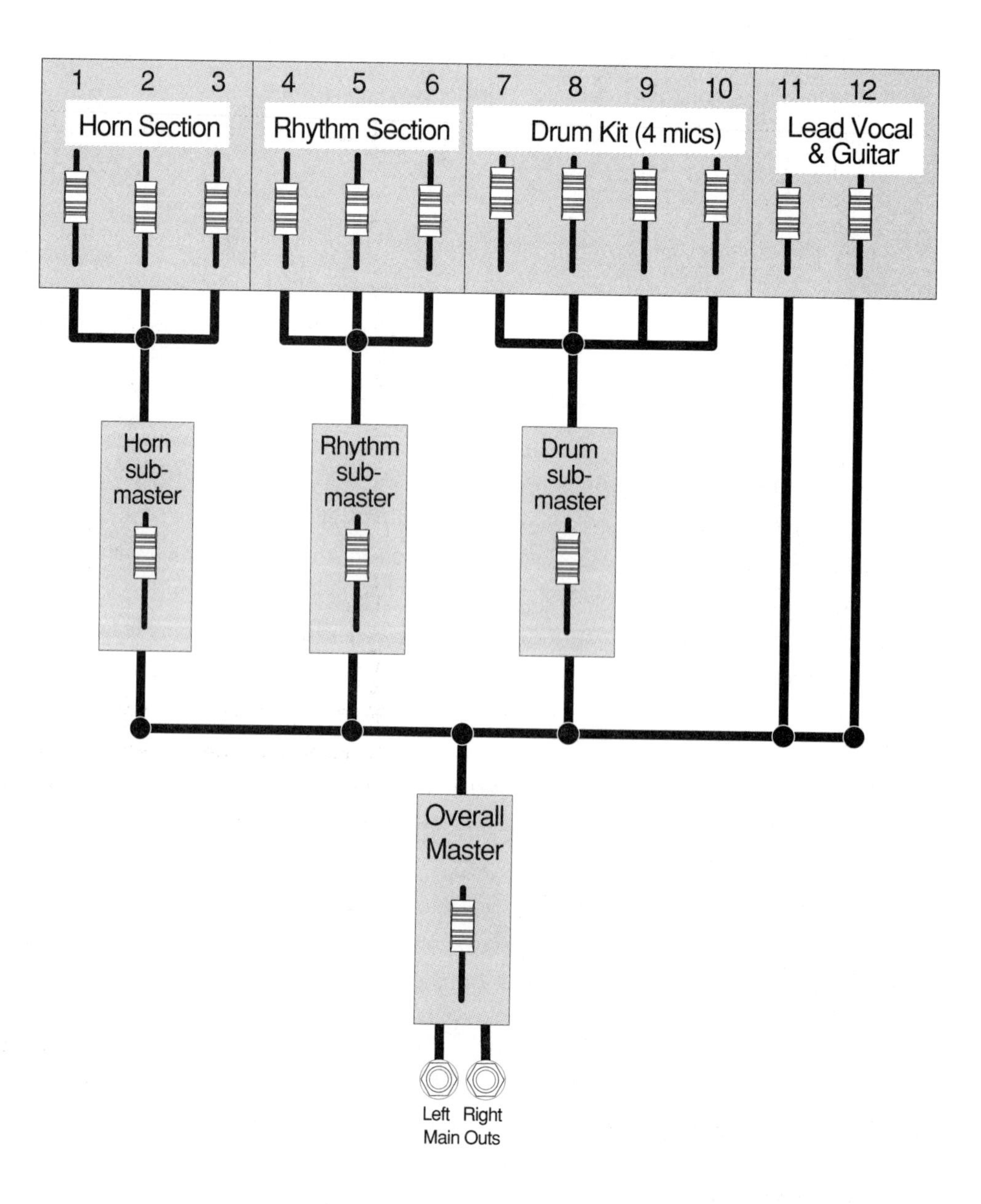

OK, let's translate the conceptual mixing setup on the opposite page into something that actually shows us which buttons get pressed.

Don't panic! This is just an extension of all the stuff we've been covering up to now. We're looking at a four-bus, 12-channel mixer. Next to each channel fader, you'll note the bus assign switches. For instance, channels one through six are assigned to 1-2, seven through ten to bus 3-4 and channels eleven and twelve directly to the stereo bus.

You'll further note that using their pan controls, channels one through three are being directed just to bus 1, while inputs four, five and six are going just to bus 2. Channels seven through ten are panned to a variety of locations. In other words, we're using buses 1 and 2 as individual mono buses, and buses 3 and 4 as a combined stereo bus.

You can also see that each bus/subgroup has its own master fader. Adjusting fader 1 would obviously affect the levels of channels one, two or three, but wouldn't have any effect on the other inputs. Adjusting the Main Mix fader would fade the entire mix.

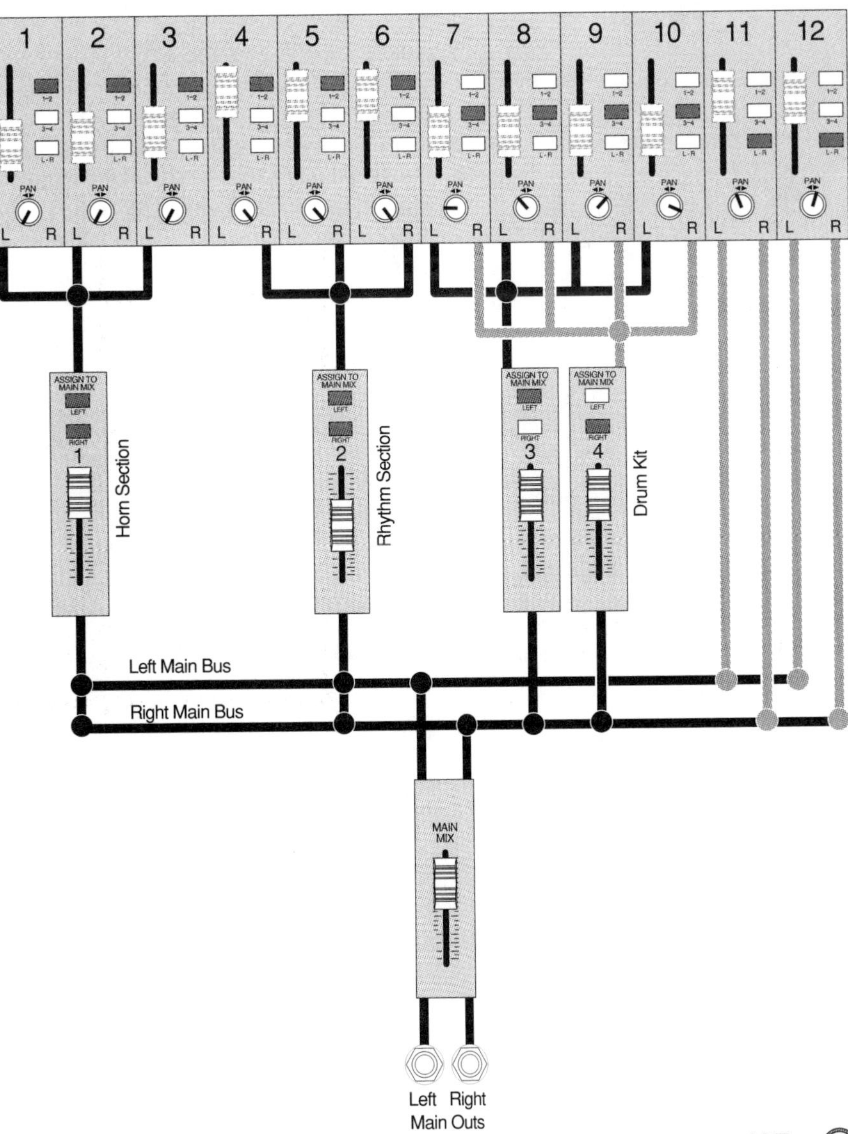

Subgroups with 4-bus mixers

Okay, so let's summarize how we set all these switches and knobs to get the desired result, working from top to bottom. Each channel in the top row is assigned to one subgroup via the switches 1-2, 3-4, or L-R. Further, each channel's Pan control determines if that channel is sent to one or both halves of the bus to which it is assigned (see channels one through six on the illustration on the previous page).

Now for a very, very important question: How does audio get from the four subgroup faders to the main output at the bottom of the drawing? You will recall that each subgroup has its own physical output jack at the back of your mixer. But we don't want to have to take patch cords and return those signals back into unused mixer inputs (assuming we had any inputs available to use!).

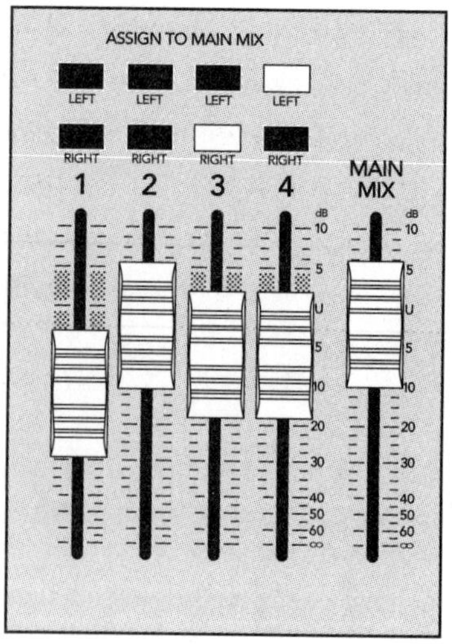

Instead, Mackie has provided a pair of switches on each subgroup master fader labeled "Assign To Main Mix." Conceptually, these are assign switches just like those on each individual channel. The difference is there are only two of them, and each is a mono assignment. Pressing both "Left" and "Right," as shown on subgroup faders 1 and 2, connects that fader's signal to both the left and right stereo buses in equal amounts. This places those signals dead-center in the stereo spectrum.

However, Groups 3-4 are being fed as a stereo bus, so we assign 3 to left and 4 to right. The result of all these assignments is that all twelve input channels make it to the main mix fader and the main left and right outputs.

Four-bus mixers use assign switches to route each bus to the left and right main mix. SR-series boards use a pan control in conjunction with a single assign button, which allows more panning flexibility, but the concept is the same. The combinations of buttons pressed above matches the hookup shown on the previous page.

Now the first ten channels of our twelve-input mix can be adjusted from just four faders. Note that inputs eleven and twelve go directly to the main mix (sorry, not available on CFX-series mixers), so they don't pass through any "submaster" volume controls. If you wanted to turn them both down at once, you'd have to adjust both their faders manually.

Subgroups with ALT 3/4

As mentioned a moment ago, it is possible to mix using subgroups on a mixer with ALT 3/4 outputs (CR-1604, 1202-VLZ and 1402-VLZ). However, you'll only be able to create *one* stereo or *two* mono subgroups. CR-1604 users must also sacrifice a pair of input channels (1202- and 1402-VLZ operators will use their mixer's control room section for this purpose).

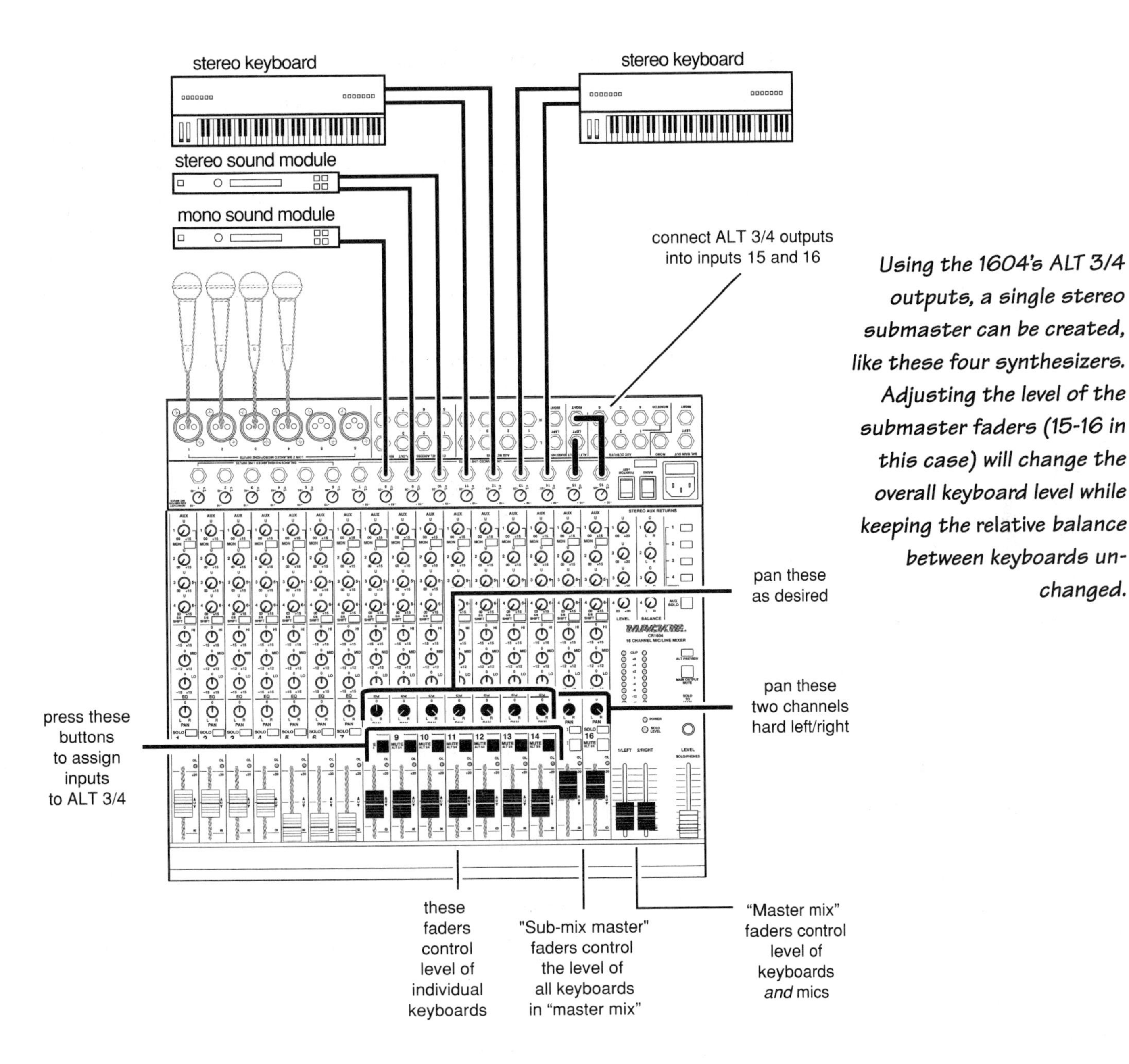

Using the 1604's ALT 3/4 outputs, a single stereo submaster can be created, like these four synthesizers. Adjusting the level of the submaster faders (15-16 in this case) will change the overall keyboard level while keeping the relative balance between keyboards unchanged.

CR-1604 SUBGROUPS

Let's begin with a CR-1604 example. In this hookup, note that the ALT 3/4 outputs connect back into unused CR-1604, 1201-VLZ or 1402-VLZ inputs. This creates a subgroup that lets you control large portions of a mix with just one or two knobs. It's like a Master level control, except it affects *part* of the total mix, but not *all* of it.

By hitting the Mute-ALT 3/4 switch on any number of input channels, you will cause those input signals to exit the mixer from the ALT 3/4 outputs, and then come back in on the channels marked "submix masters" in the illustration. (If you're finding the signal flow hard to follow, check out the simplified diagram on the following page.)

This allows you to control the level of a partial mix within your overall main stereo mix (hence the term *submix*). This lets you get the *perfect* level and pan balance between a number of related parts, and then by turning up one or two knobs, bring all parts of a submix up or down as a unit, preserving their relative balance.

To create a stereo submix, you may pan the individual members of the mix to any position, but be sure the submix masters (15 and 16 in this case) are panned hard left and hard right. You'll probably want to keep the levels of the submix masters the same.

If two independent mono submixes are desired, individual inputs must be panned either hard left or right, but the two submix masters can be panned individually to any position, even if they both end up panned center. The submix master levels don't have to be set equally.

1 2 3 4
Vocal Mics
8 9 10 11 12 13 14
Keyboards-ALT 3/4
15 16
Keys-Stereo Sub-Master
Overall Master
Left Right
Main Outs

If the CR-1604 subgroup illustration of the previous page was a bit intimidating, this simplified signal flow diagram should help you sort things out.

As an alternative to the stereo subgroup shown above, you can create a pair of mono subgroups with a CR-1604. The signal flow diagram (lower right) corresponds to the hookup shown on the page at right.

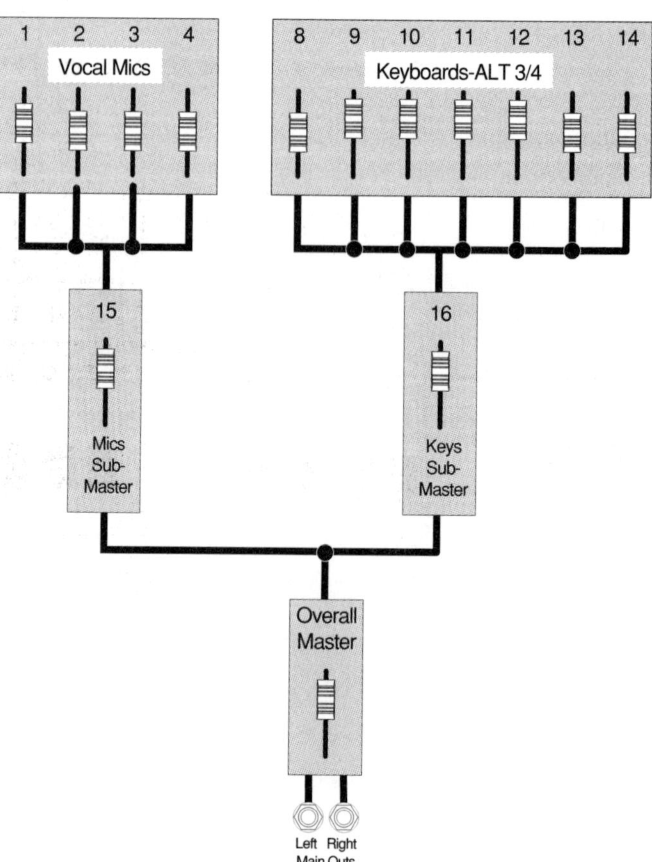

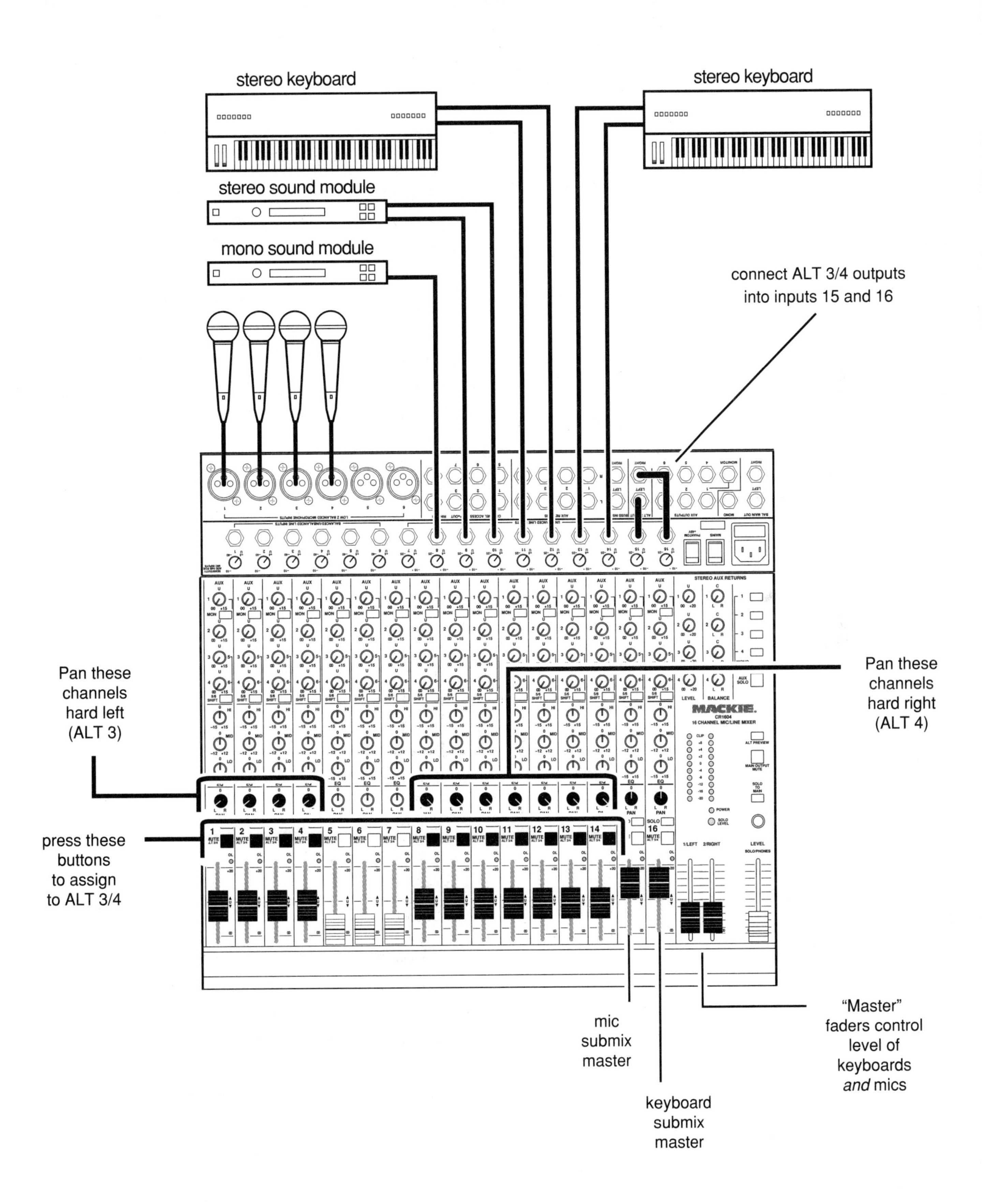

stereo keyboard
stereo keyboard
stereo sound module
mono sound module
connect ALT 3/4 outputs into inputs 15 and 16
Pan these channels hard left (ALT 3)
Pan these channels hard right (ALT 4)
press these buttons to assign to ALT 3/4
MACKIE.
CR1604
16 CHANNEL MIC/LINE MIXER
mic submix master
keyboard submix master
"Master" faders control level of keyboards *and* mics

1202- AND 1402-VLZ SUBGROUPS

1202- and 1402-VLZ users don't have to give up a pair of inputs, as is the case when using a CR-1604. Instead, the control room section is used to route signals sent to the ALT 3/4 bus to the main stereo mix.

The illustration below shows the signal flow we are shooting for: The four vocal mics have their own sub master level control. The second illustration (facing page) shows the hookup.

While the 1201- and 1402-VLZ mixers share the ALT 3/4 signal routing of the classic CR-1604, the process of setting up a submix is greatly simplified by the VLZ's control room source/assign to main circuitry. However, this approach is practical only for a single stereo submix, not two mono submixes.

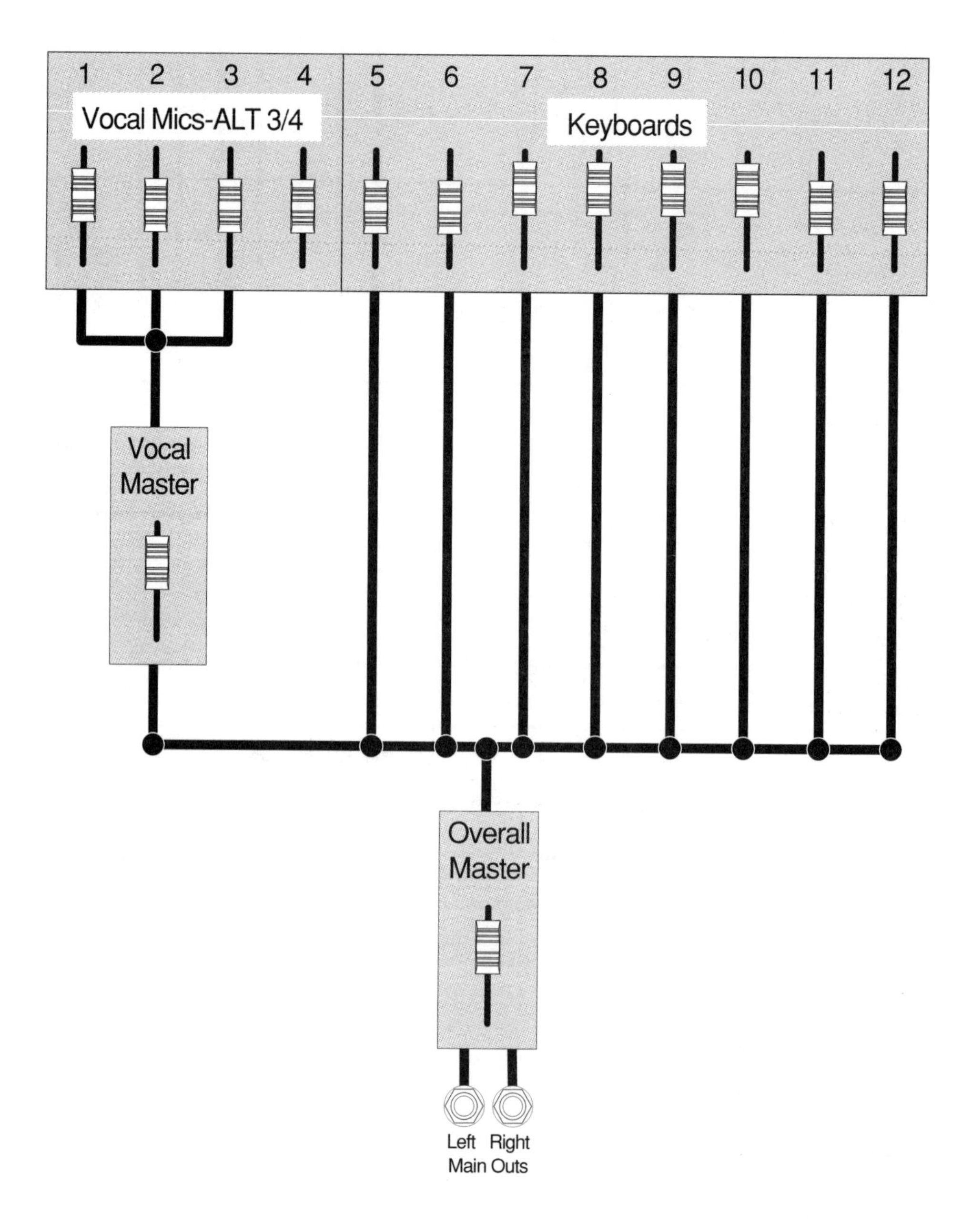

As you can see, four mics and four keyboards/sound generators are connected to our 1202-VLZ. The mic channels all have their ALT buttons pressed. This puts the mic signals out of the ALT 3/4 jacks, however, these are unconnected. Instead, the control room source is set to ALT 3/4 and finally, the control room Assign To Main button is pressed, which routes the mics back into the main mix.

As a result, you can use the control room knob to adjust the level of *all* mics when mixing, without having to individually adjust each mic channel on its own. This will help you make a quick adjustment without disturbing the relative balance between mics. The same setup applies to the 1402-VLZ, except that those users will have an extra pair of inputs to play with.

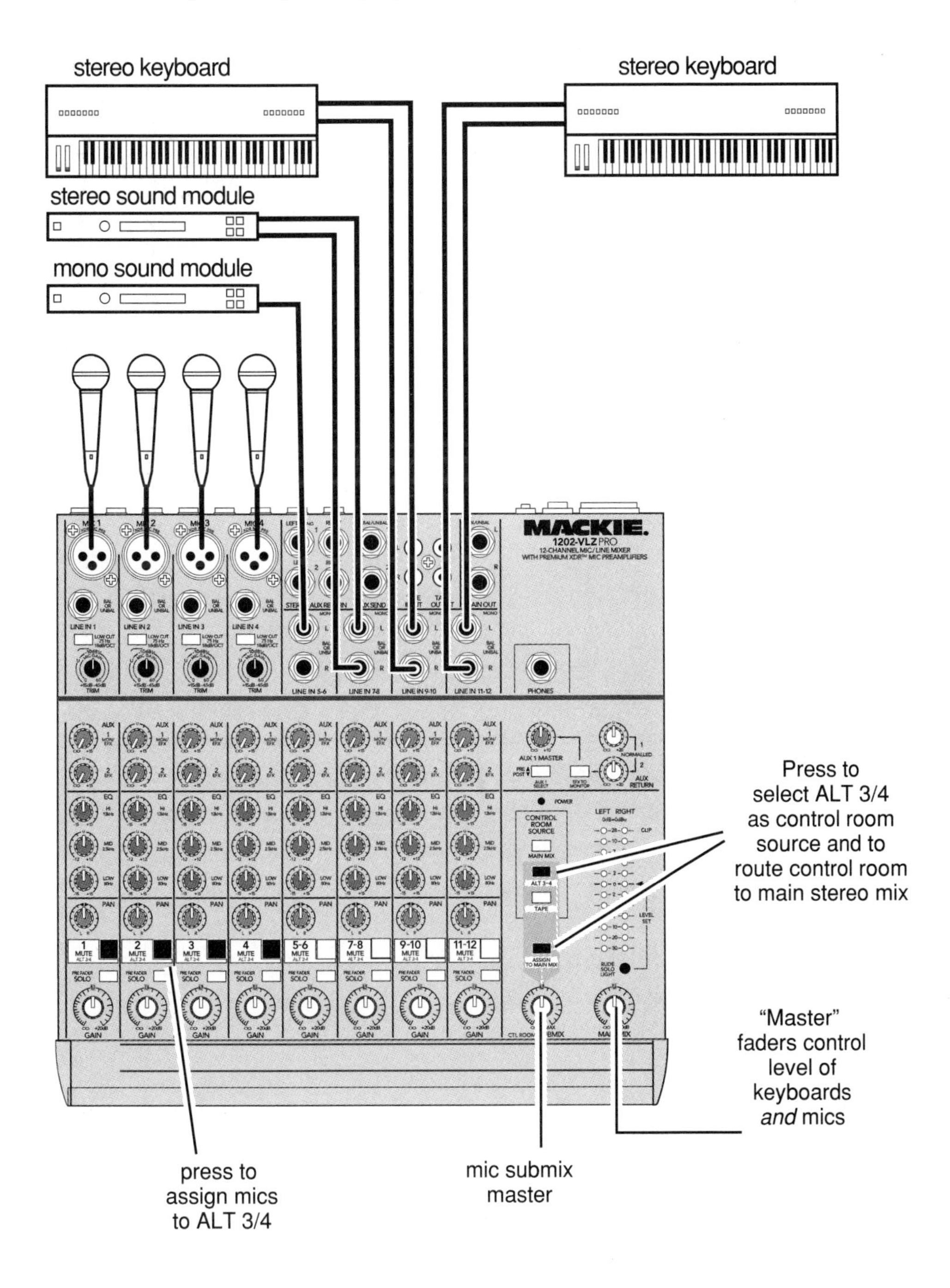

This hookup implements the signal flow diagram shown on the facing page. Use this to create a submix to simplify your mixing chores. While a 1202-VLZ is shown, a 1402-VLZ would be set up exactly the same way.

Working with Multitracks

When using your Mackie mixer with a multitrack tape recorder, you'll need to make some careful choices. These will depend on the number of tracks on your recorder, and on how many different parts you are trying to fit on those tracks.

As far as hooking things up is concerned, it doesn't matter much what kind of multitrack you are using, be it cassette, reel-to-reel, MiniDisc, tape-based digital multitrack or even a hard disk recording system.

Counting the number of available tracks is the first step in your plan. Next, you'll need to come up with a track assignment scheme that fits the parts you want to record to the available tracks. Often, you'll have more parts then tracks, which means you'll have to combine multiple parts into one mono or perhaps a stereo pair of tracks.

If you've made it through the first part of this book, you'll probably be able to make some pretty good choices about how to take the next step: Deciding how to connect your mixer to your recorder's input to record the necessary individual parts *and* mixed groups of parts to different tape tracks.

The most efficient way to record a *single* instrument to its own track is to use a direct output. This way, each channel with an insert jack can provide an independent feed to one multitrack input, leaving the rest of your mixer's resources free for other tasks.

On the other hand, recording *two or more* instruments to one track of your audio recorder can't be done using direct outs or insert outputs. Instead, you may use main mix outputs, ALT 3/4 outs, subgroup outputs or perhaps even aux send outs to route multiple inputs to a single track. (Keep in mind our earlier warnings about feedback when recorders are connected to both mixer inputs and outputs!)

Next, you need to get audio signals back from your multitrack recorder to hear what's happening while recording, and of course, when mixing. Therefore, hooking up a mixer to a multitrack is actually a two-part process—connecting mixer output to multitrack input, *and* connecting multitrack *outputs* to *mixer inputs*.

The following sections will illustrate many ways to accomplish both of these tasks. But first, a word about tape monitoring.

Monitoring "Through the Machine"

Up to this point, we've been plugging inputs into our mixer, turning up those input volumes and listening to those channels over the main stereo output.

However, when working with multitracks, a slight shift in thinking about input monitoring is required. Instead of listening to each mixer input directly, you'll listen to your input signals only after they have been passed through the multitrack, and come out of its tape track output jacks.

This is called monitoring "on input" or "through the machine." This allows you to verify exactly which signals are ending up on tape, since you can hear each track coming out of the tape recorder. You can then individually adjust track volume controls as well as solo individual tracks. Monitoring through the machine also makes it possible to match levels when "punching-in," which as you probably know, is the process of re-recording part of an existing take.

Here's how this is done: Start by routing your input signals to the multitrack, but *without sending the input channel to the mixer's main stereo output*. There are several ways to do this, described shortly.

Individual tracks are then put into Record and/or Input Monitor on the multitrack and those track outputs are connected to unused mixer inputs. Finally, *these* mixer channels are monitored over the mixer's main stereo outputs.

You can tell you're monitoring on input if you can't hear your input signals while the multitrack is playing. Stopping the deck and/or putting the appropriate tracks into record or input monitor should let you hear your live input signals again.

The diagram below illustrates the signal flow of "monitoring on input." As you can see, signal flows from the mic into Mic In 1 and out channel 1's Insert (these channel numbers are just examples—use whatever works for you).

From there, it's into the recorder's Track 1 Input, out Track 1 Output and into line input 9 of the mixer. While a video-tape based digital multitrack is shown, this and all following illustrations apply to reel-to-reel or computer-based hard disk recorders too.

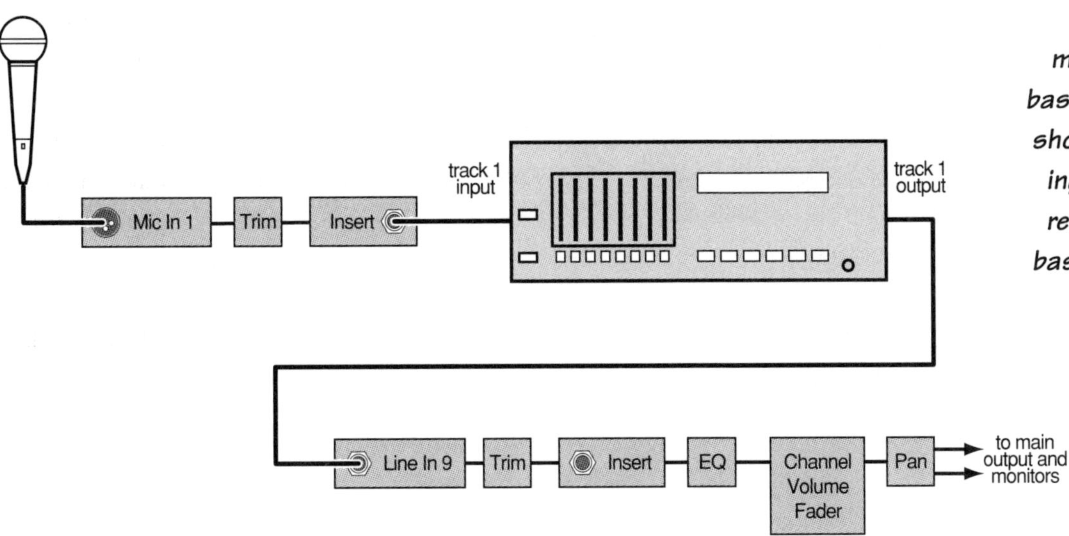

Remember that these illustrations show just one aspect of a hookup. You must combine several examples to connect your entire system.

Note that smaller mixers may not have enough inputs and outputs to fully implement this style of multitrack connection, depending on how many tracks and inputs you are working with. Still, it's important to understand this concept, even if you can't always completely achieve it. The following examples show many ways to connect your mixer to a multitrack's inputs and outputs.

To Tape

DIRECT OUTS TO TAPE INPUTS

Using a direct out to tape is the ideal way to connect a single source to one tape track. You'll almost certainly use this type of connection to your multitrack at one time or another.

You can use direct outs in conjunction with any or all of the following techniques for connecting to multitracks.

Direct outputs are the most efficient way to get individual inputs to their own tape tracks.

On most Mackies, the insert/direct out connector comes *before* the gain and EQ controls. This means that the only control you have over the signal going to tape is the trim. Depending on the input sensitivity of the multitrack, you may need to adjust the trim slightly differently than you would if you were just setting levels for your mixer alone. Alternately, if you are patching your direct out into a compressor or EQ, you can use that device's output level control as a final "signal to tape" master.

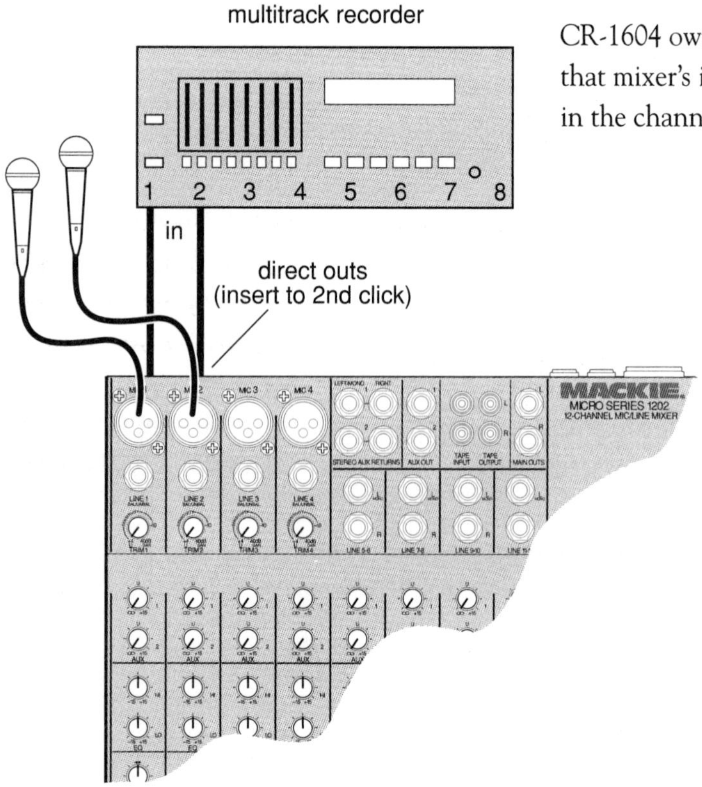

CR-1604 owners have full EQ and fader level control to tape, because that mixer's insert/direct out connector comes after the fader and EQ in the channel's signal path. 1604-VLZ and 1642-VLZ mixers have separate Direct Out connectors, in addition to their inserts. These direct outs also come after the EQ and Channel Volume Fader, so they are quite convenient for connecting individual channels to multitrack inputs.

NOTE TO CR-1604 USERS: When using a combination of direct outputs and the main stereo bus, signals in the stereo out vs. the direct outputs will be in opposite polarity. If the same signal shows up in a direct output *and* stereo mix output, the signal common to both outputs will cancel itself out if the different tracks are mixed at the same level and pan position. This is an obscure possibility, but it did actually happen to me *once*, so I thought I'd mention it.

AUX SENDS TO TAPE INPUTS

It is possible to use aux sends to feed tape track inputs. However, this is practical only with pre-fader aux sends. If you use a post-fader send, that channel's input fader must be turned up in order to get any signal from that input to an Aux Send. Unfortunately, this typically routes that channel to the main stereo mix which is a rather poor use of precious mixer resources.

Still, the only real rule in all this is that there are no real rules. There are probably a few cases where using Auxes as tape feeds makes sense. If you come up with one, feel free to use it! Yet again, you can use this type of hookup in conjunction with any or all of the other techniques described in this segment.

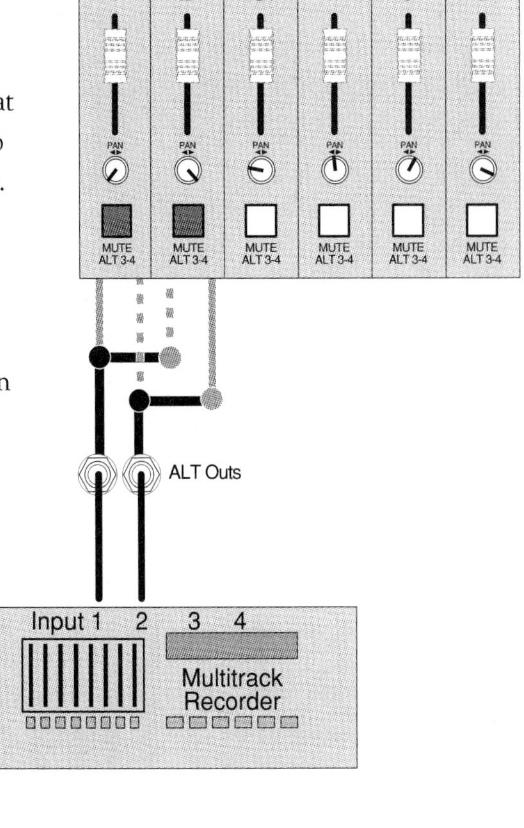

ALT 3/4 OUTPUTS TO TAPE INPUTS

If your mixer has ALT 3/4 buttons, you can connect your output 3 and 4 to the inputs of a multitrack recorder. Now, it's a simple matter to record two individual tracks each to their own channel, following the example above right.

Channels one and two have their ALT buttons depressed, and so are routed to the mixer's ALT Outs. These in turn are connected to inputs 1 and 2 of a multitrack recorder. Because the two channels are panned hard left and right, each goes to its own track. Remember that you can also use your mixer's insert jacks as direct outputs (see previous page). This is often a more efficient use of your mixer's resources because it leaves the ALT bus free for other tasks, such as the one described next.

In the second example (see left), all six inputs are routed through the mixer's ALT 3/4 bus. Obviously, there's no way each input is going to end up on its own track. Instead, all six sounds will be combined as a stereo pair on tracks 1 and 2. The pan knobs on each channel will determine that sound's position in the final stereo submix. When using this common technique, it's crucial that the relative levels of the six inputs are balanced exactly as you want them in the final mix, because there is no way to individually adjust them after they have been recorded.

An ALT 3/4 output is another handy way to get inputs to your multitrack, especially when you want to record multiple sources to the same track or pair of tracks (as seen at left).

BUS OUTPUTS TO TAPE INPUTS

Multiple output buses make multitrack routing a snap. Simply connect each of your mixer's bus outputs (left/right, 1-2 and 3-4) to the inputs of a multitrack recorder. Then you simply push assign buttons to direct channels to various tracks of your recorder. This will save you a lot of manual re-patching. (CFX users have four bus outputs, SR-series and 1604-VLZ mixers have six).

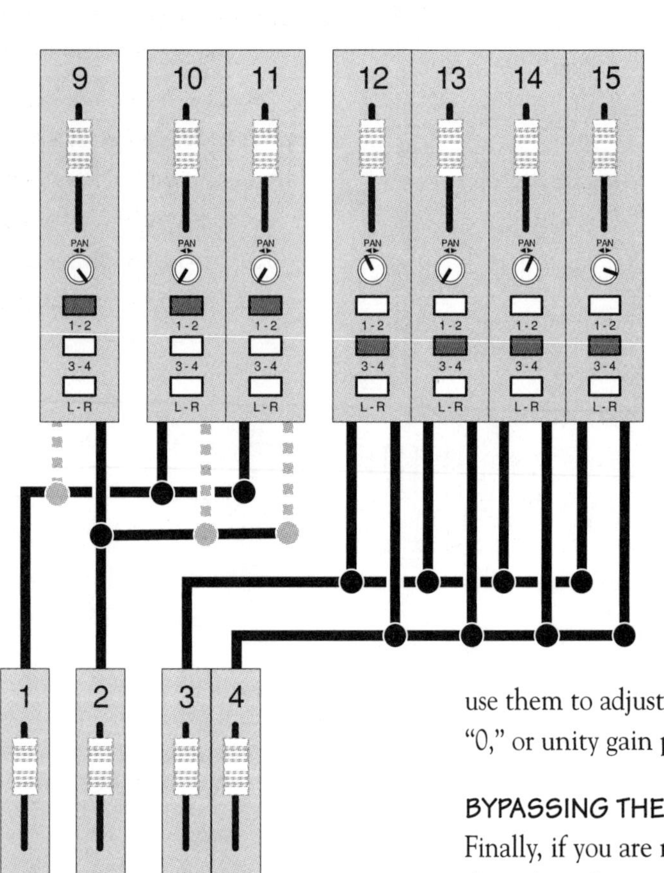

Assignable buses make routing signals to a mulitrack very simple indeed.

In the example at left, we route seven mixer channels to four different multitrack inputs. Channels nine through twelve are assigned to bus 1-2. But note the pan control—this directs nine to multitrack input 2 only, while ten and eleven go to multitrack in 1. Channels twelve through 15 are a stereo group assigned to bus 3-4. This in turn is connected to channels 3 and 4 of the multitrack. The pan knobs on each of these channels determine that sound's position in the final stereo submix. When using this common technique, it's crucial that the relative levels of the four inputs are balanced exactly as you want them in the final mix, because there is no way to individually adjust them after they have been recorded.

Note the subgroup master faders 1-4. These are in the signal path before the input of the multitrack, so you can use them to adjust overall record levels. Start with them at their nominal "0," or unity gain position and adjust them as needed.

BYPASSING THE MIXER TO TAPE

Finally, if you are recording sources with line level outputs, you could plug these directly into your multitrack's inputs and not have to tie up *any* of your mixer's resources (see facing page).

For example, electronic keyboards and drum machines can go straight to tape. In this case, you'll be using the output level control of these individual instruments to set your recording level. In some cases, these units may not have quite enough output to get optimum record levels. In this case, you may have to go back to using a mixer input channel to get enough level, or just go with slightly lower record levels, which will add a little more background hiss to the final result.

Once again, you can use this type of hook-up in conjunction with any or all of the other techniques described previously.

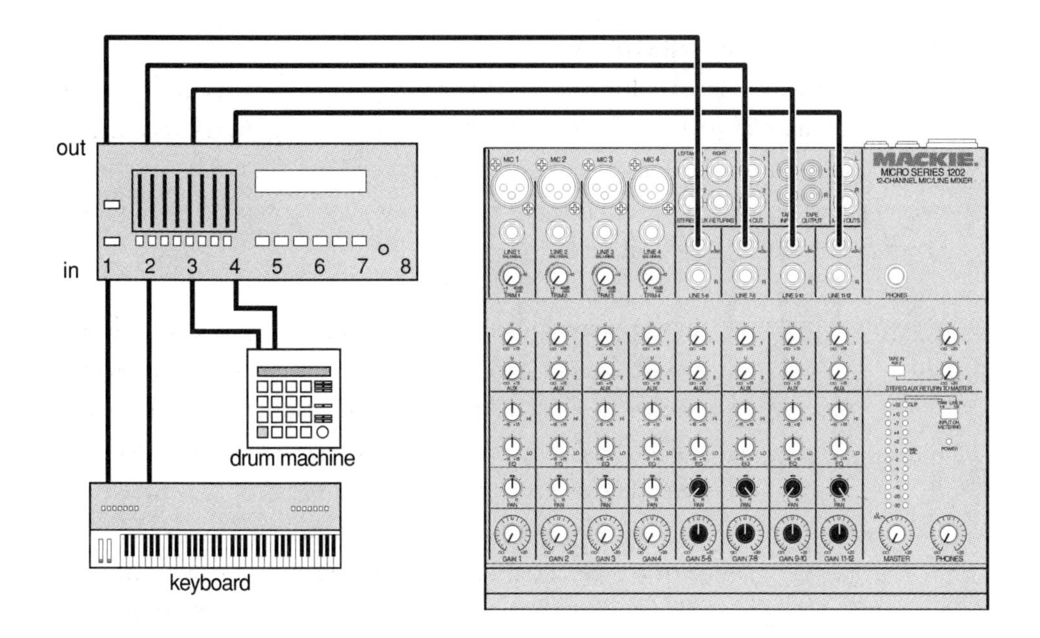

Instruments with line outputs, such as electronic keyboards or drum machines, may be connected directly to multitrack inputs, freeing mixer resources for other tasks. In this case, the 1202's first 4 channels could be used as direct outputs to send signals to other multitrack inputs.

From Tape

As mentioned earlier, you'll want to connect the outputs of your multitrack recorders (one for each track in use) back to your mixer so you can monitor through the tape machine while recording as well as when mixing. The connection of a multitrack output to a mixer input is generically known as a "tape return."

The following sections show various ways to do this. The short version is this: *Any line input on your mixer can be used as a tape return.* The method you choose will depend what mixer resources are already occupied getting signals *to* tape.

TAPE OUTPUTS TO INPUT CHANNELS

For maximum flexibility, bring tape returns into mono or stereo mixer channels (see illustration on the following page). This gives you full channel EQ and aux access so you can add effects to tape tracks during playback or monitor with effects while recording. If you return two tape outputs to a stereo line input, you'll hear one track panned hard left, the other panned hard right.

Of course, if you are using your input channels to get signals *to* tape, try the next option, multitrack outputs to stereo aux returns.

TAPE OUTPUTS TO STEREO AUX RETURNS

Stereo aux returns are another option to monitor multitrack tape outputs. If you return two tape outputs to a stereo aux return, you'll hear one track panned hard left, the other panned hard right.

Multitrack outputs can be connected to mixer line inputs or aux returns. In the illustration at right, both options are employed. Note that you will have limited independent control of the level and pan settings of tracks returned to stereo line inputs or aux returns.

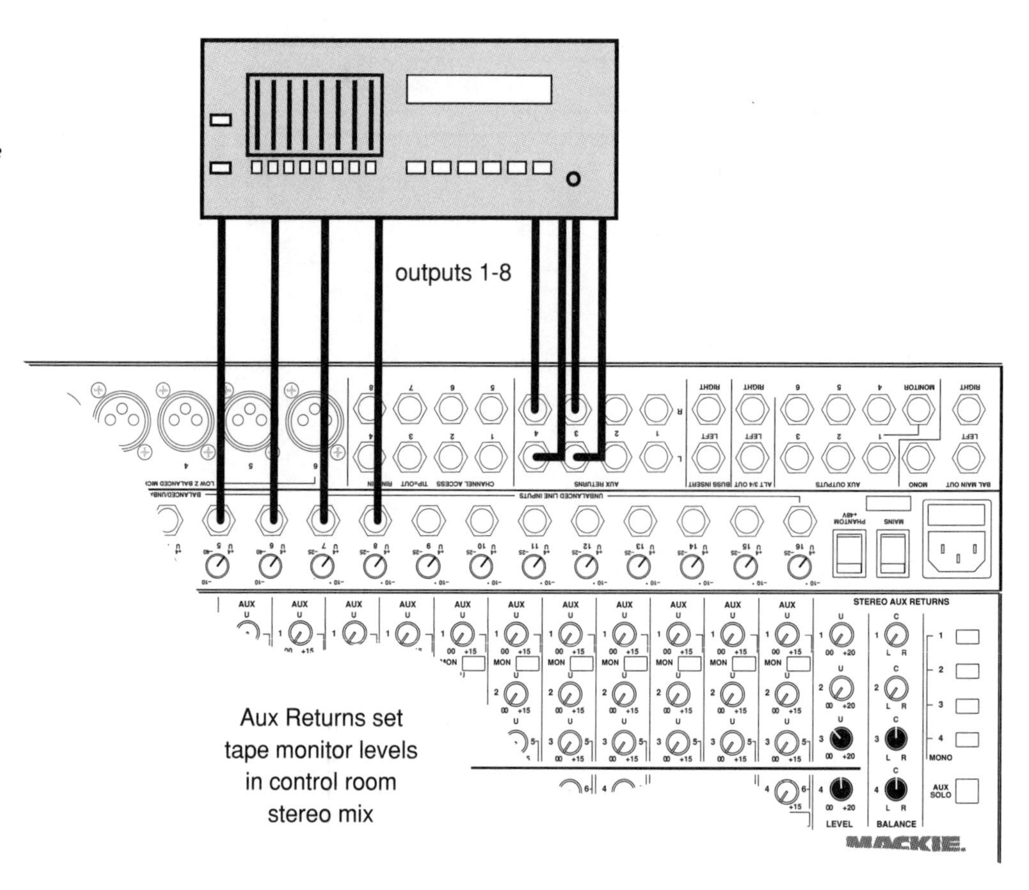

To and From Tape on the Same Channel!

One of the most elegant solutions to the multitrack connecting puzzle uses the same channel for both multitrack *input and output*! This means if you have an eight track recorder and, for instance, a 16-channel mixer, you can run your multitrack using just your first eight inputs, leaving 9-16 free for other inputs, like keyboards running off a MIDI sequencer.

There is one small catch: a pre-fader insert is required to use this technique. Therefore, CR-1604 owners will need to make an internal wiring modification before trying this trick (details on the mod are available from Mackie).

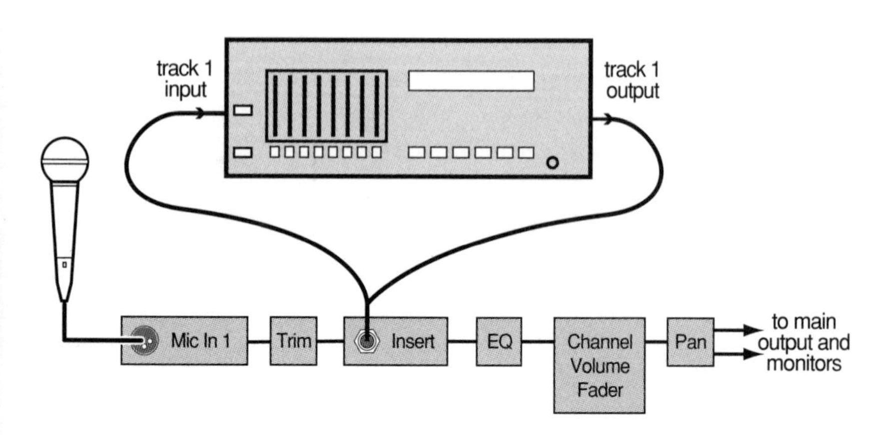

The illustration at left shows the signal flow required. You'll be using each channel's insert to divide the channel into two parts—a tape send *and* a tape return. The insert output goes to the multitrack's input, and the multitrack's output goes back to the insert return. Another way to conceptualize this is that you are "inserting a tape track" in each console input channel. The illustration below shows a hook up example using an MS1202. Any mixer with a pre-fader insert could be used the same way.

From a signal flow standpoint, only the trim control and possibly a low cut switch come between your original source and the multitrack. Therefore, only your trim control will be available to help set record levels. The rest of the channel's controls (EQ, aux sends and main volume) are *after* the channel insert, and therefore affect only the tape monitoring and playback. This means you can twiddle these controls as you wish during recording without affecting the signals going to tape.

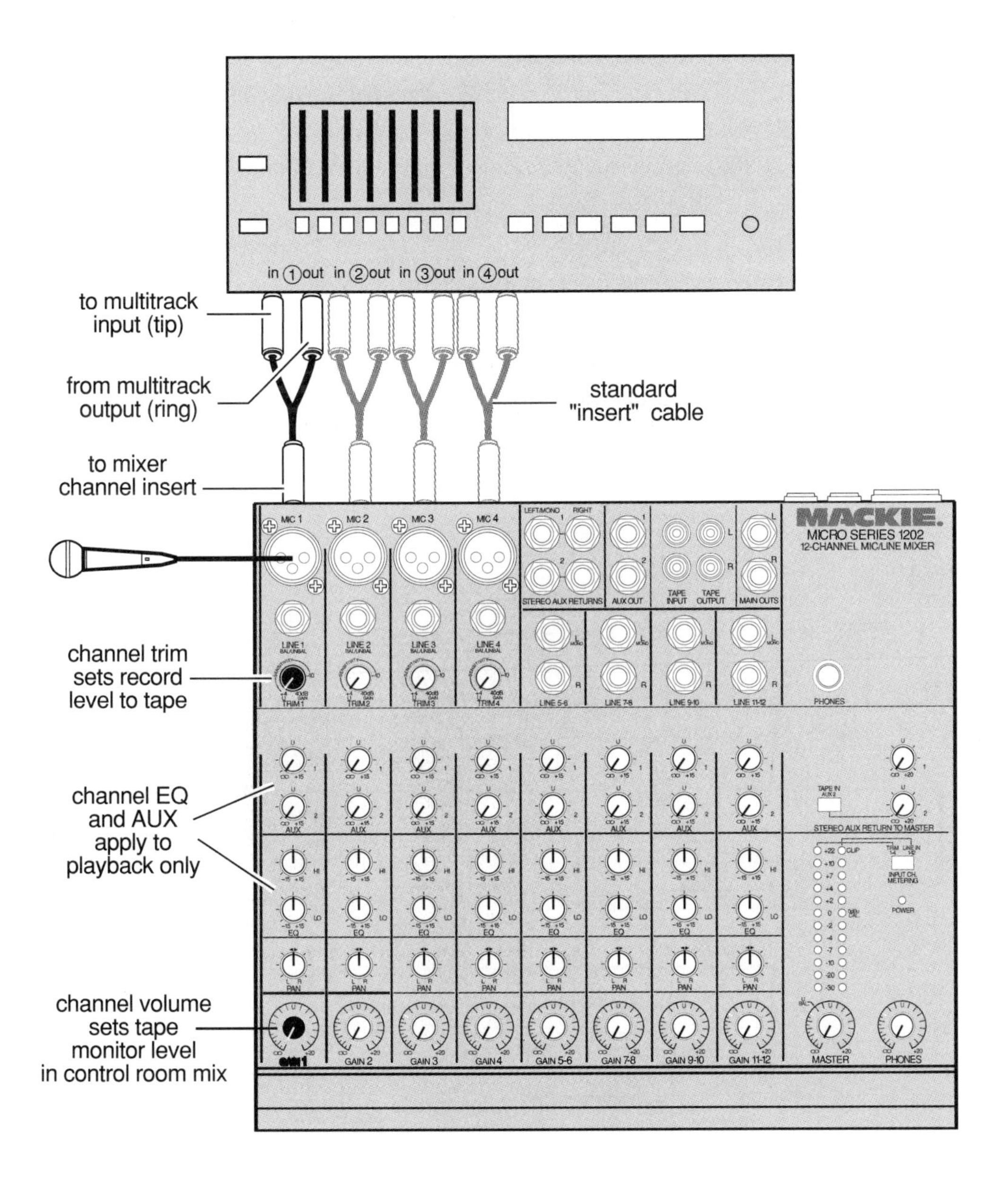

With the 1202's pre-fader insert, you can send and return signals to your multitrack on the same input channel! 1604 users will have to modify their boards to use this technique.

Recording with Effects

Where practical, most engineers prefer "tracking" (i.e. recording individual tracks) with a minimum of effects, since this allows for the most flexibility during mixing. However, this is often a luxury that smaller recording projects can't afford. For instance, you may only have one or two good effect units and therefore want to record some individual tracks with effects, freeing up the effects unit for other purposes during mixdown. Or, you may run out of tracks on your recorder and need to bounce multiple tracks down to one or two while adding effects to individual parts during the bounce. The following sections describe some of the various ways to pull this off, but remember: if you record with effects, you can't remove them later!

DIRECT OUT TO EFFECT AND TAPE

The easiest way to record *a single mic or instrument* with effects is to use a reverb or other device "in series" between a mixer (or instrument) line output and the mulitrack's input. This will often mean patching a reverb into the insert out/direct out of your mixer, and running the output of the effect into your multitrack's input. When you do this, the wet/dry mix of the effect unit must be adjusted to taste. Start with 80% wet, 20% dry and work from there.

If you do use an insert out to do this trick (as opposed to a direct out) you've got the rest of that mixer channel sitting around unused. A few pages back, we talked about a way to "insert" a tape track on an input channel. The diagram below takes the idea to its next logical conclusion: Insert an effect, then the tape track, then return the output of the multitrack back into the same channel for monitoring. A very efficient use of resources, if I do say so myself.

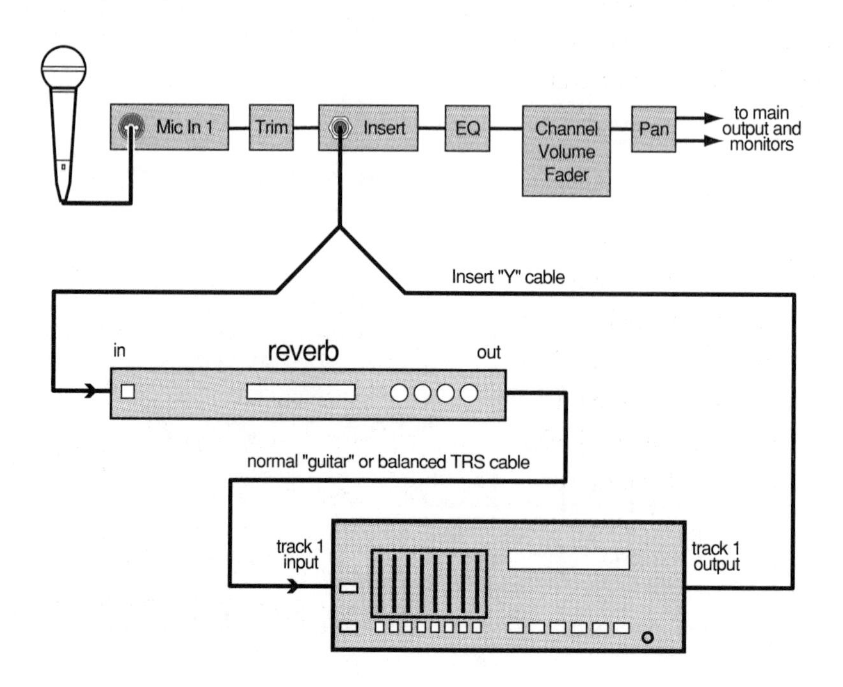

Placing an effects unit "in-series" between a mixer line out and the input of the multitrack is an easy way to record one sound to tape with effects. This diagram takes that idea a step further and returns the output of the multitrack to the same channel's insert for monitoring purposes.

FOUR-BUS MODELS—TRACKING WITH EFFECTS

The last example was great if you just wanted to record one mic to tape with effects. But what if you want to combine sounds from two or more channels, add effects to them, and put that onto a tape track? Using a four-bus mixer (CFX- or SR-series, 1604-VLZ) makes this easy (mixers with an ALT 3/4 bus will be addressed in a few pages).

No caption could reveal the true beauty of the signal flow diagram below. You're just going to have to read the text at left to reach enlightenment.

Let's say we want to record a lead and backing vocal, using two mics, onto a single tape track. We'll want the lead vocal (channel 1) to be louder and add a little extra reverb to the backing vocal (channel 2) to make it seem "further away" in the final mix. In this case, it's necessary to add the reverb while tracking; there's no way to add differing amounts of reverb to the parts later since they'll be on the same track.

The illustration at right shows the hookup we need. A CFX mixer is pictured, but the same hookup works for any four-bus board. (SR-series and 1604-VLZ models offer an additional short-cut we'll cover in a moment.)

To route our mics to track three, we simply pan them left and press the 3-4 assign button on each channel (see letter A). Next, the mixer's four bus outputs are connected to the first four multitrack inputs (B). All four buses are connected for convenience sake, although bus 3 is the only one actually used in this example. The effects unit's input is driven from Aux Send 1, while the returns are connected to a main input channel, in this case, 12 (letter C).

Now, here's what each knob does: Channel Volume Faders 1 and 2 set the relative levels of the two mics to tape (our lead vocal is pushed up a bit). The Aux 1 Sends of these channels set the amount of reverb on each mic. Note that channel 2's send is higher, adding extra reverb to the backing vocal. The EFX 1 master, below letter C, must be up in order to pass signal to the reverb.

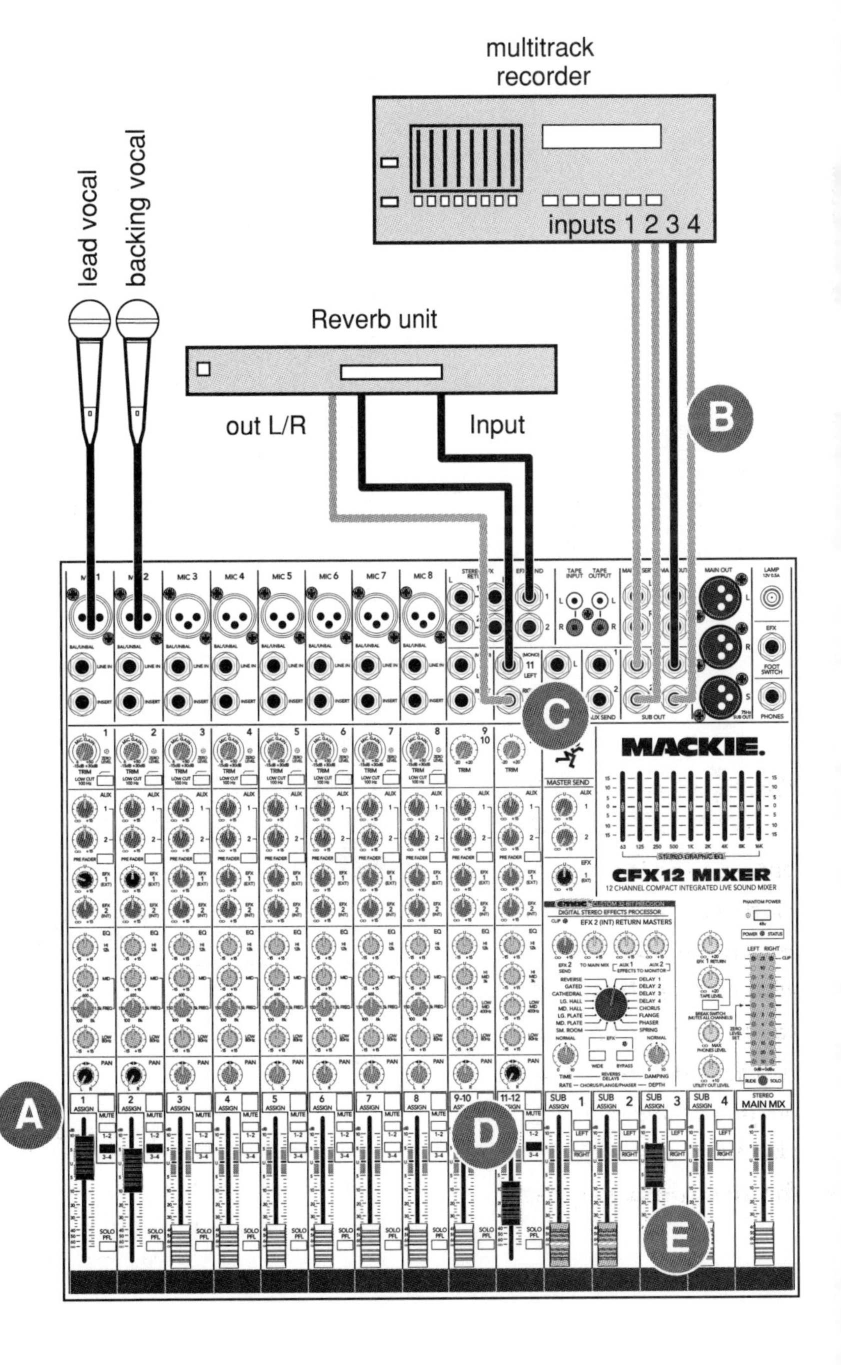

Aux send 1 exits the mixer at letter C and goes to the reverb input. (Be sure that the reverb is set to 100% wet!) The output of the reverb returns to channel 12—not the effects return! Note that stereo returns are shown, but we're only using the left channel.

Here's a crucial piece of the puzzle. At letter D, we assign channel 12 to bus 3. This means the reverb return from channel 12 and the dry mics from channels 1 and 2 are all going to the same sub output. But before they get to tape, they must pass through the sub master fader at letter E. Use this fader to set your final track 3 record levels.

Note that we are *not* assigning sub 3 to the left or right main mix. Doing so wouldn't spoil our recording, but it would mess with your ability to monitor track 3 "through the machine." While not shown in the illustration, you could run your tape returns back to unused input channels. On a 1604-VLZ or SR-series mixer, these tape return channels would be assigned to the main stereo mix. On a CFX mixer, they'd have to be assigned to 1-2, since the CFX hasn't got a dedicated assignable stereo bus.

AUX RETURNS TO SUBS (1604-VLZ, 1642-VLZ)

In the example on the previous page, we used a main mixer channel to route effects to sub outs. While this works, it does cost you a precious input channel(s). On the 1604-VLZ you can route Stereo Aux Return 3 to the bus outputs, saving an input channel or two in the process.

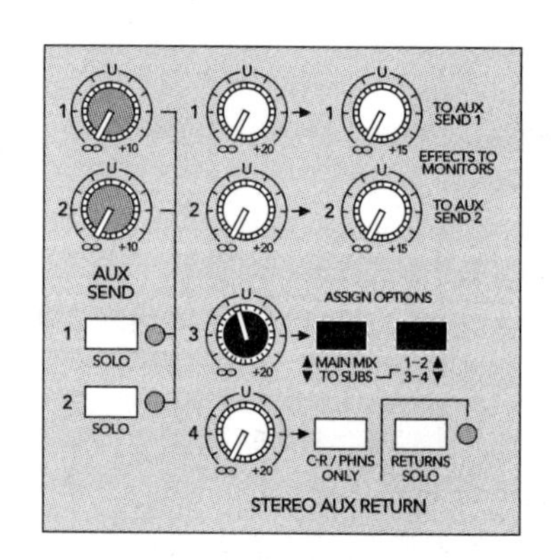

1604-VLZ and SR-series mixers can route an aux return to subgroup outputs.

To do this, connect the output of the effects unit to Aux Return 3, rather than using an input channel as shown on the previous page. Now, check out the Assign Option buttons in the illustration at left. Pressing the Main Mix/To Subs button down causes Aux Return 3 to be routed to the subs. Which subs, you ask? Direct your attention to the second button 1-2/3-4. With this button up, Aux Return 3 will be routed to bus 1-2. With the button down, as shown here, it will appear on subs 3-4. The Aux Return 3 knob (highlighted) will set the amount of effect added to the submix.

AUX RETURNS TO SUBS (SR-SERIES)

In the example on the previous page, we used a main mixer channel to route effects to sub outs. While this works, it does cost you a precious input channel(s). On a SR-series mixer, you can route Stereo Aux Return 4 to the bus outputs, saving an input channel or two in the process.

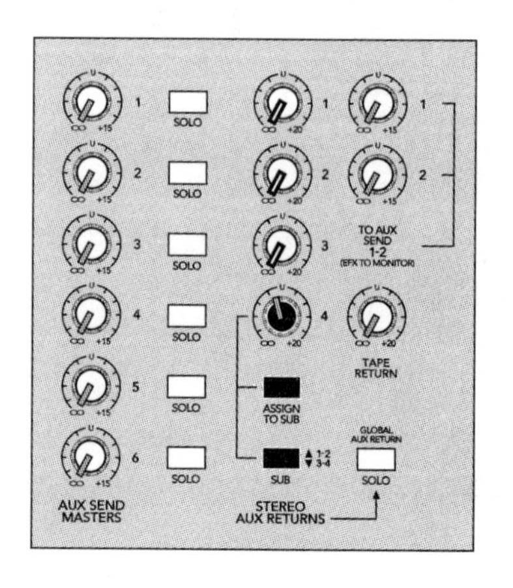

To do this, connect the output of the effects unit to Aux Return 4, rather than using an input channel as shown on the previous page. Now, check out the highlighted buttons in the illustration at left. Pressing the Assign To Sub button down causes Aux Return 4 to be routed to the subs. Which subs, you ask? Direct your attention to the second button 1-2/3-4. With this button up, Aux Return 3 will be routed to bus 1-2. With the button down, as shown here, it will appear on subs 3-4. The Aux Return 4 knob (highlighted) will set the amount of effect added to the submix.

ALT 3/4 MIXERS—TRACKING WITH EFFECTS

Users of mixers with the ALT 3/4 routing scheme can also track multiple sounds to tape with effects. However, it's a bit more complicated than the hookup used by four-bus mixers. Study the illustration at lower right. A 1202-VLZ is shown, but much the same hookup applies to a 1402-VLZ, CR-1604, etc.

Let's say we want to record a lead and backing vocal, using two mics, onto a single tape track. We'll want the lead vocal (channel 1) to be louder and add a little extra reverb to the backing vocal (channel 2) to make it seem "further away" in the final mix. In this case, it's necessary to add the reverb while tracking; there's no way to add differing amounts of reverb to the parts later since they'll be on the same track.

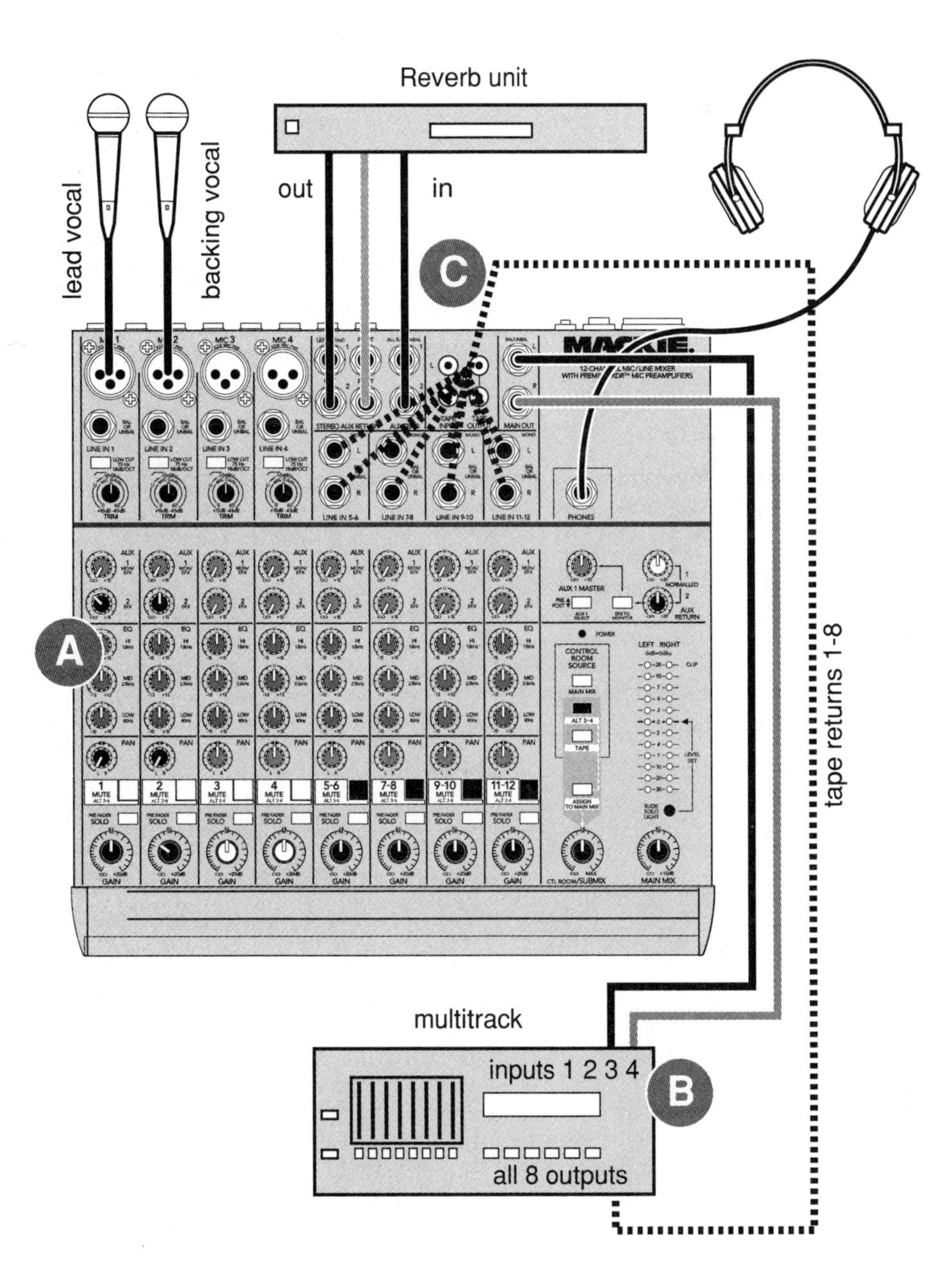

To pull this off, we're going to use the main stereo mix to feed the multitrack inputs. Our two vocal mics (see letter A) are panned left, which puts them on track 3 of the multitrack. Each vocal channel sends some signal to aux send 2, which is connected to our reverb unit near letter C. The reverb returns come back to aux return 2, in this example. Stereo returns are shown connected, but only the left reverb return will end up on tape. The Aux 2 return sets the reverb level to tape, while the Main Mix knob sets the overall record level of the vocals and reverb.

That takes care of getting the desired signals to track 3. Now let's address tape monitoring. All 8 tracks of the multitrack are returned to the stereo line input channels (see dashed lines near letter C). Each of these tape return channels *must* be assigned to ALT 3/4 or horrible feedback will result! Finally, to hear the tape returns and monitor our vocal tracks "through the machine," set the Control Room Source to ALT 3/4 and make sure Main Mix is not selected.

Note that CR-1604 users don't have the Control Room section of the later VLZ models, however, the ALT Preview button will serve the same purpose in this particular case.

ROUTING THE EMAC TO TAPE VIA SUBS

The previous example shows how to route an external effects unit through your CFX mixer's sub outputs. But wait a second—what if you want to do the same trick using the built-in emac signal processor? It can be done, but only for mono effects.

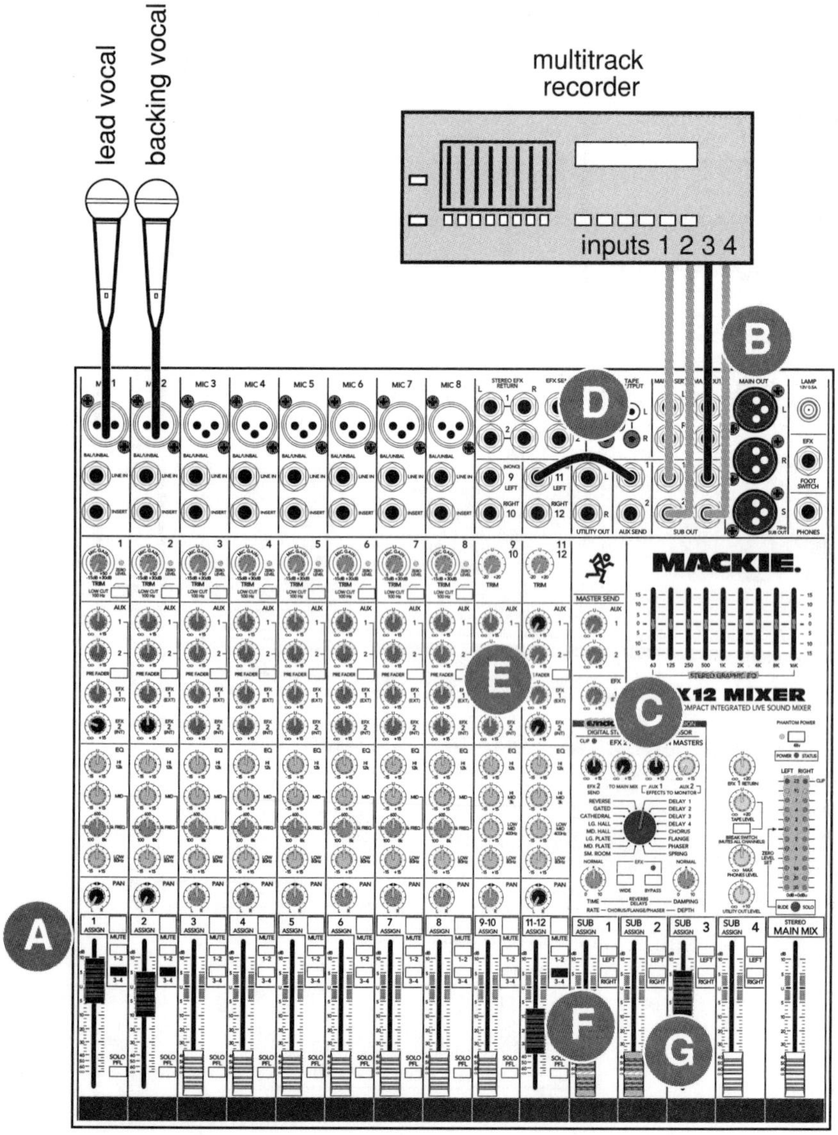

Check out the illustration at left. The mic channels (letter A) and multitrack input connections (letter B) are unchanged from our last example (recording a lead and backing vocal, with different amounts of reverb, to track 3 of our mulitrack).

Things begin to change at letter C. EFX 2 Send sets the overall signal level to the emac. The "To Main Mix" knob is off, since we want to monitor our effect levels only "through the machine." Next, Effects To Monitor, Aux 1 must be turned up as shown. This routes the output of the effects unit to the Aux Send 1 output jack.

Now comes the cable: As shown at letter D, run a cord from Aux Send 1 to Input 11, making this channel our "effects return." To avoid feedback, be sure Aux Send 1 and EFX 2 (Int) are turned off as shown (letter E). The overall effect return level is controlled by the channel 11-12 fader at letter F. Note that this channel is assigned to bus 3 thanks to the pan position and the 3-4 assign button.

This means the reverb return from channel 12 and the dry mics from channels 1 and 2 are all going to the same sub output. But before they get to tape, they must pass through the submaster fader at letter G. Use this fader to set your final track 3 record levels.

Patch Bays

Once you start buying audio equipment, your wish-list is nearly endless. New effects units, a digital recorder or extra microphones are perennial favorites. However, I'm going to make a case for one of the most mundane and under-appreciated pieces of equipment around: *The Patch Bay*.

A patch bay serves a remarkably un-exciting purpose. You plug most or all of your equipment into it. By choosing which gear gets connected to which plug at the back of the patch bay, equipment is automatically connected in the configuration you most often use. Then, when special connections are needed, plugging short cables into your patch bay creates new connections.

This will spare you endless hours crawling around behind your gear changing cables. Even better, this easy access will encourage you to experiment in ways you wouldn't otherwise attempt. It sounds like hype, but a patch bay truly helps you get the most from your entire studio!

The problem is that most people would rather spend their money on stuff that makes, or modifies, sound. The patch bay does neither, and therefore is not perceived as an important priority. All I can say is this: I have encouraged many people to spend hundreds of dollars on patchbays and additional cabling. Everyone who did so later wondered how they ever lived without it!

About Normalling

Patch bay audio connectors use a mechanical arrangement that makes an automatic connection when *no* cables are connected to its front panel inputs, while allowing other connections to be made by plugging cables in as needed. Mackie stereo inputs use the same technique.

It's a puzzle: You have a stereo input channel with two mono input jacks. Plug a cable into both and the signal on each cable will appear only in the left or right channels. But pull the right connector out and the remaining signal jumps to the position set by the Pan control.

Somehow, the mixer "knows" if there's a cable plugged into the right input. Here's how: 1/4" jacks can be constructed with extra electrical contacts that are pressed together by a little leaf-spring. These contacts are forced apart or "open" when a plug is inserted into the jack. When the plug is removed, the spring action of the metal forces the contacts back together, again closing the gap and re-establishing an audio connection.

The idea of a jack that can make its own connection without a plug is called "normalling." The term refers to a "normal" connection made automatically, or by default if nothing is plugged in. This normal connection is broken as soon as a user connection is made. If you look at the block diagram for your mixer, you'll see symbolic representations of normalling jacks.

While patchbays come with several different normalling schemes, the concept is the same: providing one useful signal path when no extra cables are connected, and the option to create a new signal flow simply by patching in a cable or two.

Finally, when you connect your new patch bay, follow the standard convention of connecting outputs to the upper row of each bay and inputs to the lower row. Think of it as a waterfall, with outputs flowing down to inputs.

CHAPTER 13

Mixing Ideas

No book can give you a cookbook solution to getting a good mix—not only are there far too many variables, but a "good" mix is a subjective judgment to begin with. Instead, this chapter introduces you to the important issues that will impact your final mix. I'll also pass along some of the habits I've formed. If they make sense to you, use them. If not, make up your own! Ultimately, *your* ears are the final judge.

Getting Sounds

Any mix is only as good as its individual parts. Time spent making sure each input is sounding its best will pay off big-time in your final mix. Before even thinking about EQ, mic selection or placement, listen carefully to the instrument. Is it in tune? Are new _______ (fill in the blank: strings, drum heads, reeds, etc.) needed? Even if the instrument is sounding at its peak, is it producing the right tone for the song? Because of the variation in harmonic content of different instruments, a given combination may complement each other well, or they may clash. No amount of audio technology can completely overcome problems like these.

Once the instruments themselves are in shape, turn your attention to the acoustical space that the performance or recording is taking place in. For example, if the room has hardwood or tile floors and lots of glass, it will be brighter and more "live" sounding (because sound reflects off of these hard surfaces). If there are carpets, drapes or upholstered furniture, the room will sound darker and more "dead," because these soft materials will absorb sound, especially higher frequencies.

Also remember that the size of the room, however furnished, will affect the length of time for a sound's reflections to fade away (this is referred to as the *reverberation time* of the room).

Next, evaluate how each instrument interacts with the room. The character of a room's sound can really enhance or detract from the sound of an instrument within it. If multiple instruments or singers are in the same room, listen to the overall blend.

Moving a mic closer to an instrument reduces the amount of "room sound" picked up by the mic. This is not because the sound of the room is getting softer! Instead, by moving the mic closer to the source, you are making the individual instrument "louder," as far as the microphone is concerned. It's the ratio between the direct instrument and the overall room sound that is changed.

By the same token, if the room sound is working to your advantage, moving the mic away from the instrument might sound fuller or more "open." Welcome to the wonderful world of microphone placement!

WHY MORE MICS DON'T SOUND BETTER

Be aware that when multiple microphones are used in a recording session or live performance, an important and serious problem appears.

When multiple microphones are in the same room, they will all "hear" some of the same sounds, even though they may be pointed at different instruments. Of course, they will most clearly pick up the sounds that they are trained on, but leakage of sound from other sources is often present.

When multiple mics are used, "leakage" of sound from one instrument into a second mic causes problems in the overall mix.

In the illustration below, each mic is getting a little bit of sound from the source the *other* microphone is trained upon. A problem is caused by the speed of sound, which is fast, but not instantaneous. The symptom appears when the signals from multiple mics are combined in your mixer.

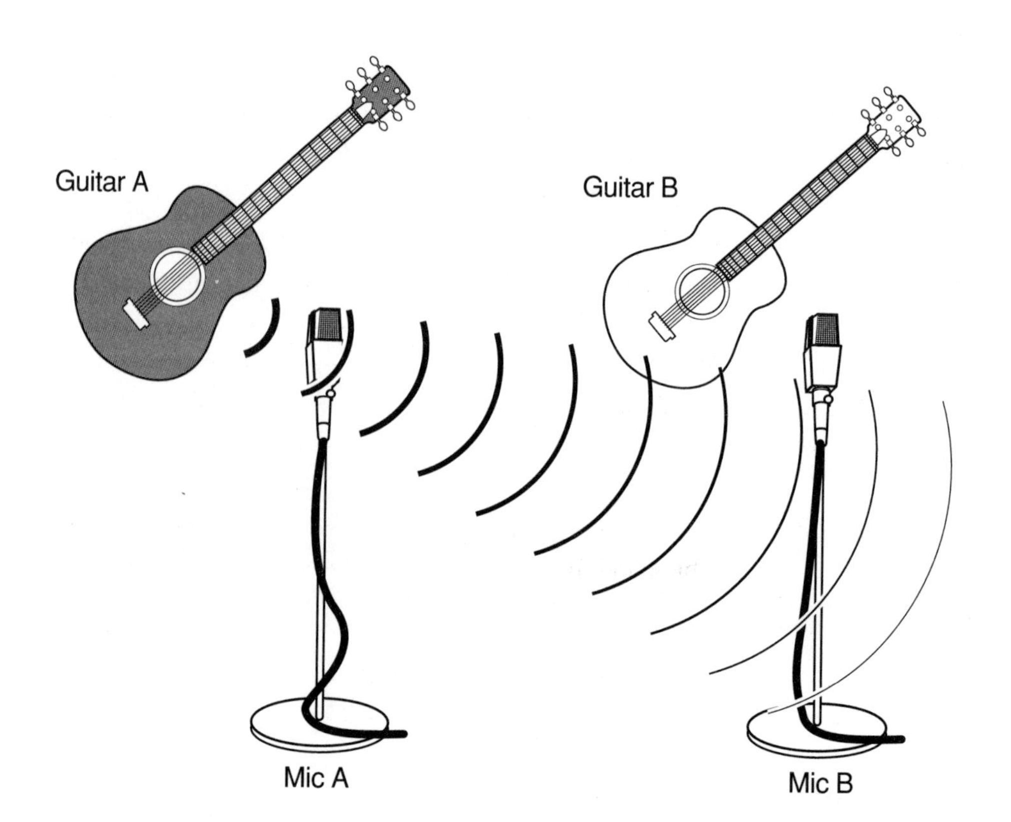

Consider Guitar A in the illustration on the previous page. In the combined mix, you will hear it nice and loud through Mic A. You will also hear a little of Guitar A's sound picked up by Mic B. This version of the sound will not only be more *ambient*, meaning it will sound more distant because it is picking up more of the sound bouncing around in the room, but it will also be delayed slightly in time, initially about one millisecond for every foot separating Mic A from Mic B. Because these time delays are so short, you won't hear them as separate events. But they do have an important effect:

When a sound is mixed with a very slightly time-delayed version of itself, an interesting thing happens. Depending on the length of this time delay, different frequencies in *both* the original and delayed sound will cancel each other out.

This cancellation is called *comb-filtering* and the resulting sound has a boxy, unnatural quality. Those familiar with effects processors that do "flanging" might recognize this sound because flanging is a similar effect created electronically.

As a result, adding additional microphones can make the resulting mix sound worse, rather than better. And the more microphones in use, the greater this problem becomes, because you have possible interaction between many different microphones, instead of just a pair.

ABOUT MICROPHONE POLARITY

Proper mic placement is the place to start when combating these problems. However, because there are logistical limits with a large, or even small group, there will often be significant distance and leakage between mics.

One simple solution experienced engineers have used for years is the reversal of *polarity* (often incorrectly referred to as "phase") between different microphones. When mixing two physically spaced microphones, reversing the polarity of one mic can improve the quality of the combined sound. This is not because the comb-filtering problem has magically gone away! Rather, by changing the polarity of one of the participating microphones, the distribution of the particular frequencies being canceled is changed. After switching polarity of one mic, the mixed mics will almost certainly sound different. If you are lucky, it will sound better and you can go on with the recording or performance. If not, change one or more mic positions and do it all over again.

A SPECIAL NOTE: when working with multiple microphones, pay close attention to the low frequencies of the sound. If turning up more microphones make a sound thinner, try reversing the polarity of (or physically moving) one of the mics involved.

OK, great. But how does one go about changing polarity of one mic? On most larger pro consoles, a polarity switch is standard issue on each input channel. Pushing this button will reverse the polarity of that mic. Unfortunately, Mackies don't have this control. But don't despair—we're going to build a solution, cheap and easy. Not only will it give you a chance to play with polarity, but with the appropriate pair of microphones, you'll be able to use this simple device to record extremely natural-sounding stereo recordings.

A VERY USEFUL CABLE

We're going to build an XLR Y cable with a twist. In a normal Y cable, each of the three connectors is wired to the same pin on the other two. The twist is that in one of the two male XLR connectors, the pin 2 and 3 connections are swapped.

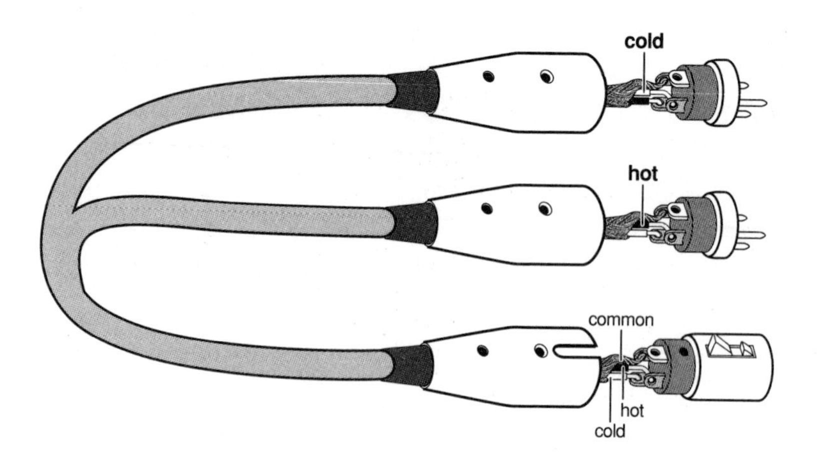

As a result, the polarity of this connector is the opposite of the other connectors. This makes the following experiment possible:

First, put a couple of mics on a snare drum, one just above the top head, the other the same distance beneath the bottom head. Listen to them both individually, then listen to them both panned to the center at roughly equal levels.

When the two mics are combined, the sound should be much thinner than when either mic is heard alone. This is because the drum's top head is moving *away* from the top mic at the same time the bottom head is moving *towards* the bottom mic. When the two mics are mixed together, this electronically cancels the main motion of the drum heads (where the low-frequency sounds are).

By connecting the Y cable to the bottom mic and connecting the two ends to a pair of adjacent Channels, you can mute or turn up and down the normal and polarity reversed bottom mic and hear the effect of the switch when mixed with the top mic. You should definitely hear the difference.

While it's a safe bet you'll want to reverse polarity on the bottom mic, the same experiment is valuable any time you have two mics picking up some of the same sounds. There's no right answer—use your ears to decide whether you want to reverse polarity on a case-by-case basis.

Here's another worthwhile experiment to try: if you have a microphone that can be set to a "figure 8" or "bi-directional" pickup pattern, use it along with a second "omni" or "cardioid" pattern mic.

The polarity-reversing Y cable is connected to the output of the figure 8 mic and brought into a pair of channels that are both turned up, but panned to the extreme left and right. Connect the omni or cardioid mic to a channel panned to the center.

The omni or cardioid mic will create a strong "center" image and bringing up the two channels of the figure 8 equally will add a startlingly natural stereo "width" to the blend. This miking technique, called "M/S" for "middle/side" can be very effective on acoustic guitar, drum overheads, group vocals and even entire ensembles.

Building Your Mix

When mixing music, I find it helpful to first listen to the rhythm section—balancing the drums, then adding the bass and so on. Once a solid foundation is established, melody and other more coloristic parts can be added to this bed. Some people find the opposite approach works better for them—put up the lead vocal or featured instrument and then build the track around that. Obviously, the type of music you're working with has an impact. Those doing dance music will probably want to start with the groove before worrying about the singing.

Whichever method you prefer, I think it's easier to build a mix part by part, as opposed to turning everything up at once and then trying to mix that.

Think Subtractivly: Take something out or turn something down before rushing to put something new in or turn something up. Is the vocal being buried? Turning down the guitar might be a better solution than turning up the vocal, only to discover that now the *drums* seem a little soft. Maybe you don't have to turn the guitar's volume down, just cut its midrange EQ a bit so it's less competitive with the singer.

Think Tone: Think in terms of frequency or tonal balance as well as working on adjusting volume levels between parts. I'm always seeking clarity and a coherent blend between sounds. Equalization is often the key. If a particular part seems too loud or soft, it may only require an EQ adjustment, not a level change, to bring it into balance with the rest of the mix. When possible, cut EQ, rather than boost it.

Don't Let It Slip Away: One key to getting a good mix is keeping your level controls somewhat near their center position, so you can turn signals up or down without running out of adjustment room. This is especially true during live performances, when you don't have the time to readjust your master volume and then change each individual level control. If most of your channel levels are in the vicinity of Unity, you're probably going to be OK.

Listen with Ears, Not Eyes: It's easy for your eyes to trick your brain into hearing things. Your eyes see your fingers nudging up a level control and you "hear" the sound getting louder. A moment later you realize that you were turning up *the wrong knob!* This is even more common with subtle adjustments, like EQ. Trust your ears, not just your eyes.

Take Artistic License: Although in many cases, a mix will try to mimic the relative level of sounds as they occur naturally, there are many cases when soft sounds are made loud and vice versa, to highlight a musical element or achieve a special mood or effect.

Mixing for a Live Audience

Although live mixing uses many of the same tools and techniques as studio mixing, the demands of live performances change one's priorities significantly both when mixing live sound for an audience through a PA/Sound Reinforcement system or mixing a live performance to a stereo tape recorder.

The key is *priorities*. In a live setting there is no substitute for getting the most important parts of the mix up right away. It's crucial that the audience be able to focus immediately on the most important element of a performance.

With a band, the most important musical element is usually the singer. Supporting the singer is especially important since the level of an unamplified voice will usually be lower than that of the supporting instruments. Once the main musical element can be heard by the audience, you can start filling in around the edges.

As you may remember, this is exactly the opposite order of events I suggested when mixing in a studio setting, where I like to begin by establishing a solid rhythm section before adding in lead instruments or vocals. If you are fortunate enough to spend time with the musicians before the crowd arrives, you can do a sound check and build your mix without worrying about the audience.

However, when the audience is present, their needs should be your first (and last!) priority. Even if you do have extra time to experiment, don't get carried away with details, like fiddling with a lot of effects. It's very important to keep a handle on the big picture, because live shows can slip quickly out of control, especially if you are distracted with a lot of button pushing.

Finally, remember that mixing sound in performance spaces is a much more problematic task than mixing in the controlled environment of the recording studio. Live mixes are often hampered by poor acoustics and loud stage volumes. This can mean that you have minimal control over the final sound, especially the overall volume. Rest assured that the person behind the mixing board will field the complaints from audience members regardless of who's at fault!

The (Over)use of Effects

Ahh, effects. "Back in my day," says Rudy, slipping into his cranky old coot voice, "we didn't have all these new fancy digital effects. No siree. All we had was wah-wah pedals and spring reverbs. And maybe a fuzz box. Why I remember the time..."

Okay, so today we've got oodles of different signal processing toys. These might be external rack units, little table-top models, foot-operated stomp-boxes or built-in units like the emac. Regardless of how they are packaged, effective use of effects calls for a flexible mixer, a good ear and above all, a modicum of restraint. It's all too easy to go overboard and end up with a sloppy wet blanket of a mix. Following are a few guidelines I keep in mind as I twiddle knobs for fun and profit.

Contrast is crucial: Don't add equal amounts of the same effect to everything in your mix. Create contrast by adding a given effect to foreground or background elements, but not both.

Be mindful of musical tempo and density: Prominent long reverbs work best with slower tempos or sparser musical arrangements. Lots of notes plus lots of reverb quickly adds up to mud. Try shorter reverb times or use a combination of reverb and delay to add ambiance without making mush.

Bright isn't always best: Listen to how your reverb responds to percussion and other strong transient sounds such as acoustic guitar or banjo. The metallic "sizzle" generated by many reverbs may be more of a distraction than a mood enhancer. Reduce the high frequency damping to "darken" the reverb's tone.

Be mindful of musical tempo and density (part 2): Long echoes with high "regeneration" or "feedback" settings, which increase the number of individual repeats, can get as cluttered as excess reverb. Also, be aware that long echoes will essentially be creating new musical events. Try making the delay time match the tempo of the song, so that echoes fall on quarter notes or some other musically relevant point. I love delay units that include a "tap" button to set the delay times. This lets you simply tap your finger on a button in tempo with the music and the unit sets its delay times to match. Don't confuse this feature with a "tapped delay line," which is a common (and useful!) type of effect that provides multiple repeats which may be set to different, arbitrary delay times.

Bright isn't always best (part 2): Most delays provide a filter to remove high frequencies from the delayed sound. This filter is often inserted within the delay feedback path, which makes each additional delay a little darker than the one before it. Filtering can help keep the echoes from being overly competitive with the original sound.

Combine delay and reverb: Using these two effects together can provide better results than either alone. For instance, I like to combine a moderate amount of reverb with an echo or two when mixing live vocals with a band. The reverb adds an almost subliminal bit of "air" around the voice, while the slightly louder (but much sparser) delay gives a sense of distance. Without the delay, a lot more reverb return would be required but by the time the reverb would be loud enough to hear, it could well be muddying up the mix.

Don't use compressors (if you don't fully understand them): A lot of people end up with compressors in their systems because someone (who, me?) told them it would impart a "professional" sound to vocals or other instruments. Or because they hoped a limiter would protect their speakers from damage. While compressors and limiters can do both these wonderful things, it is very, very easy to set their controls inappropriately, resulting in worse, rather than better, sound. Excessive compression cannot be removed from recordings, so use care when tracking. I do not recommend putting a compressor across your main stereo bus when mixing—leave that for a separate mastering step. In live performance, compression increases the risk of feedback and can rob your mix of dynamics. Learn to use your compressor by listening and experimenting on your own time, rather than during a gig or high-pressure recording session.

Know the limits of EQ: EQ is no substitute for proper microphone selection or placement. Many people hold the mistaken belief that if they could find the "magic" EQ settings they could turn the sound of a $100 mic into a $1,000 model. After all, it's just a question of different frequencies present between models, right? Wrong! While EQ lets us modify the so-called "frequency domain," it also affects the "time domain" part of the signal. The result is a "time-smearing," which is most noticeable on sounds with sharp transients, such as percussion. We have all had the experience of using EQ and finding that, while it changed the character of the sound, it failed to provide the result we had hoped. Time smearing and "ringing" of the EQ filters is one explanation for such unsatisfactory results.

In Conclusion...

Working with audio can be a fantastically rewarding experience. If you love sound and want to make it a part of your life, here are a few things to remember.

Never equate the lack of equipment with a lack of opportunity. The most important lessons can be learned with the most basic of tools. A decent microphone or two, a small mixer and a tape recorder are enough to capture meaningful and memorable sounds. Learn to use what you have.

Next, a clear understanding of signal flow concepts will serve you well, especially in unfamiliar surroundings. And you'll find the same signal flow configurations you've been introduced to in this book appear over and over in professional settings, too.

Finally, listening is paramount. If you can develop your sensitivity to sound, and learn to inspire confidence in those you work with, you'll always be in demand.

CHAPTER 14

DFX-Series Mixers

This chapter includes everything you need to know about Mackie's DFX•6 and DFX•12 compact live sound mixers. The DFX is ideal for live performance, thanks to its easy-to-use layout and big faders that make for accurate adjustments in the heat of a show. But even though the DFX is somewhat biased towards live applications, there's no reason you couldn't use one for simple recording tasks as well.

We're going to cover all the elements of the DFX-series mixers: Input Channels, Output Section, Inserts and Auxes. Remember that all the background info you'll need to understand this chapter appears in the Chapters 1-3, 5, 7 and 9.

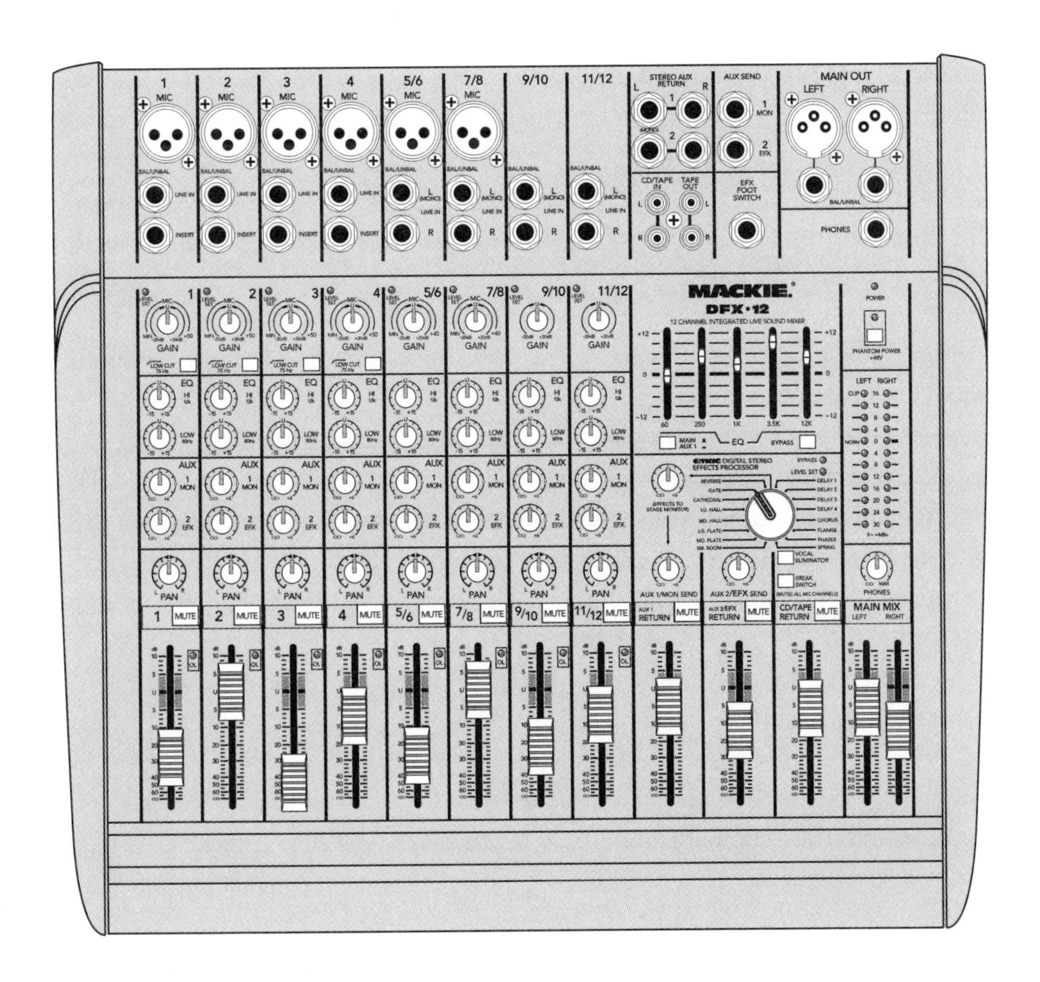

Yes! It's the Mackie DFX•12 mixer. The DFX•6 is about the same, just six fewer inputs to love.

DFX Input Sections

We'll start our tour of the DFX-Series with a look at their input channel sections. Remember that Chapter 3: Mixer Input Concepts is chock-full of important background material on this topic.

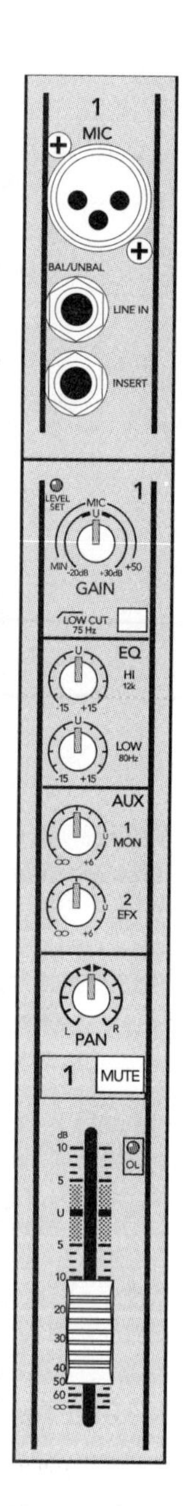

DFX-series mono mic/line input channel

The DFX-Series debuted with two models, the DFX•6 and DFX•12. The two DFX's are identical save for their number and type of input channels. The DFX•6 has four input channels—two mono and two stereo. The "six" in its model name comes from the total number of signals you can connect to its input channels, two stereo plus two mono sound sources.

The DFX•12 mixer adds two more mono mic/line inputs and two stereo line-input-only channels, for a total of 12 possible inputs—four mono and eight stereo. These inputs are spread across three slightly different types of input channel configurations: mono mic/line, stereo mic/line and stereo line-only. (Trivia: the only other Mackie with this many input channel configurations is the original CR-1604.)

Mono Mic/Line Inputs

The mono mic/line input channel is the meat-and-potatoes part of a live sound mixer. You'll find mono input channels in positions 1 and 2 on the DFX•6 and 1-4 on the DFX•12. As you've read throughout this book, I make a big deal of "signal flow," which is like a road map describing the path sounds take through your mixer. The front-panel view of a DFX mono channel is pictured at left. But how your signal gets from Point A to Point B is shown by the block diagram at the bottom of the opposite page. (Tip: The first few pages of Chapter 4 will help you understand the illustrations shown here.)

Although the mono mic/line channel has both a mic and a line input, you really should just use one or the other. Plugging both a mic and a line source into the same channel will be less than satisfactory, because the mixer offers no way to adjust levels of mic and line signals independently. This is because each channel's line and mic inputs run through the same Gain knob, as you can readily see in the diagram on the facing page. Speaking of which, this Gain knob isn't for adjusting the sound level for the audience. Instead, you should turn this knob up far enough so that the little "Level Set" LED starts blinking when you sing or play the connected instrument. Why? See page 28. (Note that most other Mackies call their Gain knobs "Trim.")

Next in line is the Insert point. You can use your DFX for years and never plug anything into this jack. But if you like, you can use the insert to connect external signal processing like compressors or equalizers to really tweak a particular channel's sound. Check out page 21 and Chapter 7 to learn more. Just remember that if you don't plug anything into the Insert, the signal automatically flows right around it to the next part of the input channel.

Following the Insert point is the Low Cut 75 Hz switch, which removes the lowest of the low frequencies from a given input channel. Depressing this button when you've connected a vocal mic to this channel will filter out low-frequency mud and stage rumble from your vocal sound. This is often the quickest way to fix a common live sound problem. See page 20 for more info.

And now, the Mute button. Pressing this interrupts the signal path, just as a dam in a river stops the flow of water. You can still have a mic or instrument banging away at the channel's input, but if the mute button is pressed, you'll never hear it. Using the mute button is a handy way to quickly silence a channel—for instance, muting before you unplug your guitar means the audience doesn't have to hear that buzzing squawk. More about muting on page 35. (The DFX's Break Switch is a handy way to mute *all* mic channels. More on it later in this chapter.)

Next, we come to the first split in our signal path. The signal is divided and sent down two different paths. One destination is the Aux 1 Monitor knob, but a copy of the signal also continues down the main path to the EQ section. Because the signals going to the Aux 1 monitor knob are taken from a point in your mixer before the equalizer, the Aux 1 mix will not be affected by channel EQ adjustments. In mixer-buzzword-speak, "the Aux 1/Monitor send is pre-EQ." In contrast, the Monitor signal comes after the Mute (post-mute!). This means Muting a channel also interrupts its Monitor send (more about this starting on page 84).

Next we come to that tone-shaping marvel, the EQ section. The DFX features high and low (i.e. treble and bass) knobs on each channel, plus a graphic EQ that can be applied overall to the entire sound mix. Leave Low and High in their 12 o'clock positions to begin with; they won't be adding or taking anything from your signal when pointing straight up. See "Understanding Equalization" on page 35 for more about EQ.

Getting from A to B: The Signal-Flow Block Diagram of a DFX-series mono mic/line input channel

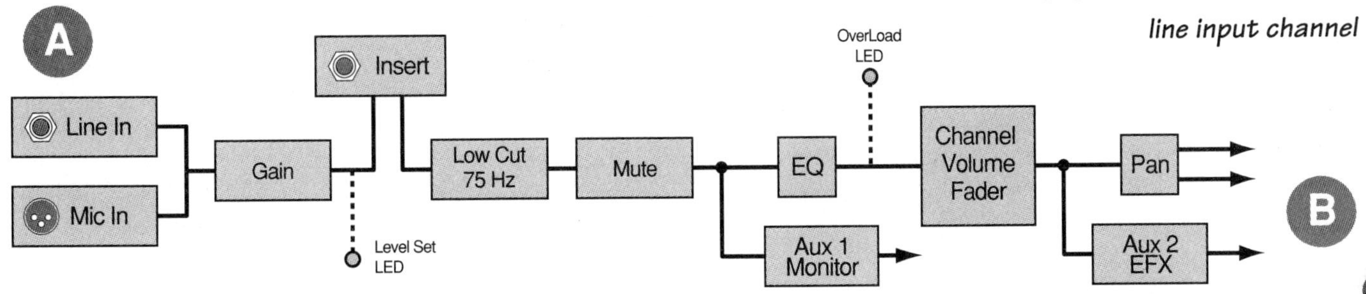

DFX-series

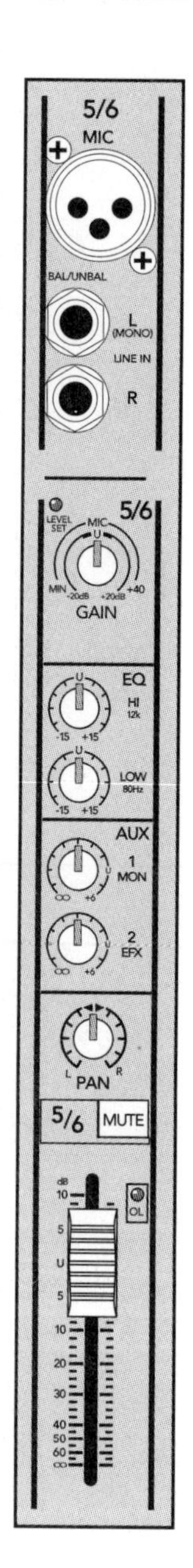

The Overload LED blinks if the signal levels in that channel are too high. It's post-EQ, because turning the Lows up a lot can push the channel into distortion. When this light blinks, turn the Gain knob down a bit. Why bother? See page 30.

Next up is the Channel Volume Fader, also known as the Big White Slider. Use this to set the relative volume of each channel to all the other sounds in your mix. You can make a channel louder by turning this slider up. But sometimes it's better to turn other channels down. Otherwise, you could find all the Channel Volume Faders at their maximums by the end of the show!

Following the Channel Volume Fader, the signal splits once again. First, a copy of this channel's sound can be sent through the Aux 2 EFX control (Auxes are also generically referred to as "sends"). Thanks to the DFX mixers' built-in effects, turning up the EFX knob on a channel is the first step in adding *echo* or *reverb* to that channel's sound. We say that "Aux 2 EFX is a post-fader send," which means that when the Channel Volume Fader is down, no signal reaches the EFX knob. Want to know more? Chapter 7 tells all.

Finally, our signal reaches the Pan control. Pan is a type of signal routing control. Check out the beginning of Chapter 5 to appreciate the full magnificence of Pan's role in the greater scheme of things. By default, leave this knob in the 12 o'clock position. Twisting it left or right "moves" the apparent position of this channel's sound back and forth in the stereo mix.

Mono Mic/Stereo Line Channels

DFX mixers also have two mono mic/stereo line input channels. On the DFX•6, these are the third and fourth faders (labeled 3/4 and 5/6). On the DFX•12 they are in the fifth and sixth fader positions, marked 5/6 and 7/8. This channel is quite similar to the mono mic/line input we just looked at. The only differences are:

- Two line inputs (Left and Right)
- No Low Cut switch
- No Insert point

A DFX-series stereo mic/line input channel and Block Diagram

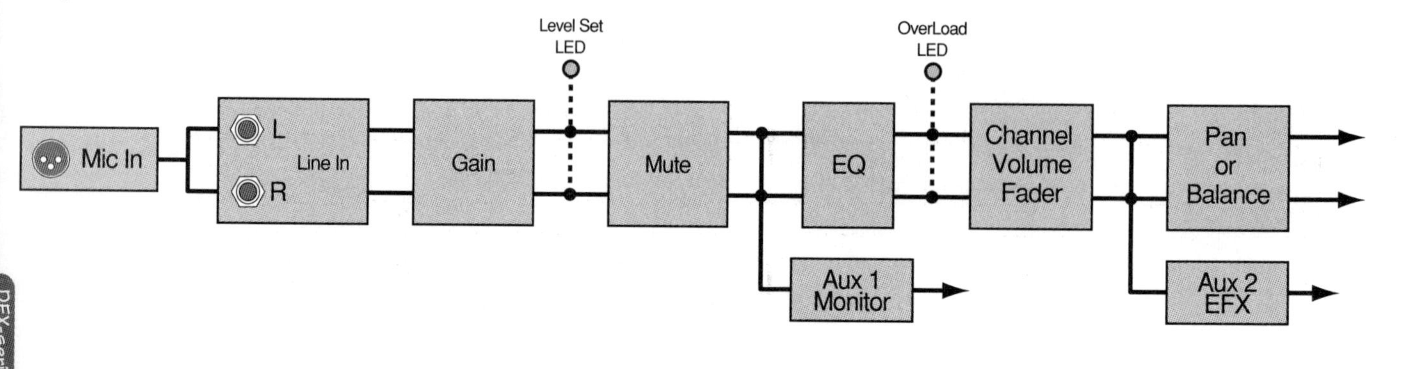

While this is a stereo channel, there is only one mic input. As you can see in the block diagram for this channel on the opposite page, any signal connected to the mic input immediately splits and is evenly balanced between left and right halves of the stereo channel.

The function of the Pan control is subtly different on a stereo channel. On a mono channel, Pan takes one signal and routes it in varying proportions between the channel's left and right output. On a stereo channel, it functions just like the "Balance" control on your home stereo. Turning it away from the 12 o'clock position changes the relative volume of the signals already present in the stereo channel's dual-signal path. Like I said, a rather subtle distinction. Just twist the knob and listen to the results. One additional item of note: The aux sends (Monitor and EFX) all tap off a mono mix of the channel's left and right channels. This means that your sends will always be mono, even if you connect a stereo input signal to a given channel.

While it's handy to have stereo channels on any mixer, especially for recording, I'm far more likely to stick to plain-old mono in live situations. This is because the geometry of a given room, the position of the loudspeakers and location of the audience all conspire to make most stereo effects a waste of time, except for the small percentage of the audience sitting *exactly* between your speakers. If you do run your PA in stereo, I suggest you avoid extreme panning of signals (see page 23) to insure a consistent sound mix for audience members on either side of the stage.

Stereo Line Input Channels (DFX•12 only)

The last two channels on the DFX•12 are stereo channels with line inputs, but no mono mic input. 'Nuff said.

A DFX-series stereo line-only input channel and corresponding signal flow

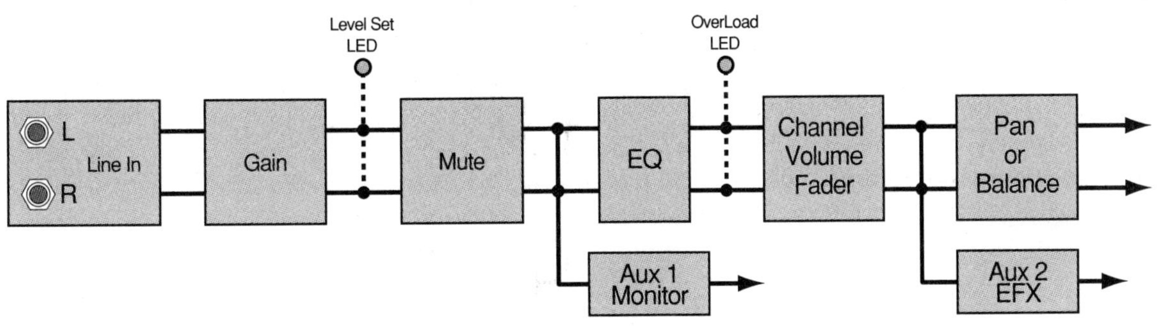

DFX-series Aux Send and Return controls. Note built-in EMAC effects processor and graphic EQ controls.

DFX Auxes

Auxes (short for auxiliary sends and returns) provide more signal routing and processing flexibility than you can shake a stick at. Before you dive into this section, be sure you've read Chapter 7 for crucial background info.

DFX-series mixers have a pair of Aux Sends. These Aux Send outputs (Monitor and EFX) are independent mixes of signals coming from any number of input channels. One of these sends (Monitor) is pre-fader, the other (EFX) is post-fader. This is smart, because there are two main ways to use auxes—stage monitoring or effects. A pre-fader send is better for stage mixes, and a post-fader send is more suitable for effects processing. Huzzah! Your DFX has one of each.

As you read in Chapter 7 (in which our hero narrowly escaped being eaten by crocodiles) Aux Sends are only half the picture. Your mixer also includes Aux Returns, which are basically stripped-down input channels. While most Mackies use little knobs to set their Aux Return levels, the DFX models have big generous faders for this task.

Aux Sends and Aux Returns work well together, but they are almost entirely independent. The one exception is the internal EMAC effects unit, which forms an internal connection between the sends and returns of your DFX. The Aux Send and Return block diagram for the DFX mixers is shown on the opposite page. While it may look a bit intimidating at first, it's actually not that difficult to sort out. Just remember that Sends are outputs and Returns are inputs.

Channel Aux Sends

Let's walk through the diagram, starting with letter A. For clarity, the components of a single mono input channel appear within the dashed rectangle (the rest of your mixer's input channels are not shown). Turning up a channel's Aux 1 Monitor knob routes some of that channel's signal to the Aux 1 bus. The most common application for this control is to create a stage monitor mix, which will help everyone on stage hear vocals or instruments clearly. Note the location of the Monitor send in the input channel—it comes before the EQ. This means EQ changes to individual channels will not affect your monitor mix.

Each channel has a second Aux send, Aux 2 EFX. Located after the EQ and the Channel Volume Fader, this knob routes part of each channel's signal to the internal EMAC effects unit. You can also use Aux 2 to send signals to an optional external effects unit (I'll tell you how later in this chapter).

Aux Send Masters

Now, looking outside the dashed box representing an individual channel, let's see where our Aux 1 and 2 signals go. The first stop for the Aux 1 bus (i.e. signals from all input channel Aux 1 sends) is the Aux 1 Monitor Send knob. If you're using Aux 1 to run stage monitors, this knob is a handy place to turn the overall monitor level up or down. Next, the signal may pass through the graphic equalizer (if the EQ's "AUX 1" button is pressed). From there, the monitor mix exits the mixer at letter B, the Aux 1 Monitor Send jack. To run a set of stage monitors, connect this output to a power amp and monitor speakers or a self-powered speaker cabinet.

DFX-series Aux Send and Return block diagram. Yeah, it looks complicated, but you'll get it.

Meanwhile, the combined mix of all channels' Aux 2/EFX sends merge together and pass through their own master volume control, the aptly named Aux 2/EFX Send Level knob (see letter D). Following this gain control, the signal splits in two

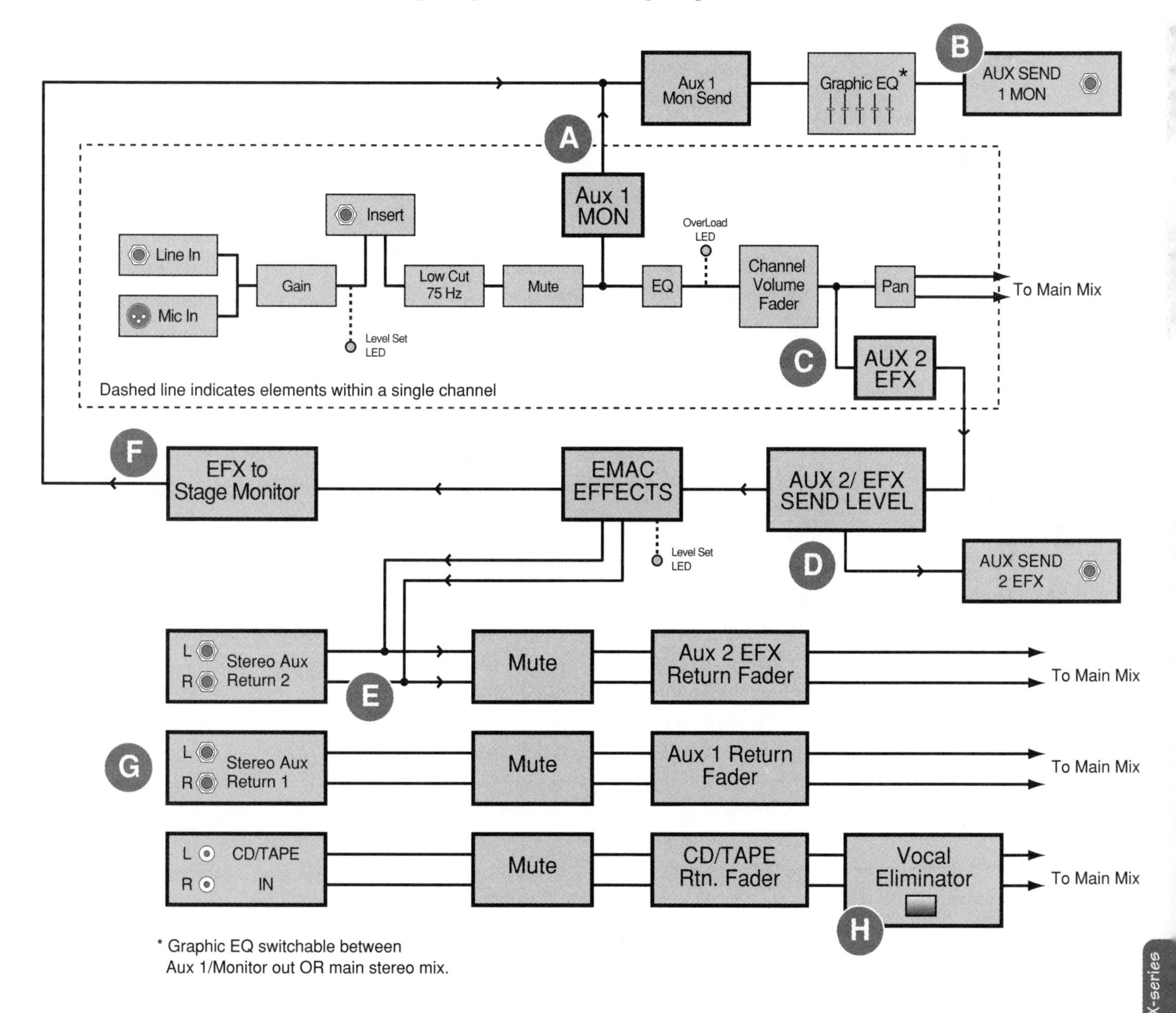

directions. Branching off to the middle-right of the diagram, the second aux mix exits the mixer at the Aux Send 2 EFX output jack. Use this connection only if you plan on connecting an external effects unit. If the internal EMAC effects meets your needs (and there's little reason why it shouldn't) you don't need to hook anything up to the Aux 2 Send output jack.

Internal EMAC Effects Routing

The other half of the signal split from the Aux 2/EFX Send Level goes to the internal EMAC effects. Keep an eye on the EMAC's Level Set LED. Adjust the Aux 2/EFX Send knob so this light blinks occasionally. If it's blinking all the time, turn it down.

Within the EMAC circuitry, your Aux 2 mix will be manipulated by the wondrous powers of digital signal processing. Twist the big preset knob and listen to the various results. What's that? You can't hear the effects? We'd better continue our way through the diagram so you can see how the effected signal gets back to the main stereo mix (and optionally, the stage monitor mix as well).

Exiting the EMAC are two versions of the effected signal. The first is a stereo-ized effect signal which is connected internally to the Aux 2 EFX Return signal path (letter E). This passes the internal effects through a Mute button and Aux 2 EFX Return fader before adding the effects to your main stereo mix. If you want the audience to hear the effects, be sure the Aux 2 Mute is not pressed, and the Aux 2 Return fader is turned up part-way.

OK, so the audience gets to hear the groovy effects. How about the band? A second, monophonic output signal exits the EMAC and feeds the EFX To Stage Monitor knob at letter F. Turning this knob up passes the effected signal up and around on the diagram to the Aux 1 Monitor Send control. Assuming your stage monitors are up and running, you'll hear the same effects the audience is enjoying. Singers always seem to love the sound of reverb or echo on their voices. Just be sure the effects are not so loud they mask musicians' intonation or timing problems!

Aux Returns

As we've mentioned before, Aux Returns are essentially stripped-down input channels. The DFX Series mixers have three such inputs: Aux 1, Aux 2 and CD/Tape Returns (see letter G). All three have Mute and Fader controls which regulate the amount of each Return's signal passed into the main stereo mix.

Aux 1 is the simplest of the three returns. It only works when an external sound source is connected to the Stereo Aux Return 1 jacks. If you're running out of input channels, use Aux 1 to connect line-level sources like keyboards or drum machines. If you're using a second mixer, you could connect its output to the Aux 1 return.

Remember that while Aux Send 1 and Aux Return 1 are both "1," they really have *nothing* else in common. For example, adjusting the Aux 1 return fader won't have any impact on your monitor levels—use the adjacent Aux 1 Monitor Send knob for that purpose.

On the other hand, the Aux 2 return fader *is* related to the Aux 2 send, thanks to the internal routing of the EMAC effects unit. Turn up the Aux 2 EFX Return fader to add effects into the main stereo mix. Of course, this will work only if you remember to turn up the Aux 2 Send in one or more channels, as well as the master Aux 2 EFX Send level control, so that signal gets to the internal effects in the first place. If you plug something into the Aux 2 return, it will be mixed with any effects signal coming out of the EMAC. If this isn't what you want, you can use an optional EFX bypass foot-switch to mute the EMAC output. Turning the Aux Send 2 master all the way down gives the same result.

The third Aux Return uses RCA-style input connectors (optional adapters can convert these to 1/4" connectors if necessary. You can hook up any line-level input to these jacks, but the most obvious choice would be a CD player. You could use this to play music during your band's break, or in case of Karaoke, use an external CD player to supply the backup music while you (or your loved ones) sing along. In that case, you might want to try using the Vocal Eliminator button, which does a reasonable job of removing most of the lead vocal from typical pop music CDs.

Connecting an External Effects Unit to your DFX

You can use an external effects unit with your DFX, but not without some head-scratching. Try this: feed the external effects unit's input from the DFX Aux Send 2 output jack. Patch the output of the external effect back into your DFX's Aux Return 1. To avoid hearing both internal and external effects at the same time, kill the EMAC output by muting Aux Return 2. Once you're happy with the sound of the external effect, you could optionally add the EMAC by un-muting the Aux Return 2 fader. This blending works best with complementary effects, such as reverb paired with an echo. Bear in mind that since both effects are being driven by the Aux 2 mix, there is no way to apply *only* echo to one channel and *only* reverb to another—everyone gets the same blend of effects. To overcome this, you'd have to use both your DFX sends, connecting Aux 1 to the external effect. Say bye-bye to yer monitor mix! Furthermore, this scheme creates another mixing problem described on page 87. Frankly, I don't recommend using Aux 1 send for effects.

DFX Inserts

Each of your DFX's mono input channels has its own insert connector (see Chapter 7 for background). Inserts let you connect an external device, such as a compressor, into the signal path of an individual input. DFX stereo inputs don't have inserts; instead, that front panel space is occupied by individual left and right line input connectors.

DFX mono channels have insert points. Use them to apply optional external signal-processing to individual channels.

From a signal flow perspective, DFX inserts come after the Gain control, but before the low-cut switch. This type of insert is called *pre-EQ*, which means that twiddling a channel's level and EQ controls won't affect the signal going out your mixer's insert, regardless of whether you are using it in the *insert, direct out,* or *split* hook-up configuration (Huh? See page 100). Remember: inserts are cool, but you don't have to use 'em.

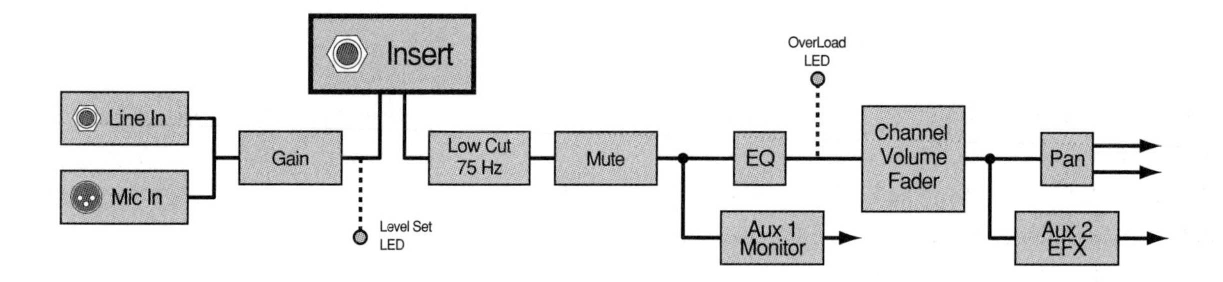

DFX Output Section

The DFX output controls, as seen from an altitude of 5 to 7 feet.

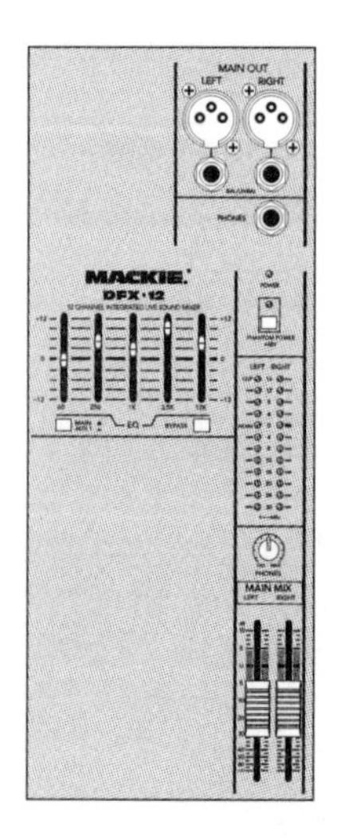

We're closing in on the end of our trip through your mixer. Last stop—the Output Section. Here's where all the signals finally end up at your mixer's main outputs, en route to the ears of your audience. Compared to the relative complexity of the Aux block diagram in the last section, understanding the output section will be a breeze. But don't forget to buzz through Chapter 9 for all the background info you need to use your mixer's output section to its fullest potential.

The Signals at the output section come from input channels and aux returns. But due to the convenient Mackie "Break Switch," they arrive via two slightly different paths. The Break Switch, as you've probably already guessed, is sort of a master mute. It's the button you press when the band takes "a fifteen minute break." Since the statistical average length of a fifteen minute break is closer to 23 minutes, it's often a nice idea to leave the audience some music to listen to while the band is goofing off. Pressing the Break Switch turns off the mics, but leaves on the CD/Tape in (DFX•6 and 12) and additionally, the stereo line inputs of the DFX•12. This explains the two pairs of stereo main mixes shown arriving at letter A and B. The stuff coming down the "A" train gets cut when you push the Break Switch, those sounds on the "B" bus are not affected.

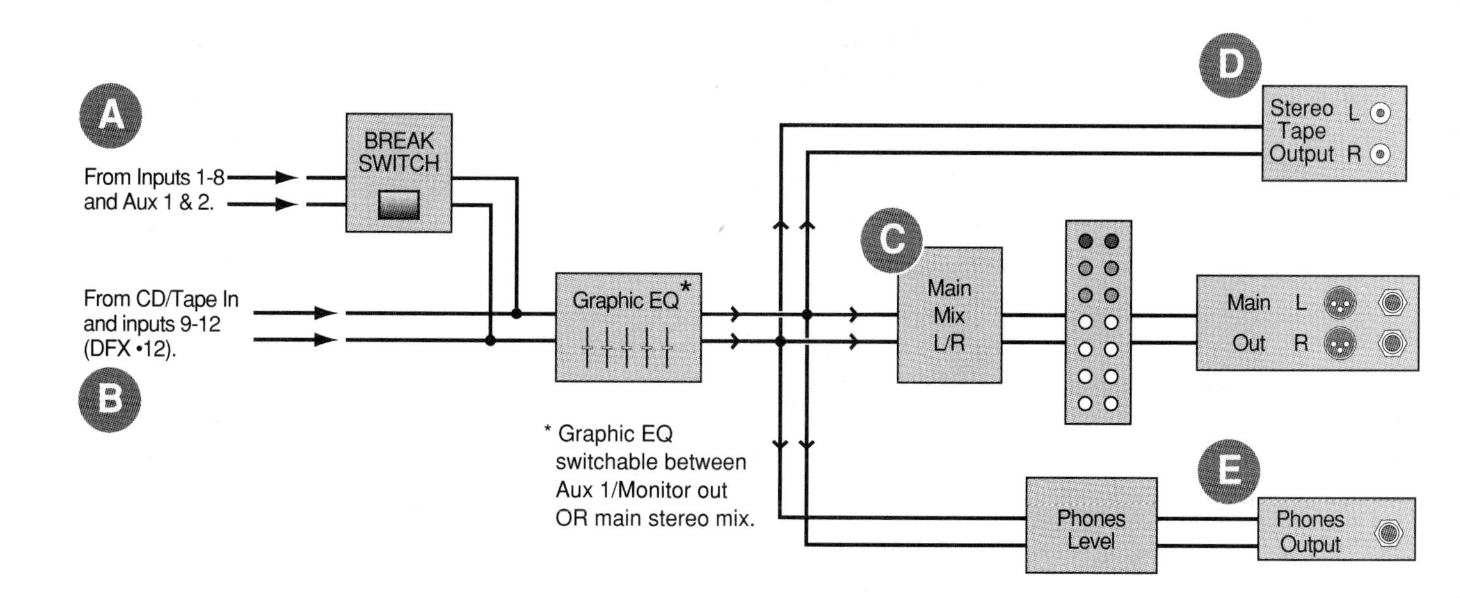

The output section of the DFX mixer shown as a signal-flow block diagram.

Next up is the Graphic Equalizer. This is applied to the main stereo mix if its Main/ Aux 1 switch is *not* pressed. Use the five band graphic for broad tonal shaping and/ or feedback suppression (see page 136). Bear in mind that effective feedback suppression often requires more than five bands (15 or even 31 bands are sometimes used), but the built-in DFX graphic is a good place to start.

Following the Graphic EQ, the signal splits in no less than three different directions. The most important is shown in the center, letter C. The signal first passes through the Main Mix left and right faders. Following this, the LED level meters do their blinky dance. From there, the signal reaches its final destination, the main outputs. Hook up the XLR or 1/4" outputs as needed. If you're using your DFX to run a live PA system, you'll connect a power amplifier or powered speakers to this point.

Main mix split number two goes to a pair of RCA jacks (see letter D). As the label suggests, this is a handy place to connect a tape recorder, recordable MiniDisc or even the line inputs of a laptop (or desktop) computer. Since this version of the main mix is tapped off upstream of the Main Mix faders, adjusting those sliders won't affect the levels on your recorder.

The third and final destination for the main mix is the headphone output, which as you can see, has its own little volume control. Slip on a pair of headphones and you can check that all your mics and instruments are working even if the Main Mix faders are turned down. Could be handy if you're playing a wedding and the ceremony is in progress at the other end of the hall.

So there you have it—your DFX mixer in a nutshell. Remember that finding an effective solution to any mixer or sound system problem usually hinges on your understanding of the underlying signal flow of your system. Refer to these pages, or the full-blown Mackie signal-flow diagram reproduced at the end of this book.

CHAPTER 15

Under The Hood

This chapter includes copies of the signal flow diagrams of each mixer described in this book. You'll find the same information in the back of your Mackie manual. Thanks to Mackie Designs for the use of these drawings and the mixer front panel artwork used throughout this book.

MS1202

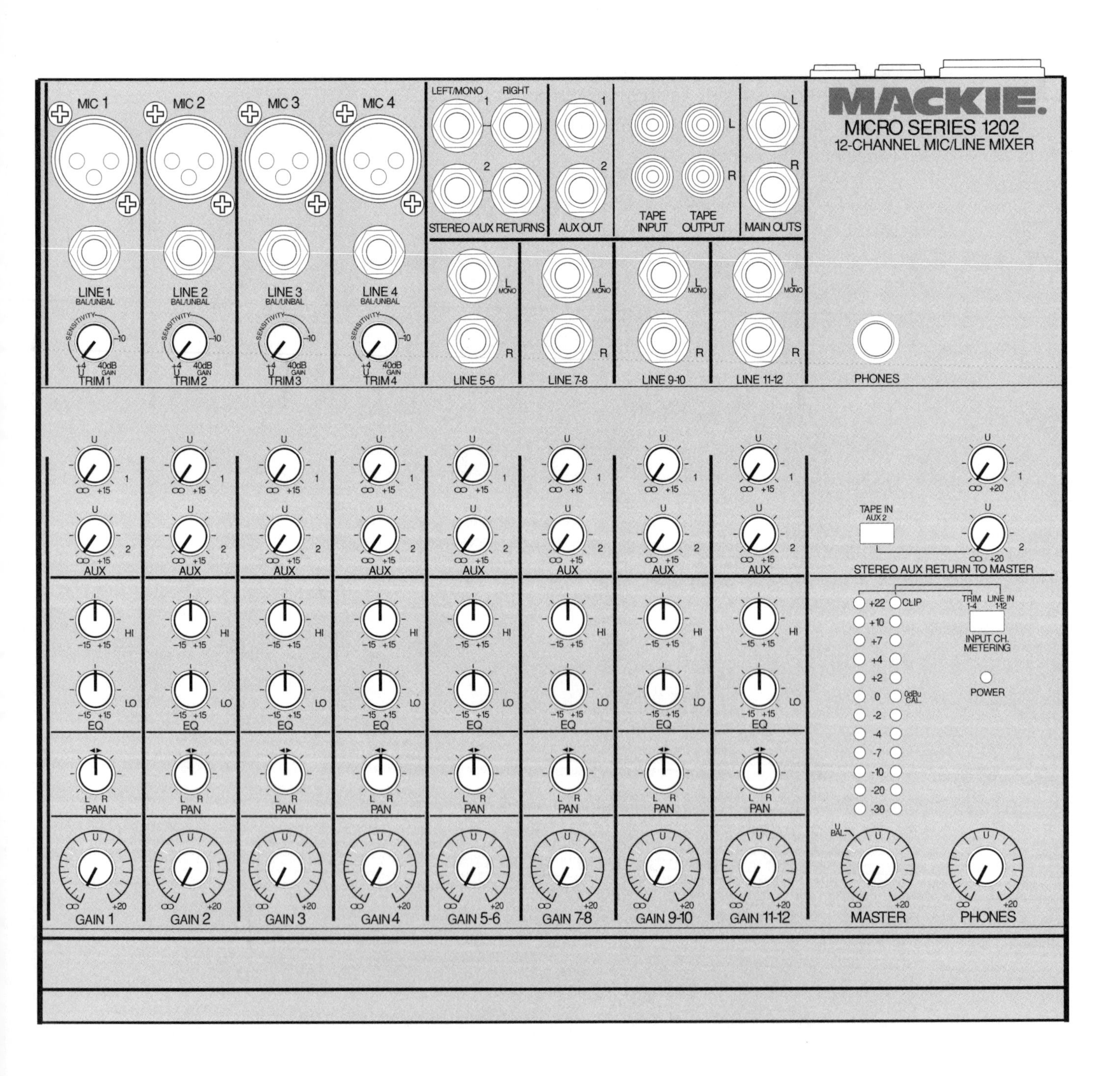
MACKIE.
MICRO SERIES 1202
12-CHANNEL MIC/LINE MIXER
MIC 1
MIC 2
MIC 3
MIC 4
LINE 1 BAL/UNBAL
LINE 2 BAL/UNBAL
LINE 3 BAL/UNBAL
LINE 4 BAL/UNBAL
SENSITIVITY
+4
-10
U
40dB GAIN
TRIM 1
TRIM 2
TRIM 3
TRIM 4
LEFT/MONO
RIGHT
STEREO AUX RETURNS
AUX OUT
TAPE INPUT
TAPE OUTPUT
MAIN OUTS
L
R
MONO
LINE 5-6
LINE 7-8
LINE 9-10
LINE 11-12
PHONES
AUX
HI
LO
EQ
PAN
TAPE IN AUX 2
STEREO AUX RETURN TO MASTER
CLIP
+22
+10
+7
+4
+2
0
-2
-4
-7
-10
-20
-30
0dBu CAL.
TRIM 1-4
LINE IN 1-12
INPUT CH. METERING
POWER
U BAL.
GAIN 1
GAIN 2
GAIN 3
GAIN 4
GAIN 5-6
GAIN 7-8
GAIN 9-10
GAIN 11-12
MASTER
PHONES

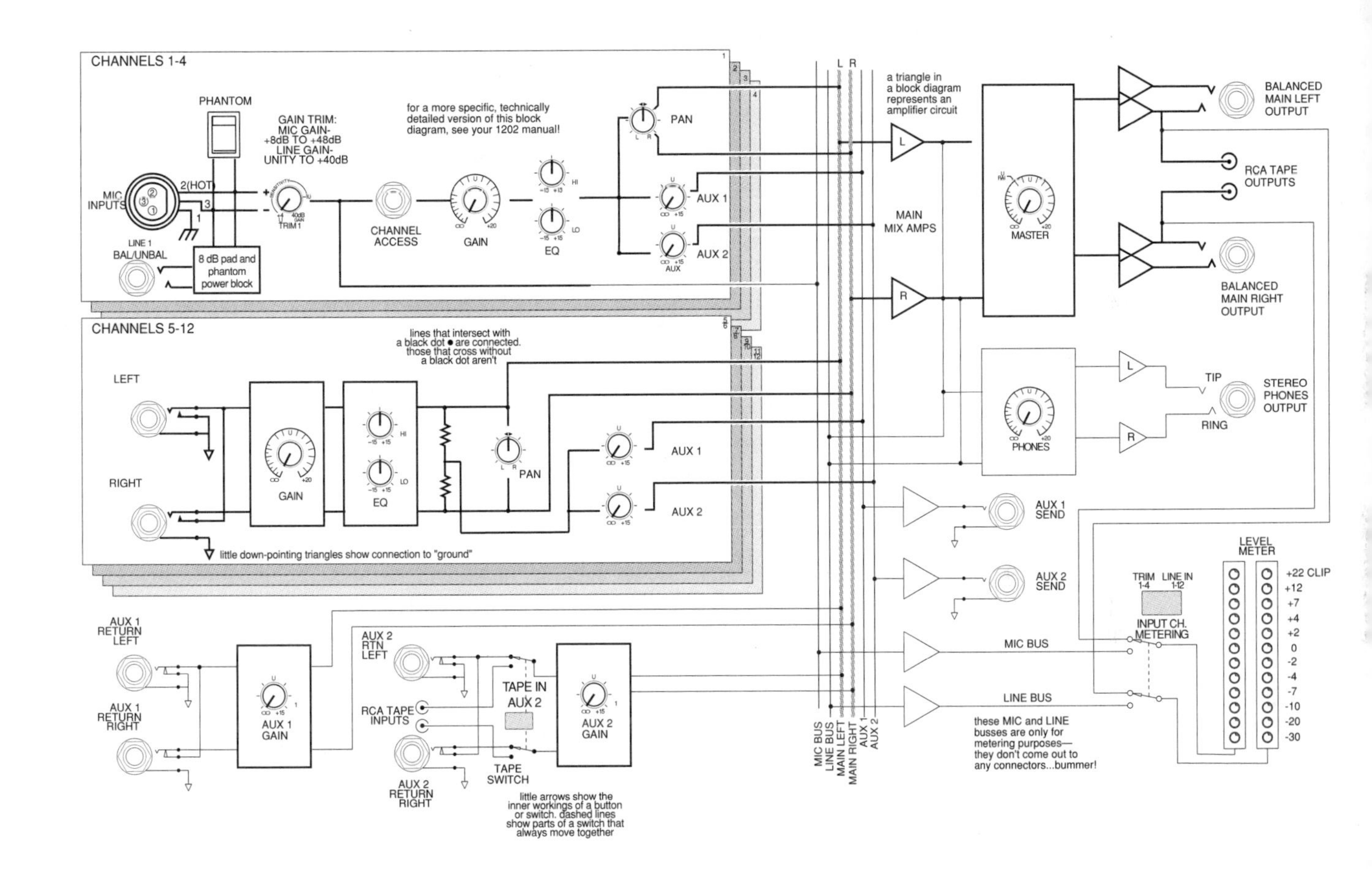
CHANNELS 1-4
PHANTOM
GAIN TRIM:
MIC GAIN-
+8dB TO +48dB
LINE GAIN-
UNITY TO +40dB
for a more specific, technically detailed version of this block diagram, see your 1202 manual!
MIC INPUTS
LINE 1 BAL/UNBAL
8 dB pad and phantom power block
CHANNEL ACCESS
GAIN
EQ
PAN
AUX 1
AUX 2
CHANNELS 5-12
lines that intersect with a black dot ● are connected. those that cross without a black dot aren't
LEFT
RIGHT
little down-pointing triangles show connection to "ground"
a triangle in a block diagram represents an amplifier circuit
MAIN MIX AMPS
MASTER
BALANCED MAIN LEFT OUTPUT
RCA TAPE OUTPUTS
BALANCED MAIN RIGHT OUTPUT
PHONES
TIP
RING
STEREO PHONES OUTPUT
AUX 1 SEND
AUX 2 SEND
LEVEL METER
TRIM 1-4
LINE IN 1-12
INPUT CH. METERING
MIC BUS
LINE BUS
these MIC and LINE busses are only for metering purposes— they don't come out to any connectors...bummer!
+22 CLIP
+12
+7
+4
+2
0
-2
-4
-7
-10
-20
-30
AUX 1 RETURN LEFT
AUX 1 RETURN RIGHT
AUX 1 GAIN
AUX 2 RTN LEFT
RCA TAPE INPUTS
AUX 2 RETURN RIGHT
TAPE IN AUX 2
TAPE SWITCH
AUX 2 GAIN
little arrows show the inner workings of a button or switch. dashed lines show parts of a switch that always move together
MIC BUS
LINE BUS
MAIN LEFT
MAIN RIGHT
AUX 1
AUX 2

1202-VLZ and VLZ PRO

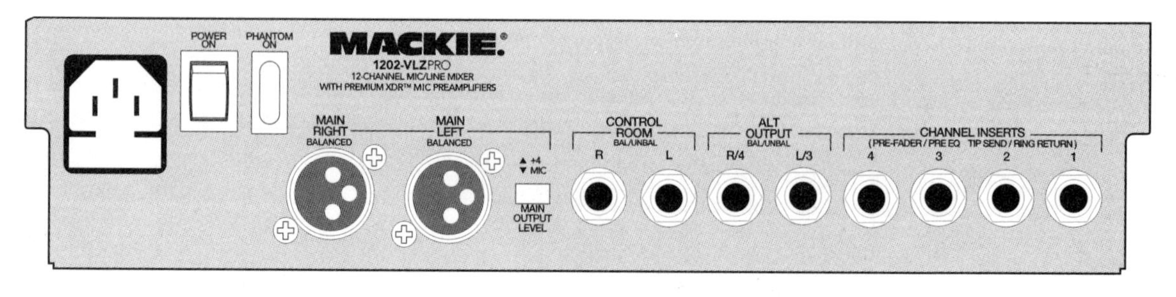

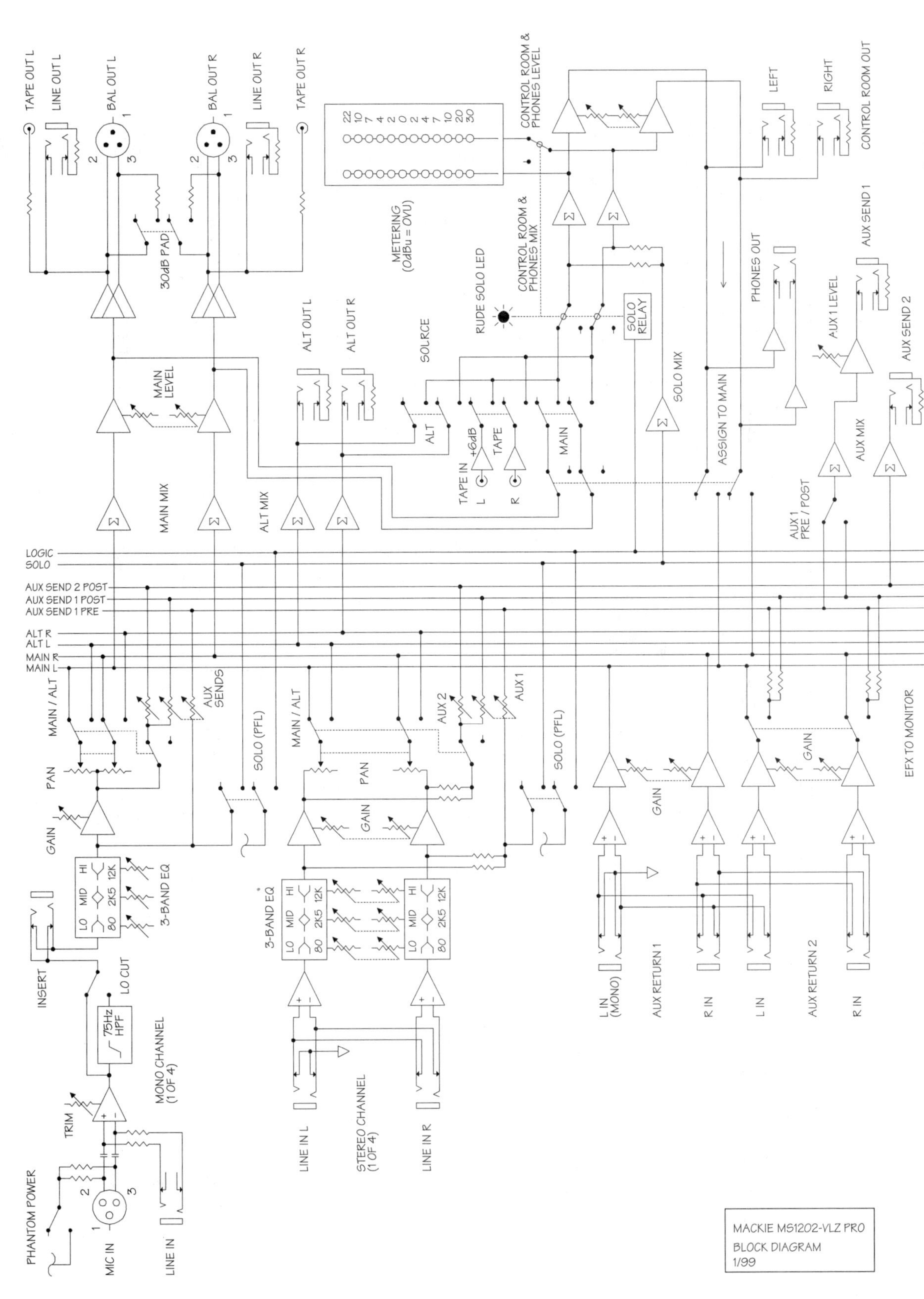
TAPE OUT L
LINE OUT L
BAL OUT L
BAL OUT R
LINE OUT R
TAPE OUT R
30dB PAD
MAIN LEVEL
MAIN MIX
ALT MIX
ALT OUT L
ALT OUT R
SOURCE
ALT
TAPE
+6dB
TAPE IN
MAIN
METERING (0dBu = 0VU)
22 10 7 4 2 0 2 4 7 10 20 30
CONTROL ROOM & PHONES LEVEL
CONTROL ROOM & PHONES MIX
RUDE SOLO LED
SOLO RELAY
SOLO MIX
ASSIGN TO MAIN
PHONES OUT
LEFT
RIGHT
CONTROL ROOM OUT
AUX SEND 1
AUX SEND 2
AUX 1 LEVEL
AUX MIX
AUX 1 PRE / POST
LOGIC
SOLO
AUX SEND 2 POST
AUX SEND 1 POST
AUX SEND 1 PRE
ALT R
ALT L
MAIN R
MAIN L
MAIN / ALT
PAN
GAIN
AUX SENDS
SOLO (PFL)
AUX 2
AUX 1
3-BAND EQ
LO MID HI
80 2K5 12K
INSERT
LO CUT
75Hz HPF
MONO CHANNEL (1 OF 4)
TRIM
PHANTOM POWER
MIC IN
LINE IN
LINE IN L
STEREO CHANNEL (1 OF 4)
LINE IN R
L IN (MONO)
AUX RETURN 1
R IN
L IN
AUX RETURN 2
EFX TO MONITOR
MACKIE MS1202-VLZ PRO
BLOCK DIAGRAM
1/99

1402-VLZ and VLZ PRO

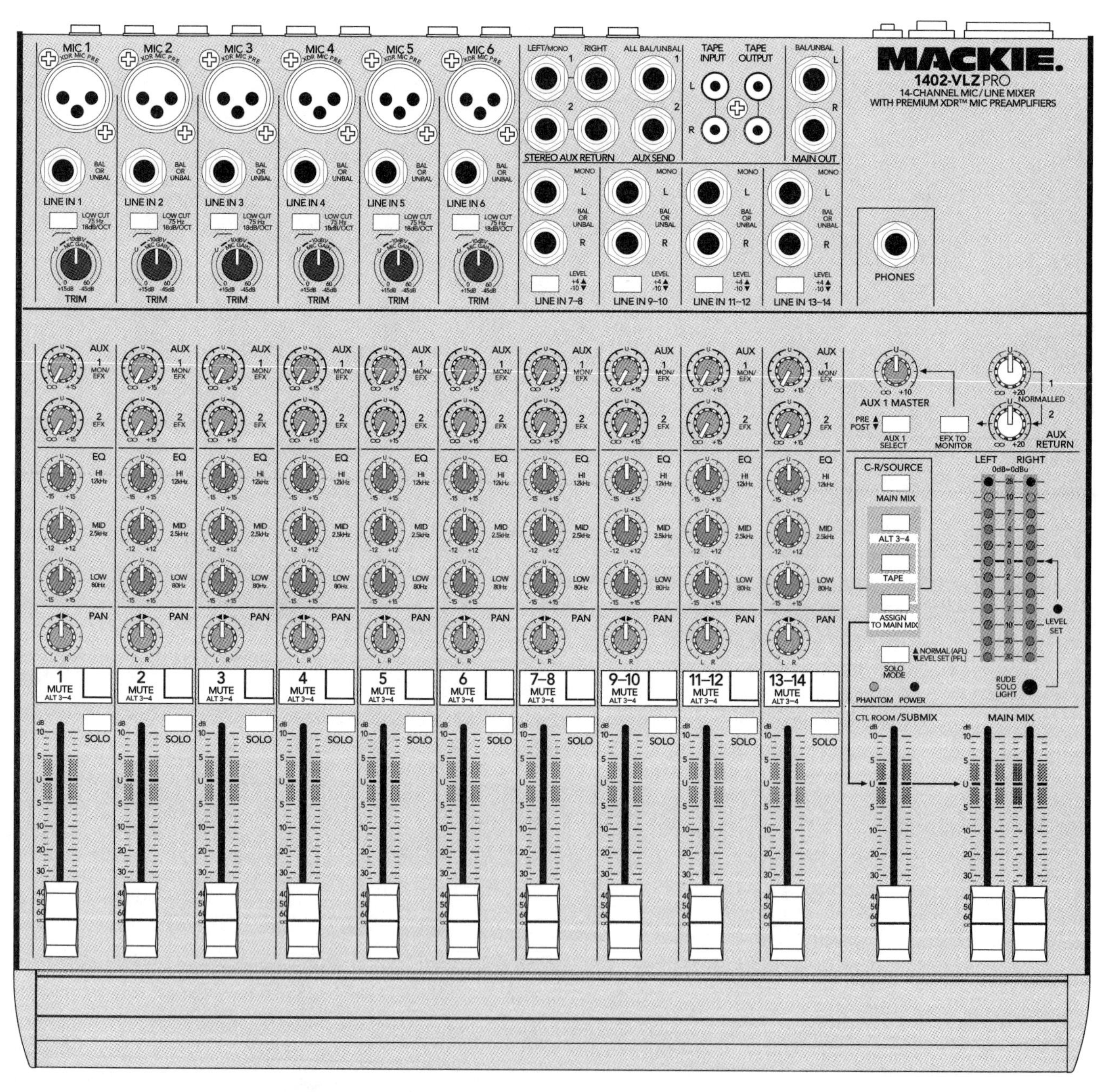

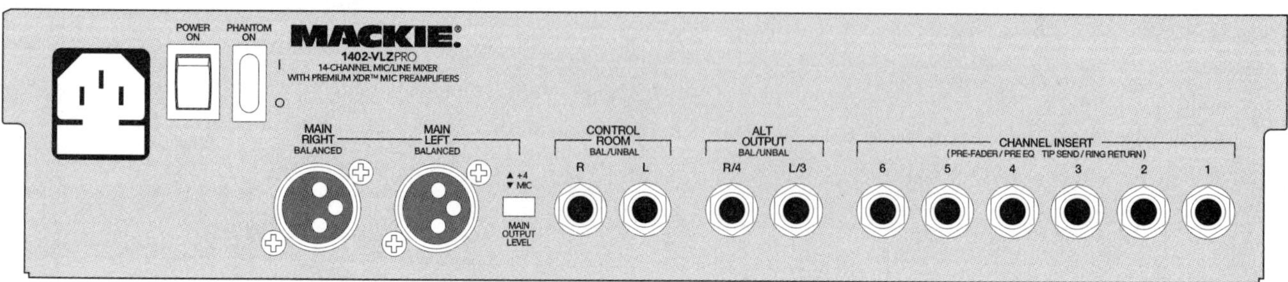

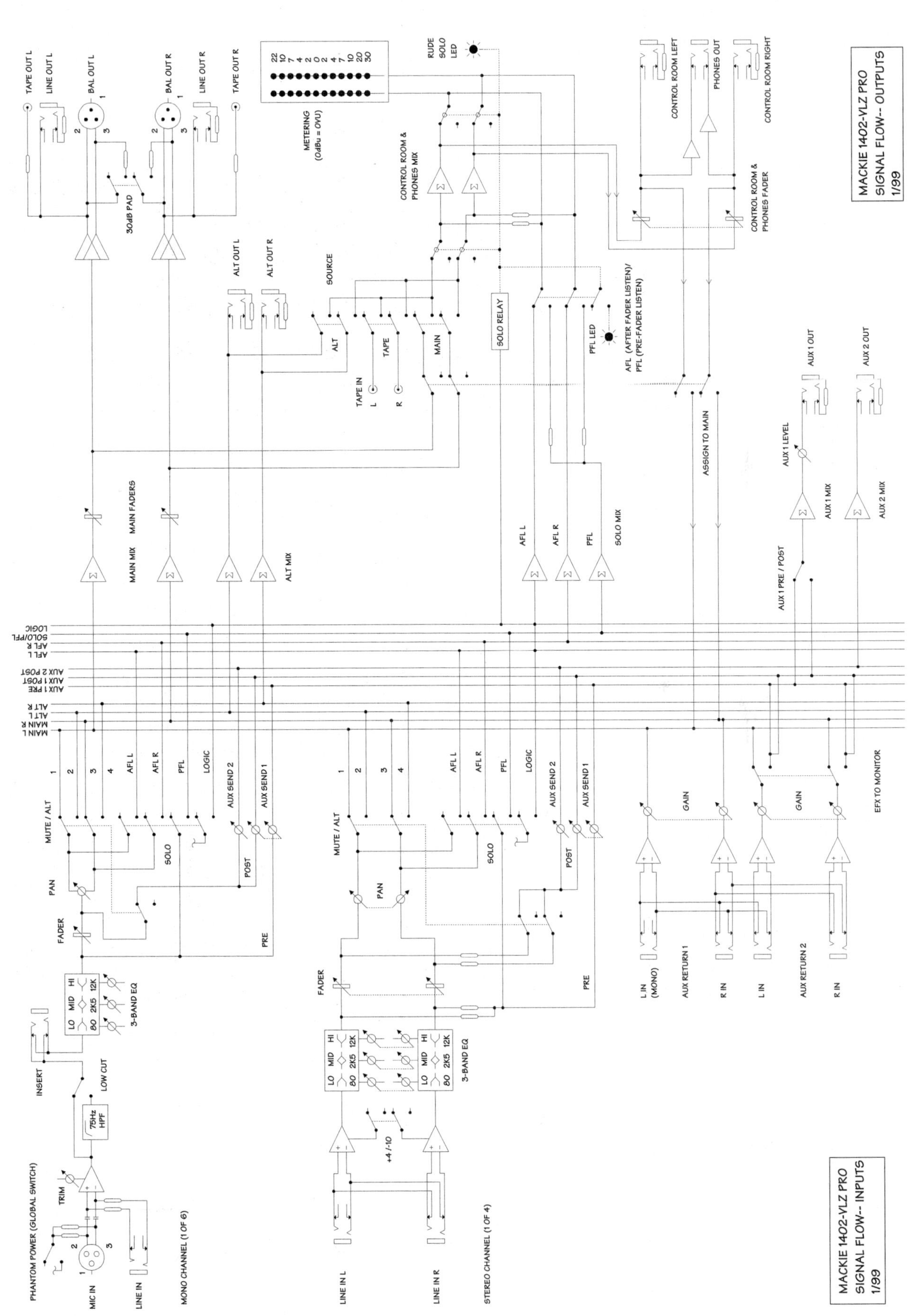
MACKIE 1402-VLZ PRO
SIGNAL FLOW-- OUTPUTS
1/99
MACKIE 1402-VLZ PRO
SIGNAL FLOW-- INPUTS
1/99
TAPE OUT L
LINE OUT L
BAL OUT L
BAL OUT R
LINE OUT R
TAPE OUT R
30dB PAD
METERING
(0dBu = 0VU)
RUDE SOLO LED
CONTROL ROOM & PHONES MIX
CONTROL ROOM LEFT
PHONES OUT
CONTROL ROOM RIGHT
CONTROL ROOM & PHONES FADER
ALT OUT L
ALT OUT R
SOURCE
ALT
TAPE
MAIN
TAPE IN
SOLO RELAY
PFL LED
AFL (AFTER FADER LISTEN)/
PFL (PRE-FADER LISTEN)
ASSIGN TO MAIN
AUX 1 OUT
AUX 2 OUT
AUX 1 LEVEL
AUX 1 MIX
AUX 2 MIX
AUX 1 PRE / POST
MAIN FADERS
MAIN MIX
ALT MIX
AFL L
AFL R
PFL
SOLO MIX
LOGIC
SOLO/PFL
AUX 2 POST
AUX 1 POST
AUX 1 PRE
ALT R
ALT L
MAIN R
MAIN L
MUTE / ALT
AUX SEND 2
AUX SEND 1
SOLO
POST
PRE
PAN
FADER
3-BAND EQ
LO MID HI
80 2K5 12K
INSERT
LOW CUT
75Hz HPF
TRIM
PHANTOM POWER (GLOBAL SWITCH)
MIC IN
LINE IN
MONO CHANNEL (1 OF 6)
+4 / -10
LINE IN L
LINE IN R
STEREO CHANNEL (1 OF 4)
GAIN
L IN (MONO)
AUX RETURN 1
R IN
L IN
AUX RETURN 2
EFX TO MONITOR

1402-VLZ

CR-1604

STEREO AUX RETURNS
LEVEL
BALANCE
MONO
AUX SOLO
MACKIE.
CR1604
16 CHANNEL MIC/LINE MIXER
ALT PREVIEW
MAIN OUTPUT MUTE
SOLO TO MAIN
POWER
SOLO LEVEL
1/LEFT
2/RIGHT
LEVEL
SOLO/PHONES
PHANTOM +48V
MAINS
AUX
MON
5/6 SHIFT
EQ
PAN
SOLO
MUTE
ALT 3/4

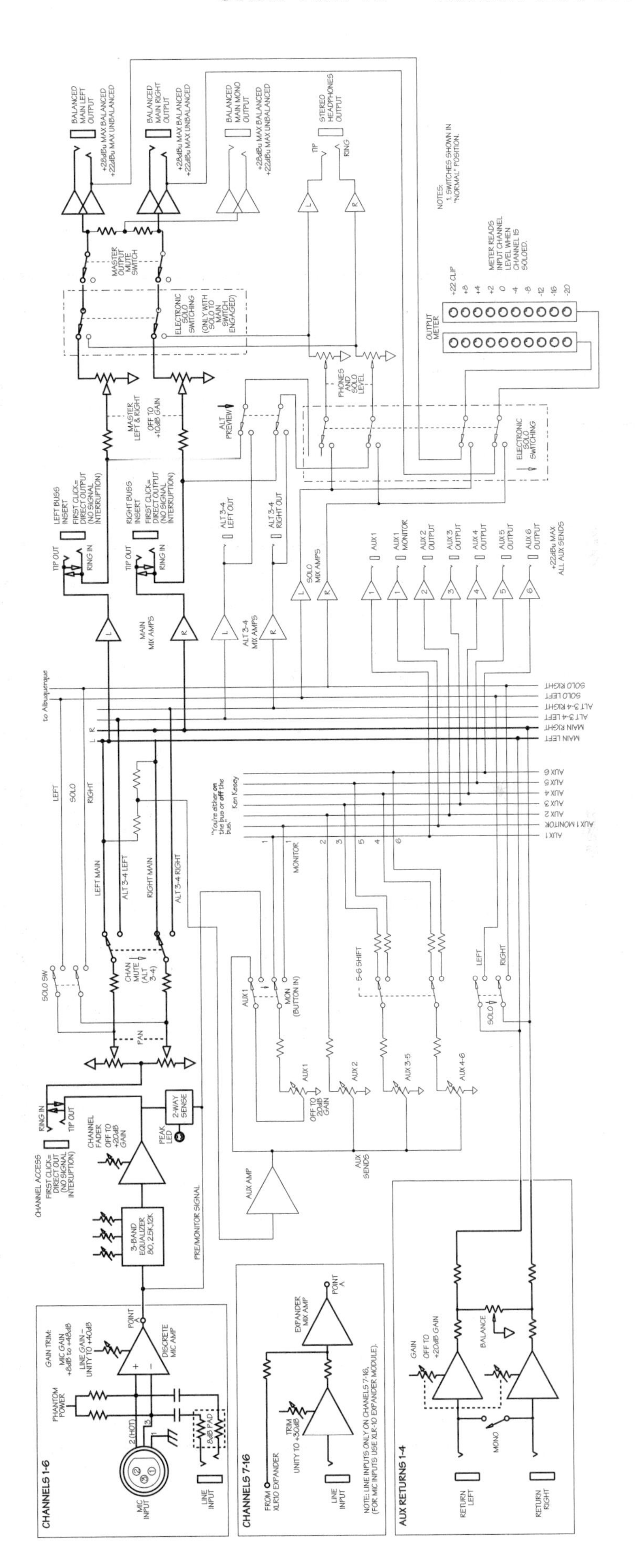

1604-VLZ, VLZ PRO

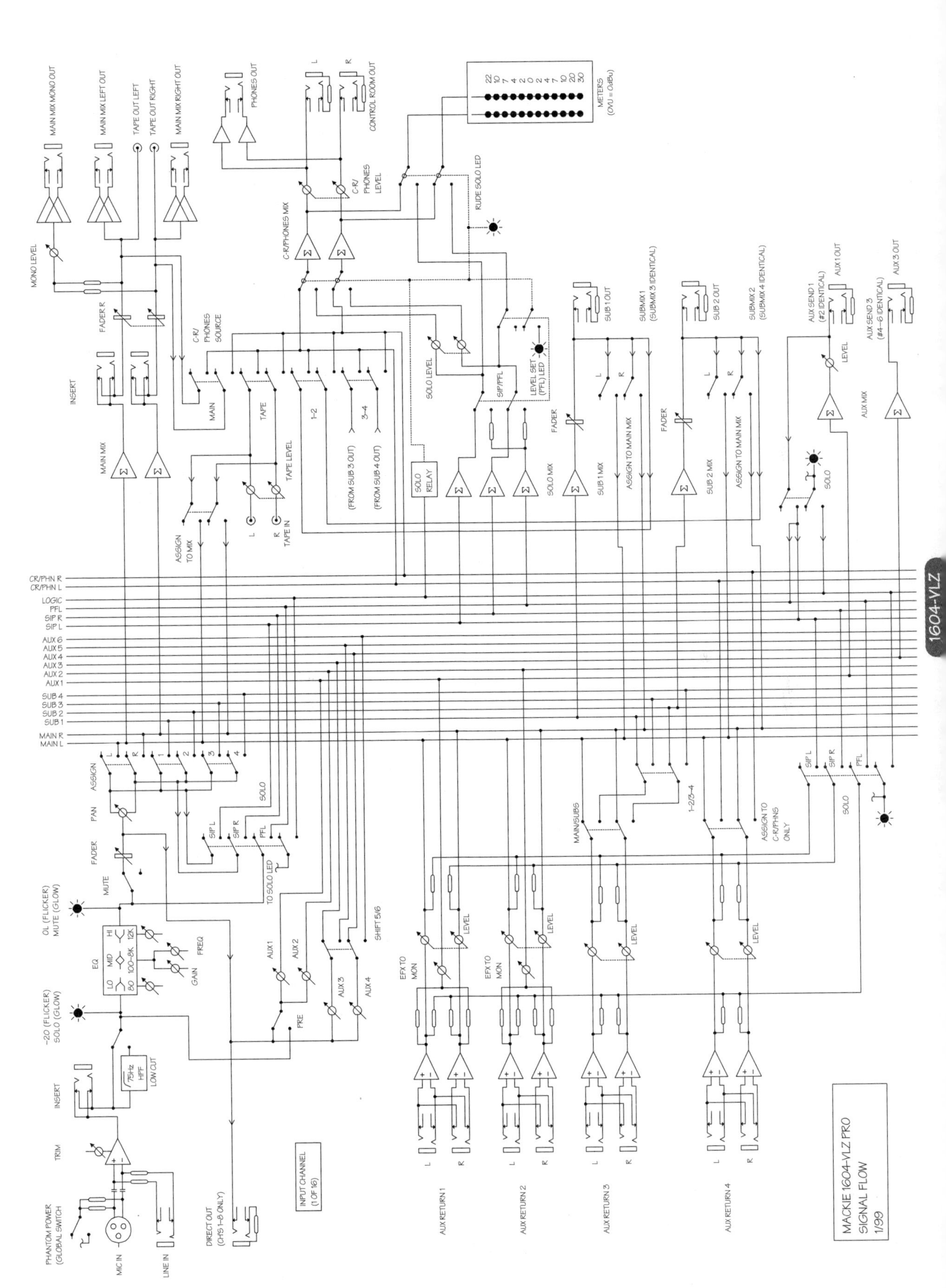
MACKIE 1604-VLZ PRO
SIGNAL FLOW
1/99
INPUT CHANNEL
(1 OF 16)
MAIN MIX MONO OUT
MAIN MIX LEFT OUT
TAPE OUT LEFT
TAPE OUT RIGHT
MAIN MIX RIGHT OUT
PHONES OUT
CONTROL ROOM OUT
METERS
(0VU = 0dBu)
AUX RETURN 1
AUX RETURN 2
AUX RETURN 3
AUX RETURN 4
MIC IN
LINE IN
DIRECT OUT
(CHS 1-8 ONLY)
PHANTOM POWER
(GLOBAL SWITCH)

1642-VLZ PRO

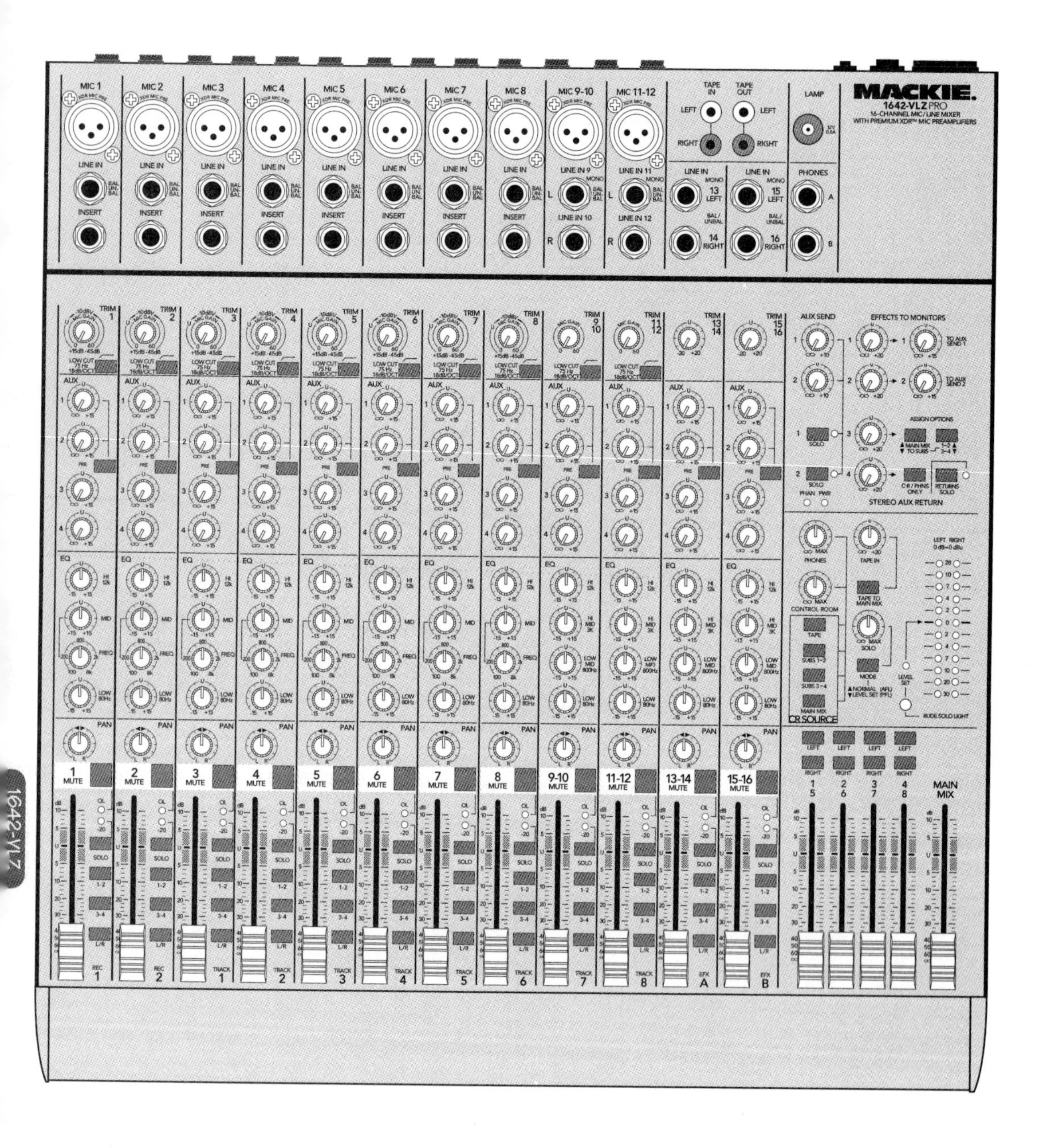
MIC 1
MIC 2
MIC 3
MIC 4
MIC 5
MIC 6
MIC 7
MIC 8
MIC 9-10
MIC 11-12
TAPE IN
TAPE OUT
LAMP
MACKIE.
1642-VLZ PRO
16-CHANNEL MIC/LINE MIXER
WITH PREMIUM XDR™ MIC PREAMPLIFIERS
LINE IN
INSERT
PHONES
TRIM
LOW CUT
AUX
EQ
PAN
MUTE
AUX SEND
EFFECTS TO MONITORS
ASSIGN OPTIONS
STEREO AUX RETURN
CONTROL ROOM
TAPE IN
TAPE TO MAIN MIX
CR SOURCE
MAIN MIX
RUDE SOLO LIGHT
MAIN MIX

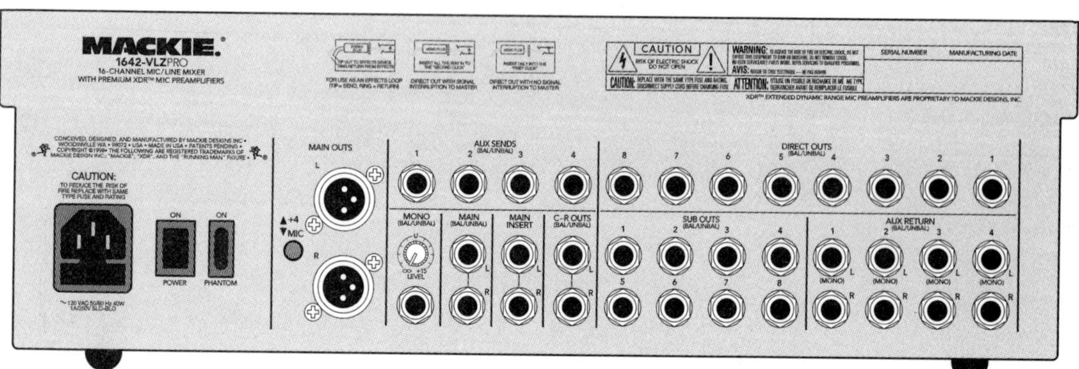
MACKIE.
1642-VLZ PRO
16-CHANNEL MIC/LINE MIXER
WITH PREMIUM XDR™ MIC PREAMPLIFIERS
CAUTION
SERIAL NUMBER
MANUFACTURING DATE
MAIN OUTS
AUX SENDS
DIRECT OUTS
MONO
MAIN
MAIN INSERT
C-R OUTS
SUB OUTS
AUX RETURN
POWER
PHANTOM

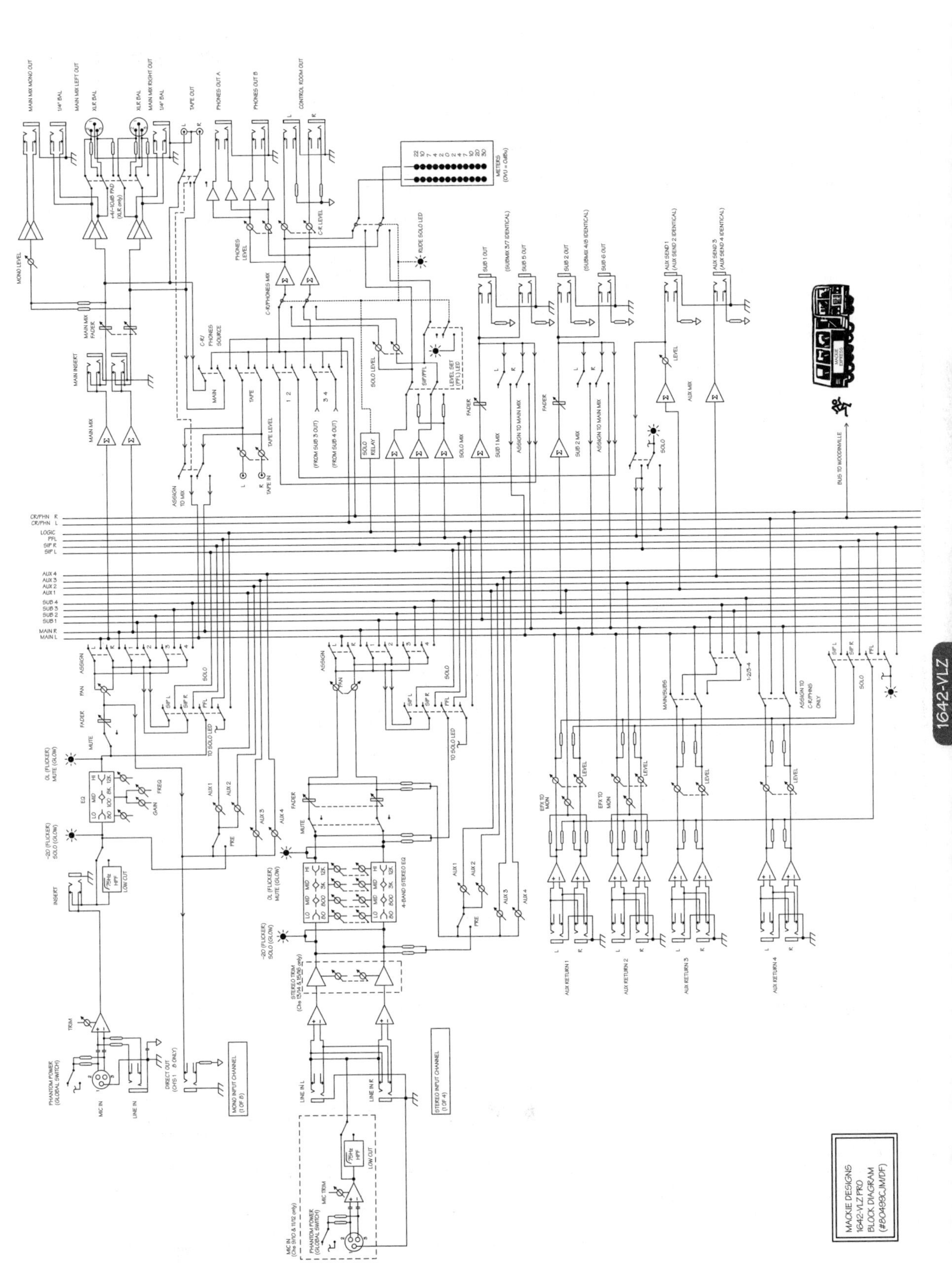
MACKIE DESIGNS
1642-VLZ PRO
BLOCK DIAGRAM

SR24•4, SR32•4

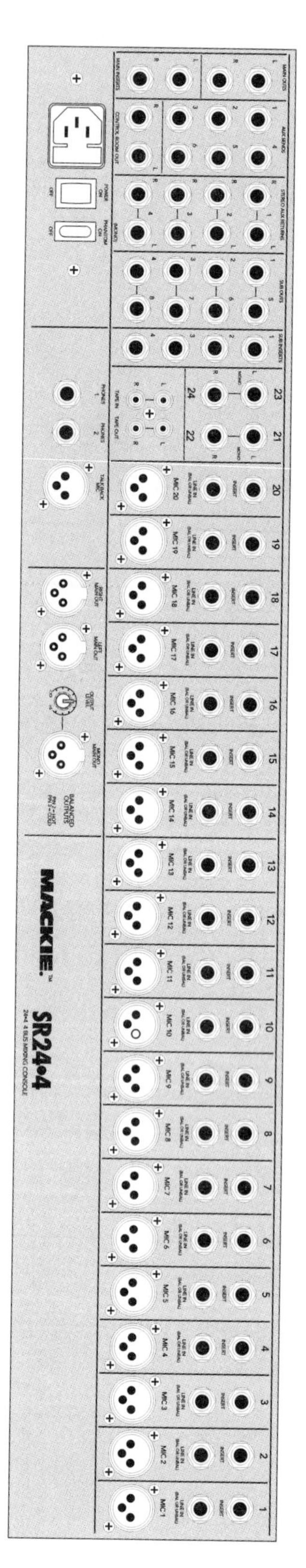

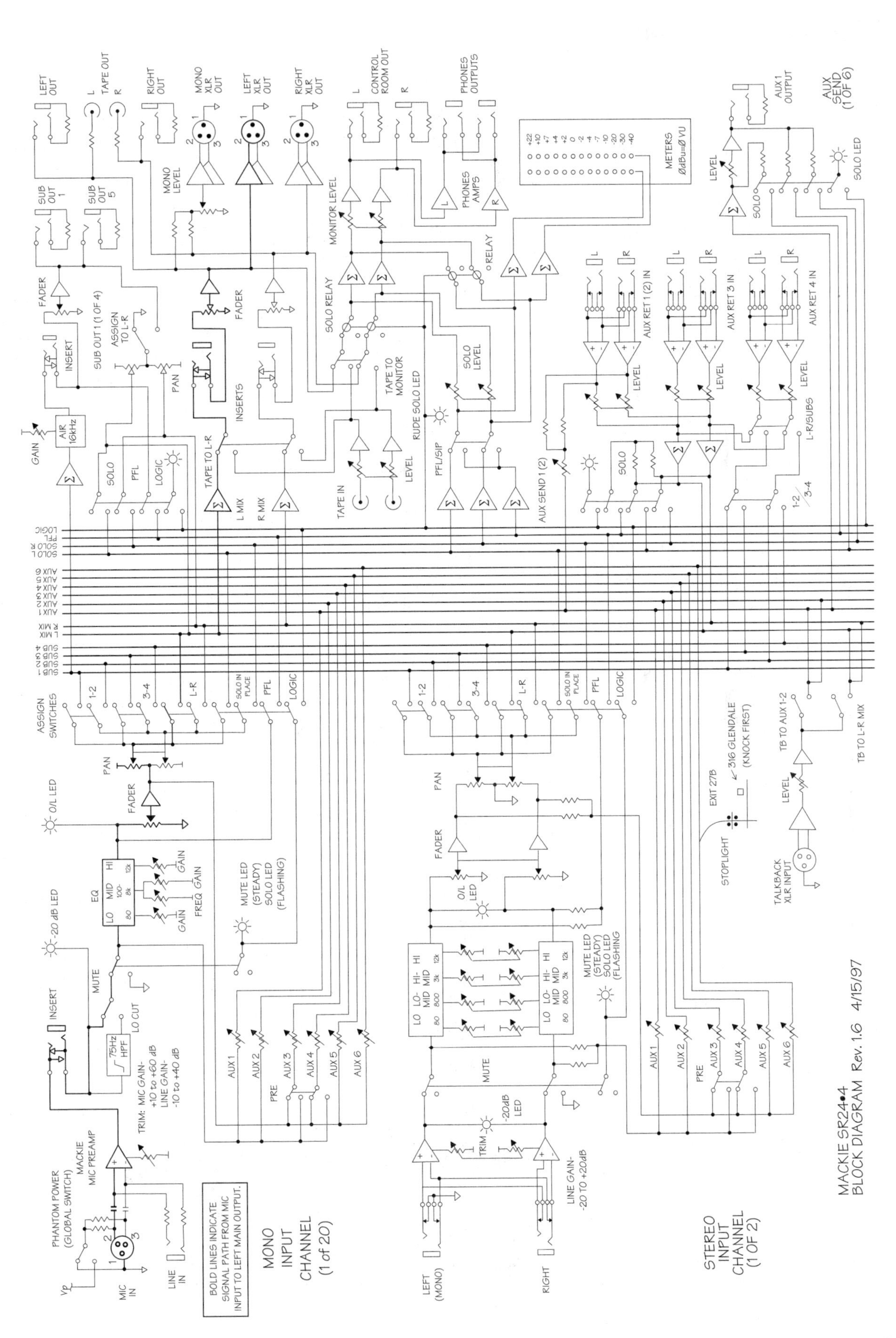
MACKIE SR24•4
BLOCK DIAGRAM Rev. 1.6 4/15/97
MONO INPUT CHANNEL (1 of 20)
STEREO INPUT CHANNEL (1 OF 2)
BOLD LINES INDICATE SIGNAL PATH FROM MIC INPUT TO LEFT MAIN OUTPUT.
SUB OUT 1 (1 OF 4)
AUX SEND (1 OF 6)
METERS Ø dBu=Ø VU
PHANTOM POWER (GLOBAL SWITCH)
MACKIE MIC PREAMP
TALKBACK XLR INPUT
STOPLIGHT
EXIT 27B
316 GLENDALE (KNOCK FIRST)
SR-series

PPM Series (mono)

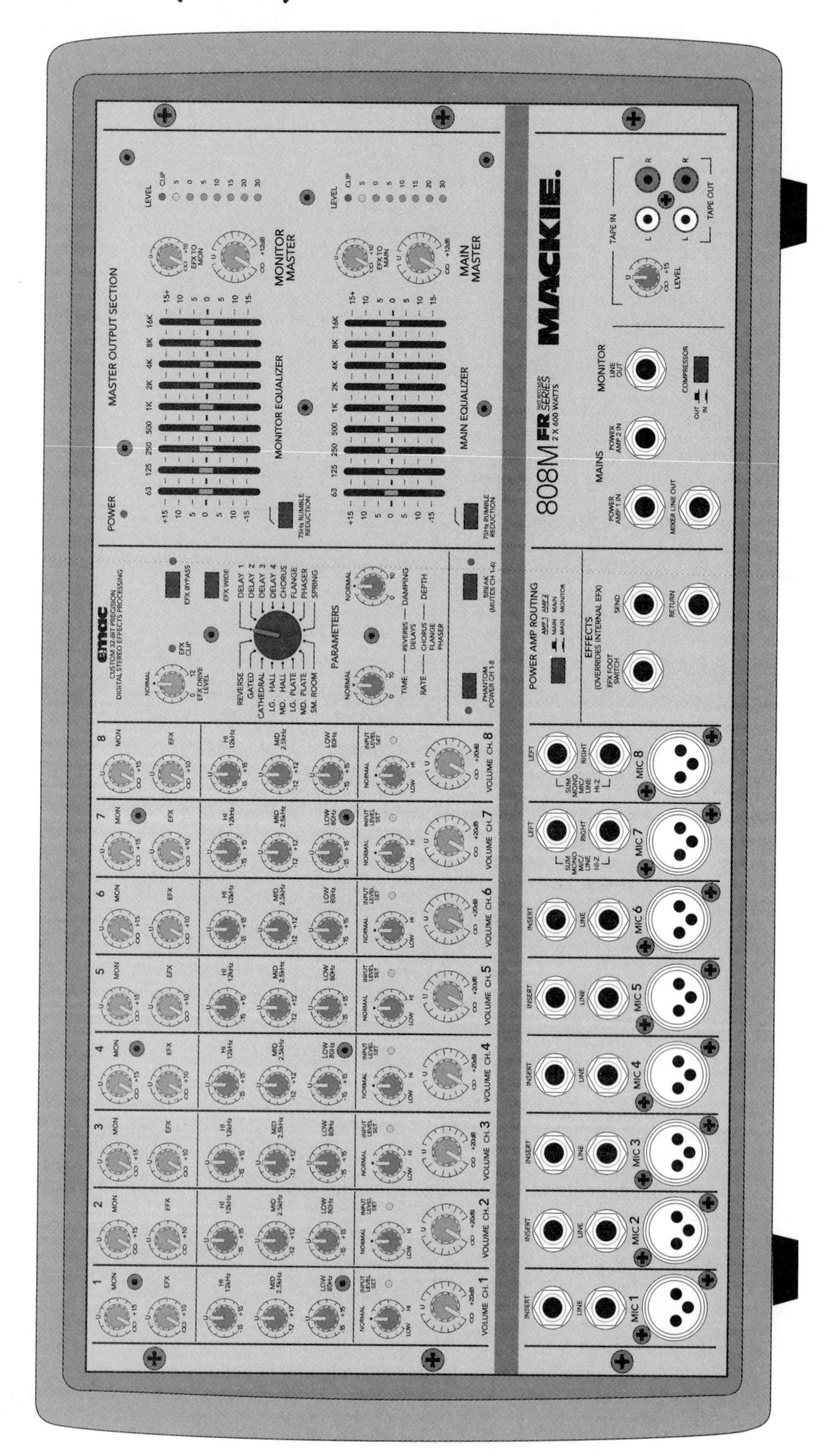

PPM-series

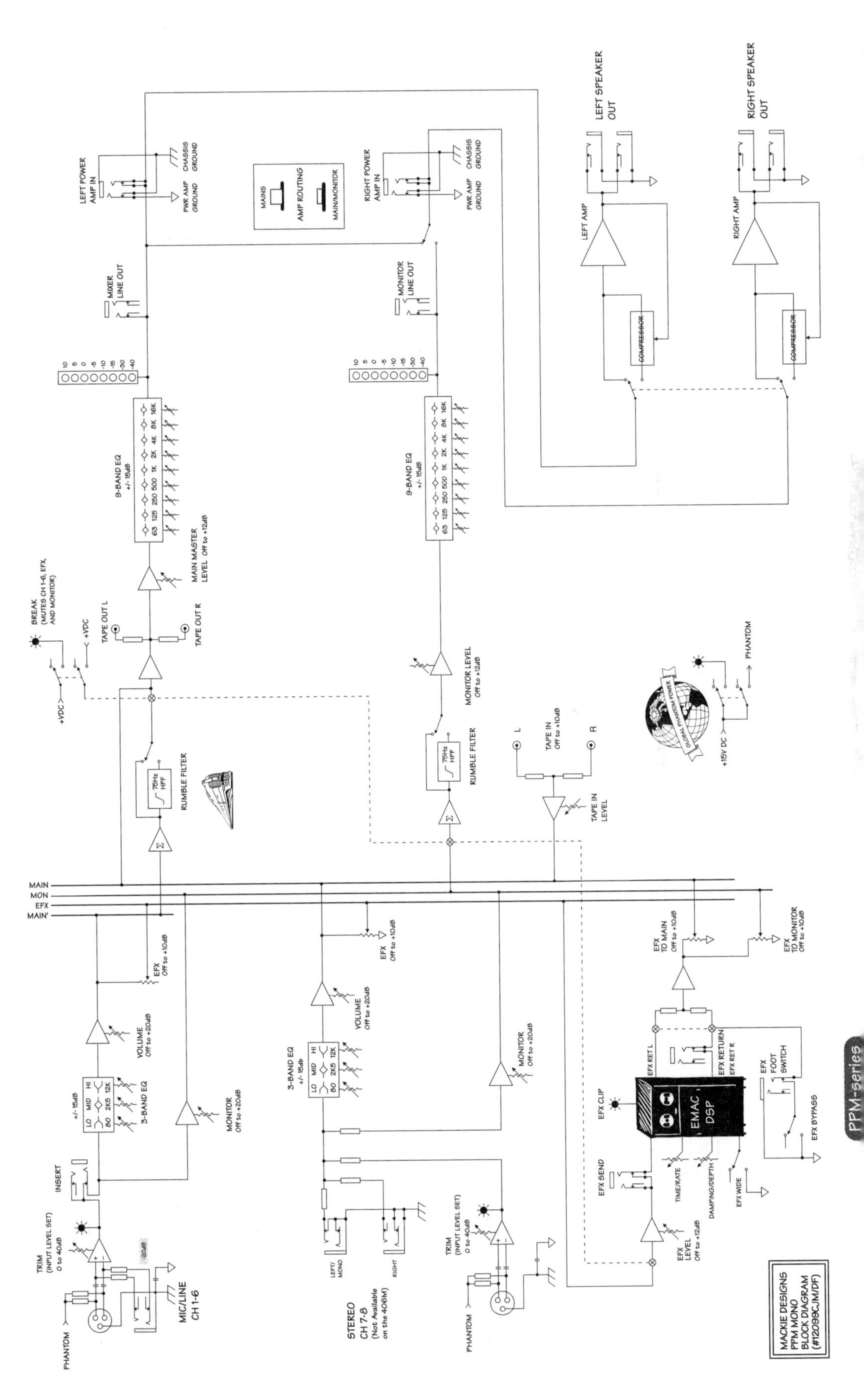
LEFT SPEAKER OUT
RIGHT SPEAKER OUT
LEFT POWER AMP IN
PWR AMP GROUND
CHASSIS GROUND
MAINS
AMP ROUTING
MAIN/MONITOR
RIGHT POWER AMP IN
LEFT AMP
RIGHT AMP
COMPRESSOR
MIXER LINE OUT
MONITOR LINE OUT
9-BAND EQ +/- 15dB
63 125 250 500 1K 2K 4K 8K 16K
MAIN MASTER LEVEL Off to +12dB
BREAK (MUTES CH 1-6, EFX, AND MONITOR)
TAPE OUT L
TAPE OUT R
+VDC
MONITOR LEVEL Off to +12dB
RUMBLE FILTER
75Hz HPF
TAPE IN Off to +10dB
TAPE IN LEVEL
PHANTOM
+15V DC
GLOBAL PHANTOM POWER
MAIN
MON
EFX
MAIN'
EFX Off to +10dB
VOLUME Off to +20dB
3-BAND EQ +/- 15dB
LO MID HI
80 2K5 12K
MONITOR Off to +20dB
INSERT
TRIM (INPUT LEVEL SET) 0 to 40dB
-20dB
MIC/LINE CH 1-6
STEREO CH 7-8 (Not Available on the 406M)
LEFT/ MONO
RIGHT
EFX TO MAIN Off to +10dB
EFX TO MONITOR Off to +10dB
EFX RET L
EFX RETURN
EFX RET R
EFX FOOT SWITCH
EFX CLIP
EMAC DSP
EFX BYPASS
EFX SEND
TIME/RATE
DAMPING/DEPTH
EFX WIDE
EFX LEVEL Off to +12dB
MACKIE DESIGNS
PPM MONO
BLOCK DIAGRAM
(#12099CJM/DF)

PPM-series

PPM Series (stereo)

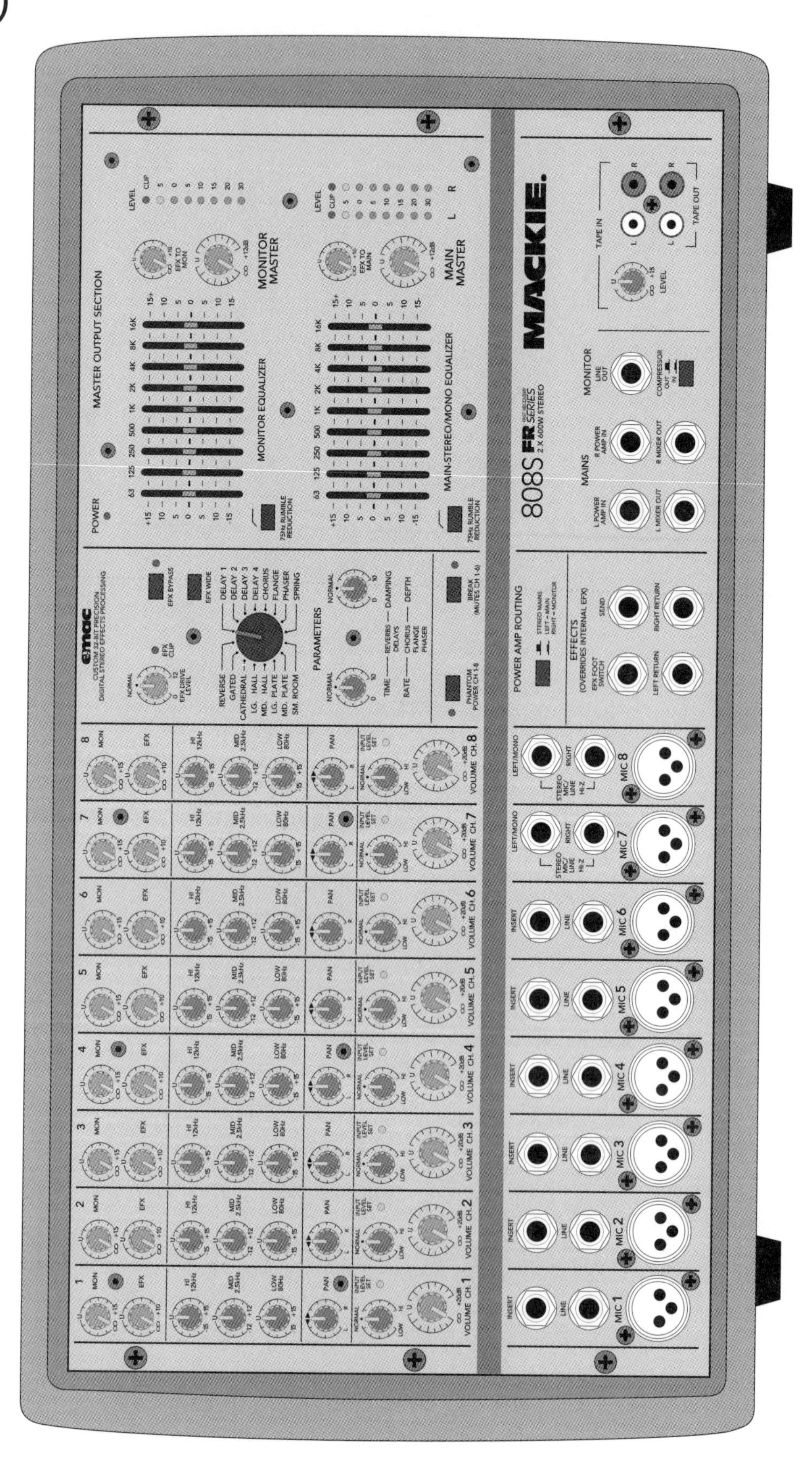

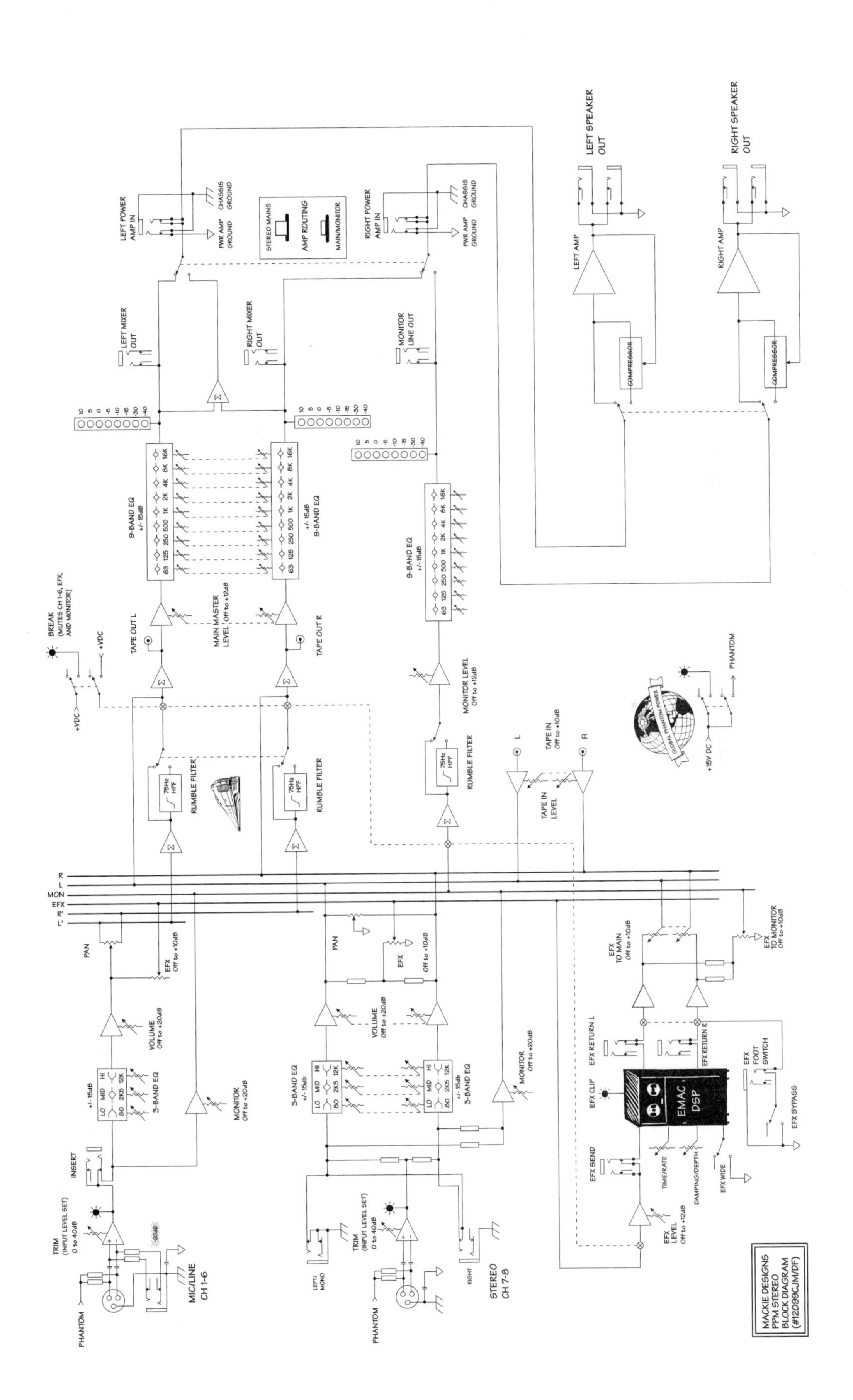
MACKIE DESIGNS
PPM STEREO
BLOCK DIAGRAM
(#12099CJM/DF)
MIC/LINE
CH 1-6
STEREO
CH 7-8
LEFT SPEAKER
OUT
RIGHT SPEAKER
OUT
LEFT AMP
RIGHT AMP
COMPRESSOR
LEFT POWER
AMP IN
RIGHT POWER
AMP IN
LEFT MIXER
OUT
RIGHT MIXER
OUT
MONITOR
LINE OUT
9-BAND EQ
+/- 15dB
3-BAND EQ
RUMBLE FILTER
TAPE OUT L
TAPE OUT R
MAIN MASTER
LEVEL Off to +12dB
MONITOR LEVEL
Off to +12dB
TAPE IN
LEVEL
EFX SEND
EFX RETURN L
EFX RETURN R
EFX CLIP
EFX BYPASS
EMAC
DSP
PHANTOM
+15V DC

CFX Series

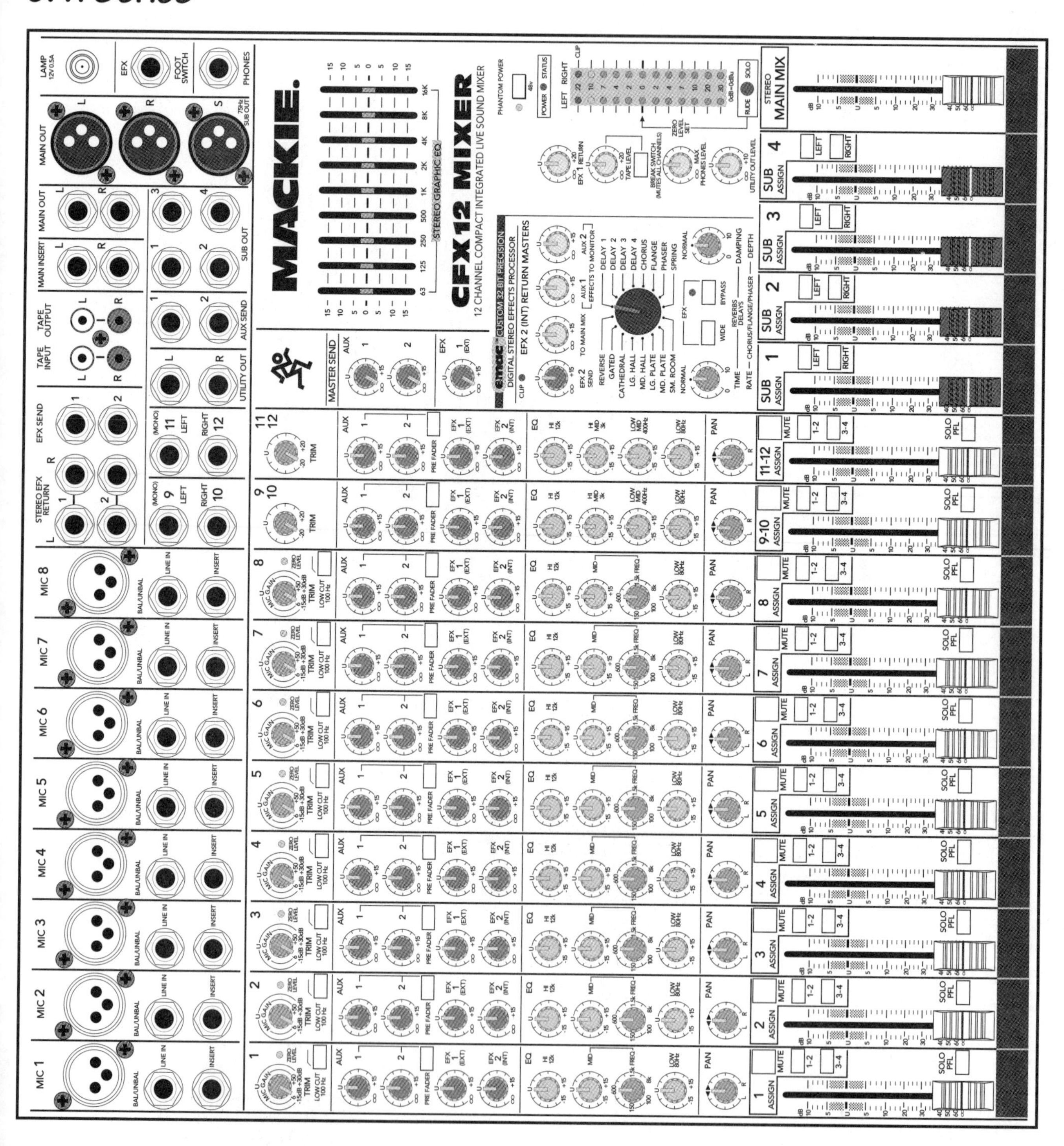
MACKIE.
CFX12 MIXER
12 CHANNEL COMPACT INTEGRATED LIVE SOUND MIXER
CUSTOM 32-BIT PRECISION
DIGITAL STEREO EFFECTS PROCESSOR
EFX 2 (INT) RETURN MASTERS
STEREO GRAPHIC EQ
MASTER SEND
STEREO MAIN MIX
MAIN OUT
MAIN INSERT
TAPE INPUT
TAPE OUTPUT
EFX SEND
STEREO EFX RETURN
SUB OUT
AUX SEND
UTILITY OUT
PHONES
MIC 1
MIC 2
MIC 3
MIC 4
MIC 5
MIC 6
MIC 7
MIC 8

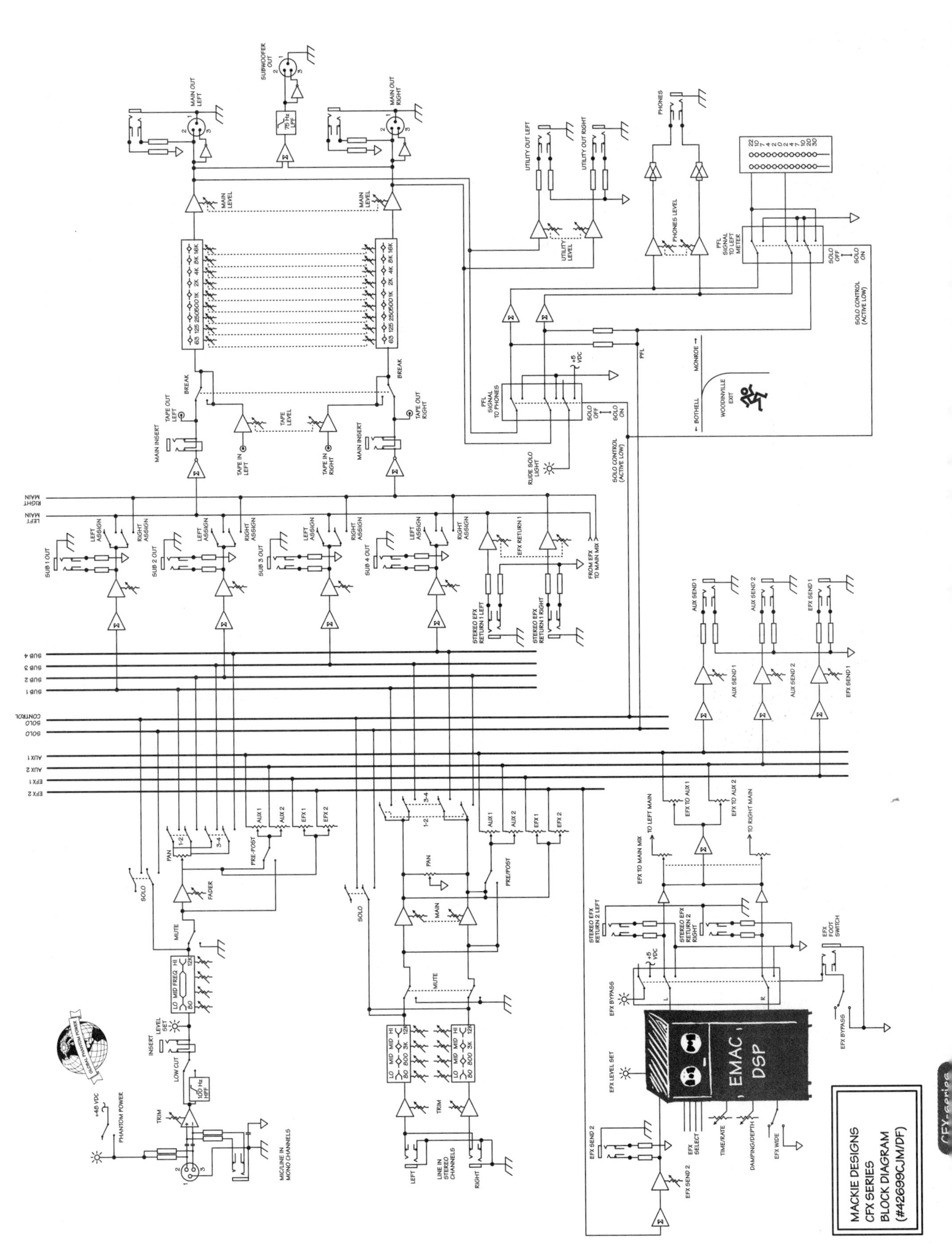
MACKIE DESIGNS
CFX SERIES
BLOCK DIAGRAM
(#42699CJM/DF)
EMAC DSP

DFX Series

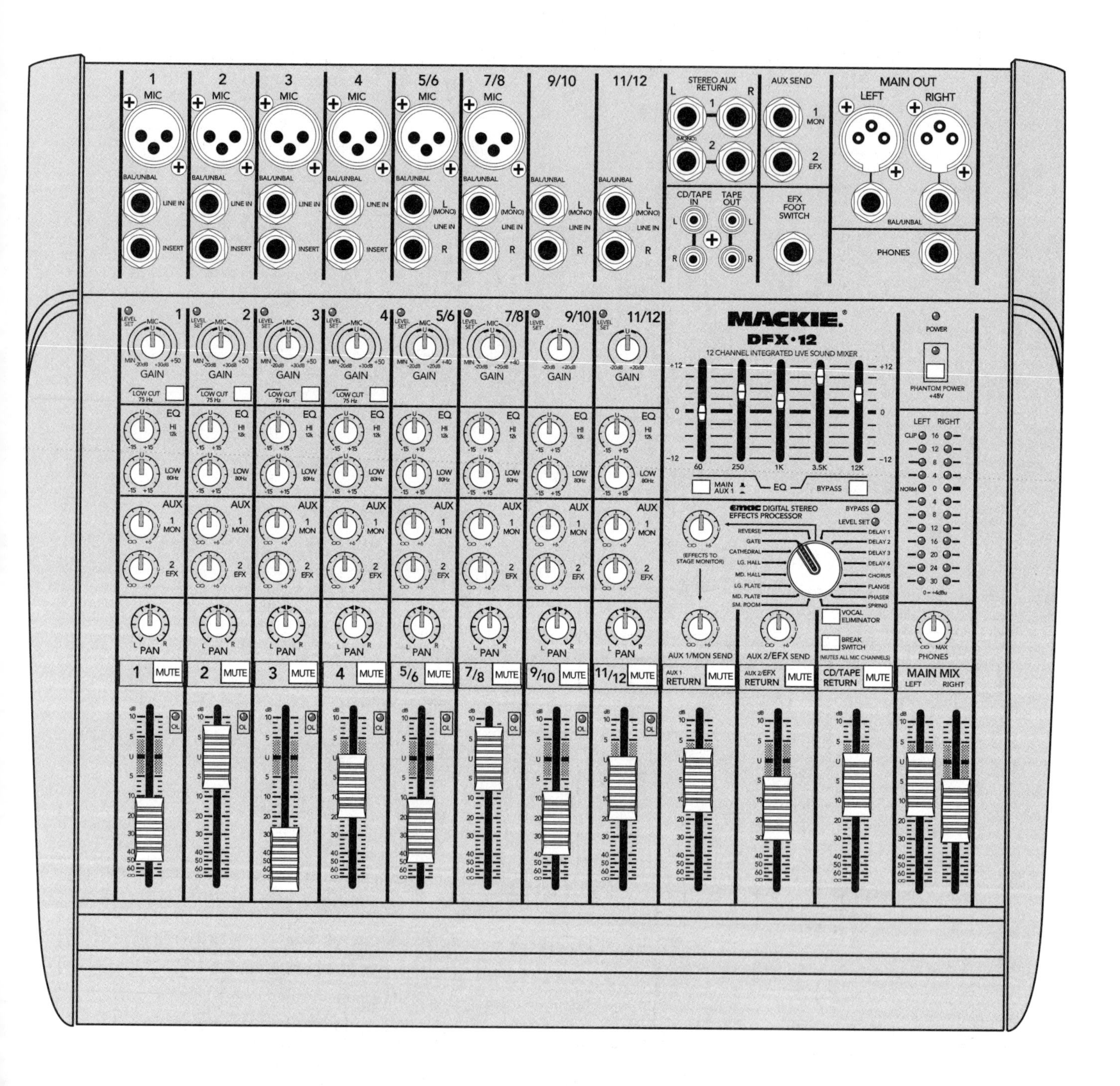
MACKIE.
DFX·12
12 CHANNEL INTEGRATED LIVE SOUND MIXER
MAIN OUT
STEREO AUX RETURN
AUX SEND
CD/TAPE IN
TAPE OUT
EFX FOOT SWITCH
PHONES
PHANTOM POWER +48V
EMAC DIGITAL STEREO EFFECTS PROCESSOR
AUX 1/MON SEND
AUX 2/EFX SEND
VOCAL ELIMINATOR
BREAK SWITCH
MAIN MIX

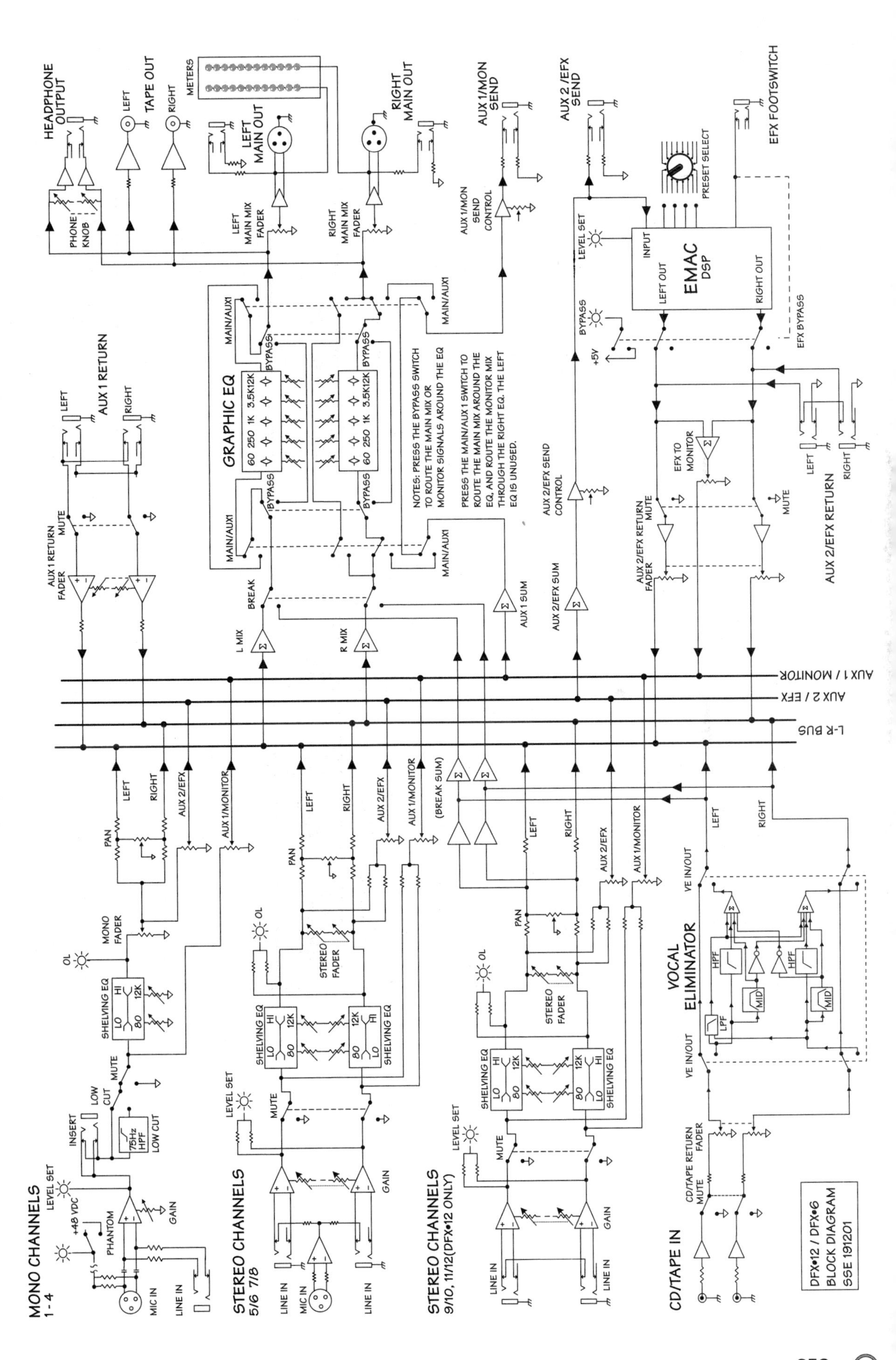
MONO CHANNELS
1 - 4
STEREO CHANNELS
5/6 7/8
STEREO CHANNELS
9/10, 11/12(DFX•12 ONLY)
CD/TAPE IN
HEADPHONE OUTPUT
TAPE OUT
METERS
LEFT MAIN OUT
RIGHT MAIN OUT
AUX 1/MON SEND
AUX 2 /EFX SEND
EFX FOOTSWITCH
PRESET SELECT
EMAC
DSP
GRAPHIC EQ
AUX 1 RETURN
AUX 2/EFX RETURN
VOCAL ELIMINATOR
AUX 1 / MONITOR
AUX 2 / EFX
L-R BUS
NOTES: PRESS THE BYPASS SWITCH TO ROUTE THE MAIN MIX OR MONITOR SIGNALS AROUND THE EQ
PRESS THE MAIN/AUX1 SWITCH TO ROUTE THE MAIN MIX AROUND THE EQ, AND ROUTE THE MONITOR MIX THROUGH THE RIGHT EQ. THE LEFT EQ IS UNUSED.
DFX•12 / DFX•6
BLOCK DIAGRAM
SSE 191201

Glossary

Amplitude
: The level of a varying signal (see *oscillation*). All else being equal, a signal with a higher amplitude will sound louder than one with a lower amplitude. When a signal's amplitude becomes too large for a circuit to pass it, *distortion* is the result.

Attenuation
: When an electronic circuit passes a signal and decreases its amplitude, the resulting signal is said to have been *attenuated.* The difference in level between the original signal and the softer one can be expressed in *decibels*, a logarithmic scale used to describe the difference in level between signals. See *gain.*

Audio Signals
: Sound is carried by the air. But mixers work with electricity, not changing air pressure. Sound can be transformed into electrical signals by microphones. We call these electrical signals *Audio signals.*

Bus
: In a mixer, there is a point where individual input signals come together and are combined, or mixed. This point is called a *mix bus.* Other names used for the mix bus are *mix output, send* or *output bus.* The term *bus* comes from the world of electrical power systems, where a "bus bar" carried the electricity for several connected circuits.

Channel
: The controls associated with one mixer input are called a channel. A channel's controls are grouped in vertical columns, seperate from the mixer's *output section.*

Clipping
: see *distortion.*

Control Room
: The room where you, your mixer and your monitor loudspeakers sit. In some mixers, there will be a set of controls just to set playback level and signal source for your "control room" monitor system. However, not all Mackies include separate controls for this function.

Direct Box
: A piece of equipment (usually a small metal box) designed to convert a 1/4" unblanced signal into an XLR signal suitable for connecting to a microphone input.

Direct Out
: An output from a single channel that is extracted from your mixer at an insert point, but doesn't get connected back to the mixer.

Distortion
: When an audio device tries to reproduce a sound that is too loud for it to handle, the result is a harsh, poor-sounding reproduction of the original. On the other hand, guitarists, myself included, think distortion can be cool. Just avoid unintentional distortion and everyone will be happy.

Equalization
: EQ for short, this is a type of mixer control that can change the *amplitude* of particular parts of a sound's *frequency* components, or *harmonics.* For example, a control that affects only the low frequencies of a sound can make the overall sound fuller, by boosting the level of the low frequencies, or thinner by cutting them.

Fader
: On mixers and other electronic control panels, a fader is a control that changes the level, or *amplitude*, of a signal. Typically, a fader is assumed to be a control that travels in a straight line, like the white-capped sliders on the 1604. In audio circles, a fader is assumed to control the volume, or *amplitude* of an audio signal. Technically, faders can be rotating knobs as well. In this book, 1202 users can assume that "fader" applies to their GAIN controls.

Feedback
: A condition where an output is connected back to its input ("feeding back") can result in a runaway situation if the signal is amplified during each successive round-trip. A microphone turned up and placed near the loudspeaker that is reproducing that mic's signal is likely to begin

Fundamental

The lowest component, or oscillation of a sound is called its fundamental. This frequency is usually interpreted to be that sound's musical pitch. Almost all sounds have additional oscillations above the fundamental, called harmonics. These harmonics give each sound its unique character.

Gain

When an electronic circuit increases the *amplitude* of a signal, it is said to have gain. The amount of gain is measured in *decibels*, a logarithmic scale used to describe the difference in level between signals. Circuits which reduce the amplitude of a passing signal are called *attenuators*. The special case when a signal is passed without changing levels is called unity gain.

Harmonics

All sounds are made up of a collection of individual frequencies. The lowest frequency component of a sound is generally heard as its musical pitch. Frequencies present above this fundamental are called harmonics. In pitched instruments, harmonics are often simple integer multiples of the fundamental, i.e. 2, 3 or 4 times the fundamental's frequency.

Insert

A connector (also called *Channel Access* or *Insert Point*) on a mixer channel that allows the channel's signal to be extracted, passed through an external signal processor and then returned back to the input channel.

Level

See *Amplitude*.

Oscillation

In the physical world and the domain of electronic circuits, oscillation is something engaged in a regular back-and-forth motion. In the real world, oscillating bodies create sound by pushing against the air.

Output Section

The portion of a mixer that controls the main stereo output volume and other mixer functions that apply to signals coming from individual mixer *channels* is called the output section.

Phantom Power

High-quality "Condenser" microphones need external power to operate. Phantom power is a voltage supplied to the mics through pins 2 and 3 of the XLR mic cable itself. Turning on the phantom switch causes 48 volts of DC (direct current) to appear at all Mackie MIC inputs.

Reverb

When a sound wave encounters a solid object, it may be partially absorbed or reflected away from it, depending on the nature of the surface. Within a confined space, reflected sound can bounce back and forth repeatedly until the sound slowly dies away. These reflections are called *reverberation*. Clap your hands in most large buildings for a demonstration!

Signal Flow

The path an audio signal takes between the input and output of a piece of audio equipment. Understanding this path makes it possible to understand and correctly predict how your sound system will work in different configurations.

Signal Processor

In general terms any electronic circuit that takes an input signal and modifies it in some way. Typically, the term is used to refer to specialized equipment separate from your mixer, like reverb or echo devices as well as equalizers, compressors and others.

TRS

An abbreviation for Tip-Ring-Sleve, this audio connector is a three conductor version of the common "guitar cord" connector, the 1/4" phone plug. Sometimes called "TRS."

Waveform

A waveform is a visual representation of a sound wave, tracing its *frequency* and *amplitude*.

Index

B

C

D

E

F

G

H

I

L

S

T

U

V

W

X

Y

AUDIO TECHNOLOGY BOOKS

FROM HAL LEONARD

Mackie Compact Mixers – Revised

Mackie Compact Mixers takes the mystery out of using your mixer. Written in a clear, musician-friendly style, this book will help you get the most from your small mixer, whatever its brand or model. Provides specific information and hook-up examples on Mackie's most popular models, including the "classic" 1202 and 1604 as well as the new 1202-, 1402-, 1604-VLZs, VLZ Pro and other models. Written by the author of *Live Sound for Musicians* and authorized by Mackie, this book explains the fundamental concepts of how mixing boards work, emphasizing how audio gets into and out of a mixer. Armed with this understanding of signal flow, you will be equipped to begin answering your own questions about how to set up and operate your mixer to best meet your needs.

00330477..$27.95

The Recording Guitarist

A Guide for Home and Studio by Jon Chappell

This is a practical, hands-on guide to a variety of recording environments, from modest home studios – where the guitarist must also act as the engineer and producer – to professional facilities outfitted with top-quality gear and staffed with audio engineers. This book will prepare guitarists for any recording situation and will help them become familiar with all facets of recording technology and procedure. Topics covered include: guitars and amps for recording; effects; mixer logic and routing strategies; synching music to moving images; and how to look and sound professional, with advice from Alex Lifeson, Carl Verheyen, Steve Lukather, Eric Johnson and others. Also includes complete info on the classic set-ups of 14 guitar greats, from Hendrix to Vai.

00330335..$19.95

Audio Made Easy – 2nd Edition

Audio Made Easy is a book about professional audio written in terms that everyone can understand. Chapters include info on mixers, microphones, amplifiers, speakers and how they all work together. New edition features a new section on wireless mics.

00330260..$12.95

Yamaha Guide to Sound Systems for Worship

The Yamaha Guide to Sound Systems for Worship is written to assist in the design, purchase, and operation of a sound system. It provides the basic information on sound systems that is most needed by ministers, members of Boards of Trustees and worship and music committees, interested members of congregations, and even employees of musical instrument dealers that sell sound systems. To be of greatest value to all, it is written to be both nondenominational and "non-brand-name."

00290243..$24.95

Live Sound for Musicians

Finally, a live sound book written for musicians, not engineers! *Live Sound for Musicians* tells you everything you need to know to keep your band's PA system working smoothly, from set-up to sound check right through performance. Author Rudy Trubitt give you all the information you need, and leaves out the unnecessary propeller-head details that would just slow you down. So if you're the player in the band who sets up the PA, this is the book you've been waiting for!

00330249..$19.95

Sound Check – The Basics of Sound and Sound Systems

Sound Check is a simplified guide to what can be a tricky subject: getting good sound. Starting with an easy-to-understand explanation of the principles and physics of sound, *Sound Check* goes on to cover amplifiers, speaker hookup, matching speakers with amps, sound reinforcement, mixers, monitor systems, grounding, and more.

00330118..$14.95

Yamaha Sound Reinforcement Handbook – 2nd Edition

Sound reinforcement is the use of audio amplification systems. This book is the first and only book of its kind to cover all aspects of designing and using such systems for public address and musical performance. The book features information on both the audio theory involved and the practical applications of that theory, explaining everything from microphones to loud speakers. This revised edition features almost 40 new pages and is even easier to follow with the addition of an index and a simplified page and chapter numbering system. New topics covered include: MIDI, synchronization, and an appendix on logarithms.

00500964..$34.95

For More Information, See Your Local Music Dealer, Or Write To:

HAL•LEONARD® CORPORATION

7777 W. Bluemound Rd. P.O. Box 13819 Milwaukee, WI 53213

www.halleonard.com

1102

Prices and availability subject to change without notice.

If you liked Mackie Compact Mixers... Check out Live Sound for Musicians!

Live Sound for Musicians shows you how to keep your PA system working smoothly, from set up and soundcheck, right through their performance, even if you've never run a PA before. This book is packed with helpful illustrations, sample setups, trouble shooting tips and it's certified 100% math free. An indispensable guide for performing musicians everywhere!

See Rudy's web site for more info—
www.trubitt.com

Topics Include:

- Sound System Basics
- Setting Up Your PA
- Soundcheck
- Troubleshooting
- Effects and Monitors
- Mixing Your Show
- Sample Setups and More!

Available Now!
by Rudy Trubitt
rudy@trubitt.com
www.trubitt.com
$19.95, 152 pages
ISBN 0-7935-6852-8